Flora of Florida, Volume V

UNIVERSITY PRESS OF FLORIDA

Florida A&M University, Tallahassee
Florida Atlantic University, Boca Raton
Florida Gulf Coast University, Ft. Myers
Florida International University, Miami
Florida State University, Tallahassee
New College of Florida, Sarasota
University of Central Florida, Orlando
University of Florida, Gainesville
University of North Florida, Jacksonville
University of South Florida, Tampa
University of West Florida, Pensacola

Flora
of ❧
Florida

VOLUME V

DICOTYLEDONS, GISEKIACEAE THROUGH
BORAGINACEAE

**Richard P. Wunderlin, Bruce F. Hansen,
and Alan R. Franck**

University Press of Florida

Gainesville · Tallahassee · Tampa · Boca Raton

Pensacola · Orlando · Miami · Jacksonville · Ft. Myers · Sarasota

23 22 21 20 19 18   6 5 4 3 2 1

Library of Congress Cataloging-in-Publication Data
Names: Wunderlin, Richard P., 1939– author. | Hansen, Bruce F., author. | Franck, Alan
R., author.
Title: Flora of Florida. Vol. 5 : Dicotyledons, Gisekiaceae through Boraginaceae /
Richard P. Wunderlin, Bruce F. Hansen, and Alan R. Franck.
Description: Gainesville : University Press of Florida, 2016. | Includes bibliographical
references and index.
Identifiers: LCCN 2017032670 | ISBN 9780813056791 (cloth : alk. paper)
Subjects: LCSH: Plants—Florida—Classification. | Plants—Florida—Identification.
Classification: LCC QK154 .W852 2016 | DDC 581.9759—dc23
LC record available at https://lccn.loc.gov/2017032670

The University Press of Florida is the scholarly publishing agency for the State
University System of Florida, comprising Florida A&M University, Florida Atlantic
University, Florida Gulf Coast University, Florida International University, Florida
State University, New College of Florida, University of Central Florida, University of
Florida, University of North Florida, University of South Florida, and University of
West Florida.

University Press of Florida
15 Northwest 15th Street
Gainesville, FL 32611-2079
http://upress.ufl.edu

To the memory of Angus Kemp Gholson (1921–2014)

and Edward Everett Terrell (1923–2011)

for their work on the Florida flora and Rubiaceae, respectively

Contents

Acknowledgments

The facilities and collections of many herbaria were utilized in preparing this volume. The courtesies extended and the loan of specimens by the curators are gratefully appreciated. These include the Florida Museum of Natural History (FLAS), Florida State University (FSU), Harvard Herbaria (A, GH), Marie Selby Botanical Garden (SEL), New York Botanical Garden (NY), University of Central Florida (FTU), University of North Carolina, Chapel Hill (NCU), and University of West Florida (UWFP). We are especially grateful to Kent Perkins (FLAS), Loran Anderson (FSU), and Austin Mast (FSU) for their continuous support. Review and helpful suggestions were kindly provided by Walter Judd (Ericaceae).

The *Flora of Florida* project has been strongly supported by the University of South Florida Institute for Systematic Botany.

Introduction

Volume 1 of the *Flora of Florida* provides background information on the physical setting, vegetation, history of botanical exploration, and systematic treatments of the pteridophytes and gymnosperms. Volumes 2 through 7 will contain the dicotyledons and volumes 8 through 10, the monocotyledons.

This volume contains the taxonomic treatments of 34 families.

ORGANIZATION OF THE FLORA

Taxa Included

Florida, with more than 4,300 taxa, has the third most diverse vascular plant flora of any state in the United States. The *Flora of Florida* is a treatment of all indigenous and naturalized vascular plant taxa currently known to occur in the state. Naturalized is defined as those nonindigenous taxa growing outside of cultivation and naturally reproducing. This includes plants that have escaped from cultivation as well as those that were intentionally or accidentally introduced by human activities in post-Columbian times. Taxa that have not been recently recollected and may no longer exist in the wild in Florida are formally treated both for historical completeness and on the premise that they may be rediscovered in the future.

A taxon is formally treated in this flora if (1) an herbarium specimen has been seen to document its occurrence in Florida, or (2) a specimen is cited from Florida in a monograph or revision whose treatment is considered sound.

Taxa Excluded

Literature reports of taxa attributed to Florida that are considered to be erroneous or highly questionable and therefore to be excluded from this flora are listed following the treatment for the genus, or in the case of genera not otherwise treated, at the end of the family. The reason for exclusion is given in each case. Most commonly, the taxon is excluded because it is based on a misidentified specimen(s), lack of documentation by means of a specimen, or it is based on a misapplied name, that is, a name correctly applied to a plant not found in Florida.

Systematic Arrangement

Recent studies have demonstrated that the traditional dicotyledons are paraphyletic and that the monophyletic monocotyledons are derived from within the dicotyledons. We believe that the arrangement as proposed by the Angiosperm Phylogeny Group (Stevens, 2016) has merit and is followed in this work with slight modifications. The linear sequence of families used here essentially follows that proposed by Haston et al. (2009). For convenience, the genera and species within each family are arranged alphabetically.

Descriptions

Descriptions are based on Florida material and are given for each family, genus, species, and infraspecific taxon.

Common Names

Non-Latinized names given for the taxa are derived from published sources as well as from our own experience. No attempt is made to list all names that have been applied to a taxon, to standardize names with a specific source, nor to supply a name for species where one is not in general usage. For plants lacking a common name, the generic name may be used as is the usual practice.

Derivation of Scientific Names

The derivation of the generic name and that of each specific and infraspecific epithet is given.

Synonymy

A full literature citation is given for each species, infraspecific taxon, and synonym. Synonyms listed are only those that have been cited for Florida in manuals, monographic treatments, and technical papers. Also included is the basionym and all homotypic synonyms of a name introduced into synonymy. The homotypic synonyms are listed in chronological order in a single paragraph, and the paragraphs of synonyms are put in chronological order according to the basionym of each. If the type of a taxon is a Florida collection and is known, this information is given. We do not attempt to lectotypify the numerous Florida taxa needing lectotypification in the belief that this is best left to monographers.

For families and genera, only the author and date of publication are given. Family and generic synonyms listed are those that have been used in the major publications pertinent to the Florida flora.

Citation of periodical literature conforms to that cited in Lawrence et al. (1968) and Bridson and Smith (1991). Other literature citations conform to that cited in Stafleu and Cowan (1976 et seq.). Author abbreviations are those listed in Brummitt and Powell (1992).

Habitat

The terminology used for plant communities generally follows that of Myers and Ewel (1990), but may vary.

Distribution

The global distribution is given for each family and genus where native and naturalized. Relative abundance in Florida (ranked as common, frequent, occasional, or rare) and the distribution is given for each species and infraspecific taxon. The format for distribution of species and infraspecific taxa is: Florida; North America (Continental United States, Canada, and Greenland); tropical America (West Indies, Mexico, Central America, and South America); Old World (Europe, Africa, Asia, Australia, and Pacific Islands). For taxa occurring in all of these areas, the phrase "nearly cosmopolitan" is used. For taxa of limited distribution in Florida, range statements by county are usually given. For taxa of wide distribution in Florida, the range is given in general terms: *panhandle*—from the Suwannee River west to Escambia County; *peninsula*—east of the Suwannee River and south of the Georgia line southward through the Florida Keys. Because of the vast floristic differences in peninsular Florida, this region is often further subdivided into northern, central, and southern regions and the keys. The northern region is east of the Suwannee River and south of the Georgia line southward through Gilchrist, Alachua, Putnam, and Flagler Counties. The central region extends from Levy, Marion, and Volusia Counties southward through Lee, Hendry, and Palm Beach Counties. The southern peninsula consists of the southernmost four counties (Collier, Broward, Monroe, and Miami-Dade). The Florida Keys consist of the chain of islands from Key Largo to the Marquesas Keys and the Dry Tortugas. Politically, they are part of Monroe County. The panhandle is subdivided into eastern, central, and western regions. The eastern region consists of the counties west of the Suwannee River west through Jefferson County, the central region extends from Leon and Wakulla Counties west through Holmes, Washington, and Bay Counties, and the western region consists of the westernmost four counties (Walton, Okaloosa, Santa Rosa, and Escambia).

Since the species distribution may change as new data are added, please refer to the *Atlas of Florida Plants* website (Wunderlin et al., 2016) for current information.

Endemic or Exotic Status

Endemic taxa are those whose global distribution is confined to the political boundary of Florida. If a taxon is a non-native, the region of nativity is given. Non-native taxa are those that are known to have become part of the flora following the arrival of Ponce de Leon in 1513. Admittedly, this is an arbitrary starting point as several species are believed to have been introduced by Paleo-Indians before 1513. Technically, these are considered as native. Another problem in interpretation arises when propagules arrive after 1513 by some means other than human activity (that is, hurricanes, storms, sea drift, or animals) and the species becomes established. Again technically, these are considered as non-natives. It is sometimes difficult to determine whether a widespread species is native or exotic, and our opinion may differ from that of others.

Reproductive Season

The sexual reproductive (flowering) season for each species and infraspecific taxon is given. The reproductive seasons are broadly defined as follows: spring—March through May; summer—June through September; fall—October through November; winter—December through February. Species "flowering out of season" are sometimes encountered.

Hybrids

Named hybrids are listed along with the putative parents, nomenclature, and usually with a comment concerning distribution in Florida.

References

Major monographs, revisions, and other pertinent literature, other than those cited in the nomenclature, are cited where appropriate in the text and listed at the end of the volume.

TAXONOMIC CONCEPTS

Taxonomic interpretations and nomenclature are generally in accord with recent monographs or revisions for the various groups except where it is believed that recent evidence necessitates a change. Citation of a monograph or revision in the text implies consideration of the work during the preparation of the treatment, but not necessarily acceptance. Where a difference of opinion exists among published treatments or the treatment in this work deviates from that of the reference cited, a discussion of alternative opinions is often provided.

Species, subspecies, and varieties are considered as entities with a high degree of population integrity. Color forms and minor morphotypes that occur within a species and that may be formally recognized as forma by other authors are accorded no formal recognition in this work.

No nomenclatural innovations are intentionally published in the *Flora*.

LITERATURE CITED

Bridson, G.D.R., and E. R. Smith. 1991. Botanico-Periodicum-Huntianum/Supplementum. Pittsburgh: Hunt Botanical Library.

Brummitt, R. K., and C. E. Powell. 1992. Authors of Plant Names. Royal Botanical Gardens, Kew. Basildon: Her Majesty's Stationery Office.

Haston, E., J. E. Richardson, P. F. Stevens, M. W. Chase, and D. J. Harris. 2009. The linear Angiosperm Phylogeny Group (LAPG) III: a linear sequence of the families in APG III. Bot. J. Linnean Soc. 161: 128—31.

Lawrence, G.H.M., A.F.G. Buchheim, G. S. Daniels, and H. Dolezal. 1968. Botanico-Periodicum-Huntianum. Pittsburgh: Hunt Botanical Library.

Myers, R. I., and J. J. Ewel, eds. 1990. Ecosystems of Florida. Gainesville: University Press of Florida.

Stafleu, F. A., and R. S. Cowan. 1976 et seq. Taxonomic Literature. Edition 2. Utrecht: Bohn, Scheltema, and Holkema.

Stevens, P. F. 2016. Angiosperm Phylogeny Website, Version 13. (http://www.mobot.org/MO-BOT/research/APweb/).

Wunderlin, R. P., B. F. Hansen, A. R. Franck, and F. B. Essig. 2016. Atlas of Florida Plants. (http://florida.plantatlas.usf.edu/).

Systematic Treatments

Keys to Major Vascular Plant Groups

1. Plant reproducing by spores .... PTERIDOPHYTES (volume I)
1. Plant reproducing by seeds.
    2. Leaves with a single midvein or with simple or sometimes dichotomously branched veins, these closely parallel and lacking secondary interconnecting cross-veinlets; seeds borne on the surface of specialized bract-scale structures aggregated into woody or fleshy cones or a single seed partly or wholly surrounded by a fleshy aril and drupelike or berrylike; perianth lacking .... GYMNOSPERMS (volume I)
    2. Leaves with parallel veins with secondary interconnecting cross-veinlets or with reticulate veins; seeds borne enclosed within specialized structures (carpels); perianth usually present.
        3. Vascular bundles occurring in a ring or in concentric cylinders; cotyledons 2; flower parts usually in other than whorls of 3 or multiples thereof; leaves usually reticulate-veined .... DICOTYLEDONS (volumes II–VII)
        3. Vascular bundles scattered (or rarely single); cotyledon 1; flower parts often in whorls of 3 or multiples thereof; leaves usually parallel-veined (sometimes with midvein only) (some plants diminutive, floating aquatics, the plant body thalloid, not differentiated into stems and leaves, rootless or with 1–few unbranched roots or plants partly or wholly submersed aquatics, the leaves, flowers, and fruits often much reduced) .... MONOCOTYLEDONS (volumes VIII–X)

Dicotyledons

GISEKIACEAE (Endl.) Nakai 1942. GISEKIA FAMILY

Herbs. Leaves opposite, simple, pinnate-veined, petiolate, estipulate. Flowers in terminal or axillary, compound dichasia, actinomorphic, bisexual, ebracteate; sepals 5, free; petals absent; stamens 5, free, alternate with the sepals; ovary superior, 5-carpellate and -loculate, the style and stigma 1. Fruit a group of achenes.

A family of 1 genus and 7 species; North America, Europe, Africa, and Asia.

The Gisekiaceae is sometimes placed in the Aizoaceae, Molluginaceae, Phytolaccaceae, or even Portulacaceae. It is anomalous in each of these and best placed in its own family.

Gisekia L. 1753.

Annual herbs. Leaves simple, pinnate-veined, petiolate, estipulate. Flowers in axillary or terminal compound dichasia, pedicellate; sepals 5, free; petals absent; stamens 5, free, alternate with the sepals, the anthers basifixed, 2-loculate, dehiscing by longitudinal slits; ovary 5-carpellate and -loculate, the style 1, terminal, the stigma capitate. Fruit a group of achenes; seed 1 per carpel, laterally compressed.

A genus of about 5 species; North America, Europe, Africa, and Asia. [Commemorates Paul Dietrich Giseke (1741–1796), German botanist, pupil of Linnaeus, and professor at Hamburg.]

Selected references: Gilbert (1993); Nienaber and Thieret (2003).

Gisekia pharnacioides L. [Resembling *Pharnaceum* (Aizoaceae).] OLDMAID.

Gisekia pharnacioides Linnaeus, Mant. Pl. 562. 1771.

Prostrate to ascending, slightly succulent annual herb; stem to 8 dm long, radiating from a central axis, glabrous, the nodes swollen. Leaves with the blade linear to oblanceolate-spatulate or elliptic, 0.5–6 cm long, 0.1–1.9 cm wide, unequal, the apex rounded to acute, sometimes minutely apiculate, the base tapering, the margin entire, sometimes revolute, the upper and lower surfaces white-flecked with bundles of raphides, the midvein evident above and below, the secondary veins inconspicuous, the petiole to 1 cm long, marginate or winged, decurrent and forming ribs on the stem. Flowers in an umbelliform, compound, congested to a lax dichasium

0.2–1.5 cm long, the peduncle 0.5 cm long, the pedicel 1–4 mm long; sepals ovate to oblong, 1–3 mm long, boat-shaped, greenish to pinkish; stamens with the filaments dilated at the base, the anthers white; carpels laterally compressed, the short style erect in bud, curving outward at anthesis. Fruit muricate, 1–1.5 mm long; seed lenticular, black lustrous, minutely punctate.

Disturbed sites. Occasional; central peninsula. Florida; Europe, Africa, and Asia. Native to Africa and Asia. Summer–fall.

AIZOACEAE Martinov 1820. MESEMBRYANTHEMUM FAMILY

Annual or perennial herbs or subshrubs. Leaves alternate or opposite, simple, pinnate-veined, petiolate or epetiolate, stipulate or estipulate. Flowers axillary or terminal, solitary or few in cymes, actinomorphic or slightly zygomorphic, bisexual or unisexual (plants polygamous), bracteate or ebracteate; sepals 4–5, basally connate, equal or unequal; petals 100–200 (including petaloid staminodes) or absent; stamens (1)3–80(600), free or connate in groups, sometimes adnate to the sepal lobes, the anthers 2-loculate, longitudinally dehiscent; ovary superior, inferior, or half inferior, (1)3- to 12-carpellate and -loculate, the styles as many as the carpels, free or basally connate. Fruits loculicidal or circumscissile dehiscent capsules or indehiscent and nutlike or a berry; seeds 1–many, arillate or exarillate.

A family of about 130 genera and about 2,500 species; nearly cosmopolitan.

Tetragoniaceae Link, 1831.

Selected references: Boetsch (2002).

1. Leaves basally connate-perfoliate ....**Carpobrotus**
1. Leaves sessile or short-petiolate.
    2. Ovary inferior or half-inferior.
        3. Leaves alternate; flowers greenish or yellowish; fruit a nutlet .... **Tetragonia**
        3. Leaves opposite; flowers pink to purple; fruit a capsule...**Aptenia**
    2. Ovary superior.
        4. Leaves alternate; fruit a 3- to 5-valved capsule; seed 1 in each carpel................................**Galenia**
        4. Leaves opposite; fruit a circumscissile dehiscent capsule; seeds few to many in each carpel.
            5. Leaves in equal or subequal pairs; estipulate... **Sesuvium**
            5. Leaves in unequal pairs; stipulate.
                6. Sepals appendaged; stamens 5–10; seeds few; stipules scarious, entire or nearly so ....
                ....**Trianthema**
                6. Sepals unappendaged; stamens 1–3; seeds numerous; stipules hyaline, laciniate...............
                ....**Cypselea**

Aptenia N. E. Br. 1925.

Subshrubs. Leaves opposite, simple, in subequal pairs, pinnate-veined, petiolate, estipulate. Flowers solitary, axillary, pedunculate, ebracteate; sepals 4; petals (including petaloid staminodes) numerous, basally connate into a short tube; stamens 80; ovary inferior, 4-carpellate and -loculate. Fruit a 4-valved, loculicidal dehiscent, fleshy capsule; seeds numerous, exarillate.

A genus of 2 species; North America, Europe, Africa, and Australia. [From the Greek *a*, without, and *pteron*, wing, in reference to the capsules lacking wings.]

Selected reference: Vivrette (2003c).

Aptenia cordifolia (L. f.) Schwantes [*Cord*, heart, and *folia*, leaf, in reference to the heart-shaped leaves.] BABY SUN ROSE; HEARTLEAF ICEPLANT.

> *Mesembryanthemum cordifolium* Linnaeus f., Suppl. Pl. 260. 1782 ("1781"). *Litocarpus cordifolius* (Linnaeus f.) L. Bolus, Fl. Pl. South Africa 7: t. 261(11–12). 1927. *Aptenia cordifolia* (Linnaeus f.) Schwantes, Gartenflora 77: 69. 1928. *Tetracoilanthus cordifolius* (Linnaeus f.) Rappa & Camerone, Lav. Ist. Bot. Giard. Colon. Palermo 14: 64. 1954.

Prostrate succulent subshrub; stem to 6 dm long, sometimes rooting at the nodes, minutely papillate. Leaves with the blade cordate-ovate, 1–3 cm long, 0.5–2.5 cm wide, the apex acute, the base cuneate, the margin entire, the upper and lower surfaces minutely papillate, the petiole ca. 1 cm long. Flowers solitary, the peduncle 0.5–1.5 cm long; hypanthium 6–7 mm long, papillate; sepal lobes, unequal, the 2 larger ones foliaceous, 4–7 mm long, the 2 shorter ones narrower; petals linear, 3–4 mm long, pink to purple, reflexed; petaloid staminodes linear, ca. 3 mm long, white. Fruit obconic, 1.3–1.5 cm long, ca. 1 cm wide; seeds ca. 800, obovoid, compressed, brownish black, tuberculate.

Dry disturbed sites. Rare; St. Lucie County. Escaped from cultivation. Florida, Oregon, and California; Europe, Africa, and Australia. Native to Africa. Summer–fall.

Carpobrotus N. E. Br. 1925. HOTTENTOT FIG

Perennial herbs. Leaves opposite, simple, in equal pairs, pinnate-veined, basally connate-perfoliate, estipulate. Flowers solitary, terminal on short branchlets, pedunculate, ebracteate; sepals 5; petals (including petaloid staminodes) numerous, free; stamens in 4(7) series of ca. 100 each; ovary inferior, 8- to 12-carpellate and -loculate, the style absent, the stigmas 8–12. Fruit a berry; seeds numerous, exarillate.

A genus of 13 species; North America, Mexico, South America, Europe, Africa, Australia, and Pacific Islands. [From the Greek *karpos*, fruit, and *brota*, edible.]

Selected reference: Vivrette (2003d).

Carpobrotus edulis (L.) L. Bolus [Edible.] COMMON HOTTENTOT FIG.

> *Mesembryanthemum edule* Linnaeus, Syst. Nat., ed. 10. 1060. 1759. *Carpobrotus edulis* (Linnaeus) L. Bolus, Fl. Pl. South Africa 7: sub t. 247 1927.

Prostrate, succulent suffruticose herb; stem to 3 m long, glabrous. Leaves with the blade linear, 5–13 cm long, 0.5–1 cm wide, the apex acute, basally connate-perfoliate, sharply 3-angled in cross-section, somewhat curved, serrate on the lower angle, the upper and lower surfaces glabrous. Flowers terminal on a short, erect branchlet, the peduncle 2–6 cm long; hypanthium turbinate, 2–3 cm long; sepal lobes unequal, the outer 2 lobes 1–6 cm long, foliaceous, sharply 3-angled in cross-section, the lower angle margin serrate near the apex, the inner 2 lobes smaller, the margin entire, membranaceous; petals (including petaloid staminodes) linear, 3–4 cm long, ca. 2 mm wide, the apex acute or obtuse, yellow, turning pink in age; stamens with the anthers yellow; styles 8–15 mm long, radiating horizontally. Fruit clavate to subglobose, 2–3.5 cm long, the apex depressed, yellowish; seeds ca. 1000, obovoid, compressed, brown, slightly tuberculate.

Dunes. Rare; Volusia, Martin, Franklin, and Escambia Counties. Escaped from cultivation. Florida and California; Mexico and South America; Europe, Africa, Australia, and Pacific Islands. Native to Africa. Summer.

Cypselea Turpin 1806.

Annual or weakly perennial herbs. Leaves opposite, simple, in unequal pairs, pinnate-veined, petiolate, stipulate. Flowers solitary or 2–3 in cymes, terminal on a short branchlet and appearing axillary, bracteate; sepals 4–5; petals and petaloid staminodes absent; stamens (1)3; ovary superior, 2-carpellate, 1-loculate. Fruit a circumscissile dehiscent capsule; seeds numerous, arillate.

A genus of 8 species; North America, West Indies, Mexico, and South America. [From the Greek *kypsele*, a hollow vessel, in reference to the shape of the circumscissile capsule.]

Selected reference: Ferren (2003b).

Cypselea humifusa Turpin [*Humi*, ground, and *fus*, pour out, in reference to its prostrate habit.] PANAL.

Cypselea humifusa Turpin, Ann. Mus. Hist. Nat. 7: 219, pl. 12(5). 1806.

Prostrate succulent herb; stem to 5 cm long, glabrous. Leaves with the blade elliptic to elliptic-oblong or ovate, the larger ones 5–10 mm long, the smaller ones 2–6 mm long, the apex obtuse, the base broadly cuneate, the margin entire, the upper and lower surfaces glabrous, the petiole subequaling the blade, dilated below, sometimes connate and sheathing the node; stipules scarious, laciniate. Flowers solitary, terminal on a short, lateral branchlet and appearing axillary, short-pedunculate; bracts 2, scarious; hypanthium turbinate, 1.5–2 mm long; calyx lobes ovate to oval, ca. 1.5 mm long, erect, the apex obtuse, the margin scarious, the outer surface green or purple-tinged, the inner surface white to reddish; stamens adnate to the sepal lobes; styles erect, slightly basally connate. Fruit subglobose, ca. 2.5 mm long and wide, membranaceous, circumscissile below the middle; seeds numerous, subreniform, brown, smooth.

Pineland depressions and ditches. Occasional; central and southern peninsula. Florida, Alabama, Louisiana, Arizona, Nevada, and California; West Indies and South America. Native to West Indies. Spring–summer.

Galenia L. 1753.

Suffrutescent perennial herbs or subshrubs. Leaves alternate, simple, pinnate-veined, epetiolate, estipulate. Flowers solitary, axillary, sessile, ebracteate; sepals 5; petals and petaloid staminodes absent; stamens 10, in pairs alternate with the calyx lobes; ovary superior, 2- to 5-carpellate and -loculate, the styles 4(5). Fruit a loculicidal dehiscent capsule; seeds 4(5), exarillate.

A genus of 27 species; North America, Africa, and Australia. [Commemorates Claudius Galenius (ca. AD 130–200), Roman naturalist and physician.]

Selected reference: Vivrette (2003b).

Galenia secunda (L. f.) Sonder [Turned toward one side, in reference to the leaves.] ONESIDED GALENIA.

> *Aizoon secundum* Linnaeus f., Suppl. Pl. 261. 1782 ("1781"). *Galenia secunda* (Linnaeus f.) Sonder, in Harvey, Fl. Cap. 2: 474. 1862.

Prostrate, suffrutescent perennial herb or subshrub; stem to 3 dm long, coarsely villous with loosely appressed, grayish trichomes. Leaves with the blade, rhombic to obovate or spatulate, 1.5–2.2 cm long, 7–9 mm wide, folded inwardly, the apex obtuse and recurved, the base cuneate, the margin entire, the upper and lower surfaces grayish white-papillose. Flowers usually terminal on a short, lateral branch; sepal lobes oblong or oblong-lanceolate, ca. 3 mm long, white to yellow, erect, the apex incurved, ciliate, the outer surface pubescent; stamens 10, of unequal length, in pairs, shorter than the calyx lobes. Fruit 4(5)-angled, leathery, the apex depressed; seeds reniform, blackish brown, shiny, tuberculate-striate.

Disturbed sites. Rare; Escambia County. Known only from a single 1901 collection (*Curtiss 6869*, GH, NY). Florida and New Jersey; Africa and Australia. Native to Africa. Summer–fall.

Sesuvium L. 1759. SEAPURSLANE

Annual or perennial herbs. Leaves opposite, simple, in equal or subequal pairs, pinnate-veined, petiolate or epetiolate, estipulate. Flowers solitary or few in cymes, these terminal, but appearing axillary, sessile or pedicellate, ebracteate; sepals 5; stamens 5 or 30; petals absent; ovary superior, 2- to 5-carpellate and -loculate. Fruit a circumscissile dehiscent capsule; seeds numerous, arillate.

A genus of about 8 species; nearly cosmopolitan. [For *Sesuvium*, land of the Sesuvii, a coastal Gallic tribe.]

Selected reference: Ferren (2003a).

1. Flower or fruit sessile or on a pedicel to 1 mm long; stamens 5 .... **S. maritimum**
1. Flower or fruit on a pedicel 3 mm long or longer; stamens 30 .... **S. portulacastrum**

Sesuvium maritimum (Walter) Britton et al. [Growing by the sea]. SLENDER SEAPURSLANE.

> *Pharnaceum maritimum* Walter, Fl. Carol. 117. 1788. *Sesuvium pentandrum* Elliott, Sketch Bot. S. Carolina 1: 556. 1821, nom. illegit. *Mollugo maritima* (Walter) Seringe, in de Candolle, Prodr. 1: 393. 1824. *Squibbia maritima* (Walter) Rafinesque, New Fl. 4: 16. 1838 ("1836"). *Sesuvium maritimum* (Walter) Britton et al., Prelim. Cat. 20. 1888. *Halimus maritima* (Walter) Kuntze, Revis. Gen. Pl. 1: 263. 1891.

Prostrate to ascending, succulent annual herb; stem to 4(10) dm long, glabrous. Leaves with the blade spatulate to narrowly oblanceolate or obovate, 1–3.5 cm long, 1–10 mm wide, the apex rounded to obtuse, the base narrowly cuneate, clasping the stem, the margin entire, the upper and lower surfaces glabrous. Flowers usually solitary, sessile or subsessile (pedicel to 1 mm long); sepal lobes ovate to ovate-oblong, 2–4 mm long, the apex obtuse, with a subapical, dorsal, prolonged appendage, the inner surface pink or purplish; stamens 5; ovary 2- or 3-carpellate and -loculate. Fruit ovoid, 4–5 mm long; seeds 30–50, reniform, ca. 1 mm long, brownish black, smooth, iridescent.

Coastal dunes, brackish marshes, and salt flats. Occasional; peninsula west to central panhandle. New York south to Florida, west to Texas, also Kansas and Oklahoma; West Indies. All year.

Sesuvium portulacastrum (L.) L. [*Portulaca* (Portulacaceae) and the Latin *astrum*, in reference to an incomplete likeness to that genus.] SHORELINE SEAPURSLANE.

> *Portulaca portulacastrum* Linnaeus, Sp. Pl. 446. 1753. *Sesuvium portulacastrum* (Linnaeus) Linnaeus, Syst. Nat., ed. 10. 1058. 1759. *Halimus portulacastrum* (Linnaeus) Kuntze, Revis. Gen. Pl. 1: 263. 1891.

Prostrate, succulent perennial herb; stem to 1 m long, glabrous, often rooting at the nodes. Leaves with the blade elliptic to obovate or oblanceolate to 4–6 cm long, 1–2.5 cm wide, the apex obtuse-rounded to acute, the base narrowly cuneate, clasping the stem, the margin entire, the upper and lower surfaces glabrous. Flowers solitary, the pedicel 3–20 mm long; sepal lobes broadly ovate-lanceolate to lanceolate, 0.5–1 cm long, 4–6 mm wide, the apex obtuse, with a subapical, dorsal, prolonged appendage, the inner surface pink or purplish; stamens 30, in fascicles; ovary 5-carpellate and -loculate. Fruit conic, ca. 1 cm long; seeds 30–60, reniform, 1–1.5 mm long, black, smooth, shiny.

Coastal dunes, brackish marshes, and salt flats. Frequent; peninsula, central and western panhandle. Pennsylvania, North Carolina south to Florida, west to Texas; West Indies, Mexico, Central America, and South America; Europe, Africa, Asia, and Australia. Native to North America, tropical America, Africa, Asia, and Australia. All year.

Tetragonia L. 1753.

Annual or perennial herbs. Leaves alternate, simple, pinnate-veined, petiolate, estipulate. Flowers solitary, axillary, pedunculate, bisexual or sometimes unisexual (plants polygamous), ebracteate; sepals 4–5; petals absent; stamens 10–13; ovary half-inferior, 2- to 8-carpellate and -loculate, the styles 2–8. Fruit indehiscent, nutlike; seeds 1–10, exarillate.

A genus of 60 species; North America, West Indies, Mexico, Central America, and South America; Africa, Asia, Australia, and Pacific Islands. [From the Greek *tetra*, four, and *gonia*, angle, in reference to the shape of the fruit.]

Selected reference: Vivrette (2003a).

Tetragonia tetragonioides (Pall.) Kuntze [To resemble *Tetragonia.*] NEW ZEALAND SPINACH.

Demidovia tetragonioides Pallas, Enum. Hort. Demidof 150, t. 1. 1781. *Tetragonia quadricornis* Stokes, Bot. Mat. Med. 3: 127. 1812, nom. illegit. *Tetragonia tetragonioides* (Pallas) Kuntze, Revis. Gen. Pl. 1: 264. 1891.
Tetragonia expansa Murray, Commentat. Soc. Regiae Sci. Gott. 6: 13. 1783.

Prostrate, succulent annual or perennial or suffrutescent herb; stem to 1.4 m long, glabrous. Leaves with the blade ovate-rhombic to triangular, 2–6 cm long, 2–5 cm wide, the apex acute to obtuse, the base cuneate to truncate, the margin entire, the upper surface papillate, the lower surface glabrous, the petiole 1–3 cm long, winged. Flowers solitary, the peduncle ca. 2 mm long; sepal lobes triangular or ovate, 1–2 mm long, the inner surface yellow; stamens in fascicles or separate. Fruit obconic, 8–12 mm long, nutlike, with 4–5 hornlike protuberances; seeds pyriform to subreniform, 1–1.5 mm long, light brown, smooth.

Disturbed coastal sites. Rare; Duval, Levy, Hillsborough, and Manatee Counties. Escaped from cultivation. New York and Massachusetts south to Florida, also Wisconsin, North Dakota, Washington, Oregon, and California; West Indies, Mexico, Central America, and South America; Asia, Australia, and Pacific Islands. Native to Asia, Australia, and Pacific Islands. Spring–fall.

Trianthema L. 1753. HORSEPURSLANE

Annuals. Leaves opposite, simple, pinnate-veined, in unequal pairs, petiolate, stipulate. Flowers solitary or 2–3 in cymes, axillary, pedunculate, bracteate; sepals 5; petals and petaloid staminodes absent; stamens 5–10; ovary superior, 1- to 2-carpellate and -loculate. Fruit a circumscissile dehiscent capsule; seeds 1–12, arillate.

A genus of 20 species; North America, West Indies, Mexico, Central America, and South America; Africa, Asia, and Australia. [From the Greek *treis*, three, and *anthemon*, flower.]

Selected reference: Ferren (2003c).

Trianthema portulacastrum L. [*Portulaca* (Portulacaceae) and the Latin *astrum*, indicating an incomplete likeness to that genus.] DESERT HORSEPURSLANE.

Trianthema portulacastrum Linnaeus, Sp. Pl. 223. 1753. *Trianthema monogyna* Linnaeus, Mant. Pl. 69. 1767, nom. illegit.

Prostrate or decumbent, succulent annual herb; stem to 1 m long, the young ones with lines of trichomes decurrent below the petiolar sheaths. Leaves in unequal pairs, one much larger than the other, the blade suborbicular to obovate or subrhombic, 2–4 cm long, 1.5–3 cm wide, the apex rounded, truncate, or retuse, often short-apiculate, the base broadly cuneate, the margin entire, the upper and lower surfaces glabrous, the petiole subequaling the blade, dilated proximally; stipules winged, scarious, entire, those of a pair sheathing the nodes. Flowers 1–3 in the leaf axil, sessile; bracts 2, adnate to the hypanthium; hypanthium 1–2 cm long; sepal lobes ovate-lanceolate to lanceolate, ca. 2.5 mm long, concave, with a hornlike protuberance on the back just below the apex, the outer surface green, the inner surface pink-purple to whitish. Fruit short-cylindric or turbinate, somewhat curved, the apex truncate, with winged appendages, circumscissile at or a little below the middle; seeds reniform, ca. 2 mm long, reddish brown to black, papillate.

Open, often disturbed sites, usually coastal. Occasional; central and southern peninsula. New Jersey south to Florida, west to California; West Indies, Mexico, Central America, and South America; Africa, Asia, and Australia. Native to North America, tropical America, Africa, and Asia. All year.

PHYTOLACCACEAE R. Br. 1818. POKEWEED FAMILY

Herbs, shrubs, or vines. Leaves alternate, simple, pinnate-veined, petiolate, estipulate. Flowers in terminal or axillary spikes, panicles, or racemes, actinomorphic or zygomorphic, bisexual, bracteate, bracteolate; sepals 4–5, free; petals absent; stamens 4–20; ovary inferior or superior, 1- to 12-carpellate, the carpels completely or partly connate, the styles 1 or as many as the carpels. Fruit a 1- to many-seeded berry or an achene.

A family of 18 genera and about 135 species; nearly cosmopolitan.

The Phytolaccaceae is treated here in the broad sense, conforming to that of Nowicke (1968), Rogers (1985), and Nienaber and Thieret (2003), and with the exception of our placing *Gisekia* in the Gisekiaceae.

Agdestidaceae (Heimerl) Nakai, 1942; *Petiveriaceae* C. Agardh, 1824.

Selected references: Nienaber and Thieret (2003); Rogers (1985).

1. Vine; ovary inferior; fruit winged with enlarged, persistent sepals... **Agdestis**
1. Herb or shrub; ovary superior; fruit not winged.
    2. Inflorescence spicate; fruit an elongate achene, apically 2-lobed, each lobe tipped with 1–3 sharply reflexed spines; plant with a strong garlic- or skunk-like odor when fresh ....**Petiveria**
    2. Inflorescence racemose; fruit a subglobose or somewhat flattened berry; plant without a strong odor when fresh.
        3. Ovary 2- to 12-carpellate, the styles 2–many .... **Phytolacca**
        3. Ovary 1-carpellate, the style 1.
            4. Stamens 4; stigma capitate; fruit red or orange (rarely yellow) ....**Rivina**
            4. Stamens 8–16; stigma penicillate; fruit purple-black ....**Trichostigma**

Agdestis Moc. & Sessé ex DC. 1817.

Perennial vines. Leaves simple, pinnate-veined, petiolate, estipulate. Flowers in terminal or axillary simple or compound panicles, rarely solitary, bract 1, bracteoles 2; sepals 4(5), free; petals absent; stamens 15–20, unequal, the anthers cleft at both ends; ovary (3)4-loculate, but 1-loculate by abortion of the septa, the styles 4(5). Fruit an achene, crowned by the enlarged, persistent, winglike perianth segments, 1-seeded.

A genus of 1 species; North America, West Indies, Mexico, Central America, and South America. [From *Agdistis*, a mythical, hermaphroditic monstrosity, the genus being anomalous (a "monstrosity") in the Menispermaceae where it was first placed, being the only genus in that family with bisexual flowers.]

Agdestis clematidea Moc. & Sessé ex DC. [Resembling *Clematis* (Ranunculaceae).] ROCKROOT.

Agdestis clematidea Moçiño & Sessé y Lacasta ex de Candolle, Syst. Nat. 1: 543. 1817 ("1818").

Perennial subwoody vine, to 15 m or more; stem puberulent to glabrate. Leaves with the blade ovate to reniform, 2–9 cm long and wide, the apex obtuse to rounded, often mucronate, the base cordate, the margin entire, the upper and lower surfaces punctate with oxalate crystals, the petiole 2–3(5) cm long. Flowers usually axillary, sometimes terminal, in a lax panicle 4–15 cm long, sometimes bearing a single flower in the axil of a bract, or bearing a branch resembling the main axis, the peduncle 2–10 cm long; bract subulate to lanceolate, 1–2 mm long; bracteoles inconspicuous; pedicel 3–10 mm long; sepals elliptic to oblong or obovate, 4–6.5 mm long, white to greenish white, reticulate-veined; stamens slender, unequal, shorter than to slightly longer than the sepal segments, inserted irregularly, the filaments filiform; stigmas recurved. Fruit obconical, ca. 2.5 mm long and broad, 1-seeded by abortion, crowned by the enlarged, greenish, spreading, winglike, prominently veined sepals, these ca. 6 mm long, ca. 3 mm wide.

Hammocks and disturbed sites. Rare; central and southern peninsula. Escaped from cultivation. Florida and Texas; West Indies, Mexico, Central America, and South America. Native to Mexico and Central America. Spring–fall.

Petiveria L. 1753.

Perennial herbs or subshrubs. Leaves simple, pinnate-veined, petiolate, estipulate. Flowers in elongate terminal or axillary spiciform racemes, bract 1, bracteoles 2; sepals 4, free, the lobes subequal; stamens 4–6(8), inserted irregularly on a hypogynous disk at the ovary base, the filaments of varying lengths, the anthers linear, cleft at both ends, dorsifixed; ovary 1-carpellate and -loculate, the style absent, the stigma sessile, penicillate. Fruit an achene, the apex 2-lobed, armed with 3–6(13) reflexed spines on the lobes.

A genus of 1 species; North America, West Indies, Mexico, Central America, and South America. [Commemorates James Petiver (1658–1718), British botanist, entomologist, and apothecary.]

Petiveria alliacea L. [Resembling *Allium* (Alliaceae), in reference to the garlic-like odor.] GUINEA HEN WEED.

Petiveria alliacea Linnaeus, Sp. Pl. 342. 1753. *Petiveria foetida* Salisbury, Prodr. Stirp. Chap. Allerton 214. 1796, nom. illegit.

Erect or ascending perennial herb or subshrub, to 2 m; stem simple or sparsely branched, sometimes angled, puberulent to glabrate. Leaf with the blade elliptic to obovate, 3–20 cm long, 1.5–7 cm wide, the apex acuminate to acute, obtuse, or rounded, the base acute to cuneate, the margin entire, the upper and lower surfaces puberulent to glabrate, the petiole 0.4–1.5 cm long. Flowers 8–30 in a slender, spiciform raceme 8–40 cm long, often drooping, the peduncle 1–4 cm long, sessile or the pedicel to 1 mm long; bract triangular, ca. 1.5 mm long, the bracteoles lanceolate, to 1 mm long; sepals linear-lanceolate to linear-oblong, 3–5 mm long, subequal, white or greenish to pinkish, often pubescent basally; stamens ca. 3 mm long; ovary tomentose, with 4 reflexed bristles at the apex. Fruit elongate-cuneate, 8–12 mm long, striate, apically 2-lobed, each lobe tipped with 1–3 sharply reflexed spines 3–5 mm long.

Hammocks and disturbed sites. Occasional; peninsula. Florida and Texas; West Indies, Mexico, Central America, and South America. All year.

Phytolacca L. 1753. POKEWEED

Perennial herbs. Leaves simple, pinnate-veined, petiolate, estipulate. Flowers in terminal or axillary, leaf-opposed racemes, bracts 1, bracteoles 2; sepals 5; petals absent; stamens (9)10(12), inserted in a whorl at the base of the calyx on a hypogynous disk, the filaments free, the anthers 2-thecal, opening by longitudinal slits; the ovary superior, 6- to 12-carpellate and -loculate, the carpels united at least in the proximal half, the styles as many as the carpels, usually connivent. Fruit a berry; seeds several per locule, lenticular.

A genus of about 25 species; nearly cosmopolitan. [From the Greek *phyton*, plant, and the Latin *lacca*, varnish, in reference to the purple pigment of the berries.]

Selected reference: Caulkins and Wyatt (1990).

Phytolacca americana L. [Of America.] AMERICAN POKEWEED.

Phytolacca americana Linnaeus, Sp. Pl. 441. 1753. *Phytolacca decandra* Linnaeus, Sp. Pl., ed. 2. 631. 1762, nom. illegit.

Phytolacca rigida Small, Bull. New York Bot. Gard. 3: 422. 1905. *Phytolacca americana* Linnaeus var. *rigida* (Small) Caulkins & Wyatt, Bull. Torrey Bot. Club 117: 366. 1990. TYPE: FLORIDA: Miami-Dade Co.: Miami, 5–21 May 1904, *Small & Wilson 1893* (holotype: NY).

Erect perennial herb, to 3 m; stem glabrous. Leaves with the blade lanceolate to elliptic-lanceolate or ovate, 10–30 cm long, 3–18 cm wide, the apex acuminate, the base rounded to cuneate, sometimes oblique, the margin entire, the upper and lower surfaces glabrous, the petiole 1–6 cm long. Flowers several to many in an erect to drooping raceme, 2–30 cm long; peduncle 3–8 cm long; bract 1–2 mm long; bracteoles minute; sepals ovate to suborbicular, 2.5–3.3 mm long, white or greenish white to pinkish or purplish; carpels with the stigmas connivent at anthesis, separating and persisting in a ring around the apex of the fruit, the pedicel 3–12 mm long. Fruit oblate, 6–11 mm wide, purple-black; seeds ca. 3 mm long, black, shiny.

Hammock margins and disturbed sites. Common; nearly throughout. Quebec to Florida, west to Washington, Oregon, and California; Mexico; Europe, Africa, and Asia. Native to North America. Spring–fall, all year in southern counties.

Authors have disagreed on the status of plants with erect as opposed to drooping racemes and pedicels that are shorter than the berries. These have been treated as a distinct species (*P.*

rigida) or considered conspecific with *P. americana*. This problem was reviewed by Caulkins and Wyatt (1990), who concluded it best to treat these plants at the varietal level. Because characters used to separate the taxa are not always correlated and occur throughout the range of the species, we elect to follow Nowicke (1968) and others, who treat *P. americana* as a single variable taxon.

Rivina L. 1753.

Perennial herbs or subshrubs. Leaves simple, pinnate-veined, petiolate, estipulate. Flowers in terminal or axillary racemes, bract 1, bracteoles 2; sepals 4, free, subequal; stamens 4, the anther dorsifixed; ovary superior, 1-carpellate and -loculate, the style 1, the stigma capitate. Fruit a berry; seed 1.

A genus of 1 species; North America, West Indies, Mexico, Central America, South America, Asia, Australia, and Pacific Islands. [Commemorates August Quirinus Rivinus (Bachmann) (1652–1723), professor of botany, medicine, and chemistry at Leipzig.]

Rivina humilis L. [Low growing.] ROUGEPLANT.

> *Rivina humilis* Linnaeus, Sp. Pl. 121. 1753. *Solanoides pubescens* Moench, Methodus 307. 1794, nom. illegit. *Rivina pallida* Salisbury, Prodr. Stirp. Chap. Allerton 67. 1796, nom. illegit. *Rivina laevis* Linnaeus var. *pubescens* Grisebach, Fl. Brit. W.I. 59. 1864. *Tithonia humilis* (Linnaeus) Kuntze, Revis. Gen. Pl. 2: 552. 1891. *Rivina laevis* Linnaeus forma *humilis* (Linnaeus) Voss, Vilm. Blumengärtn., ed. 3. 1: 876. 1895.
>
> *Rivina humilis* Linnaeus var. *glabra* Linnaeus, Sp. Pl. 122. 1753. *Rivina laevis* Linnaeus, Mant. Pl. 41. 1767. *Piercea glabra* (Linnaeus) Miller, Gard. Dict., ed. 8. 1768, nom. illegit. *Solanoides laevis* (Linnaeus) Moench, Methodus 307. 1794. *Rivina gracilis* Salisbury, Prodr. Stirp. Chap. Allerton 67. 1796, nom. illegit. *Tithonia humilis* (L.) Kuntze var. *glabra* (Linnaeus) Kuntze, Revis. Gen. Pl. 2: 552. 1891. *Rivina humilis* Linnaeus var. *laevis* (Linnaeus) Millspaugh, Publ. Field Columb. Mus., Bot. Ser. 2: 41. 1900, nom. illegit.
>
> *Rivina acuminata* Rafinesque, New Fl. 4: 13. 1838 ("1836"); non Kunth, 1817. *Piercea acuminata* Rafinesque, New Fl. 4: 13. 1838 ("1836"), nom. alt. TYPE: "Florida to Arkanzas."
>
> *Rivina obliquata* Rafinesque, New Fl. 4: 13. 1838 ("1836"). *Piercea obliquata* Rafinesque, New Fl. 4: 13. 1838 ("1836"), nom. alt. TYPE: FLORIDA.

Erect, straggling, or vinelike perennial herb or subshrub, to 1 m; stem glabrate to pubescent. Leaves with the blade lanceolate, elliptic, or oblong to deltoid or ovate, 2.5–15 cm long, 1.7–6 cm wide, the apex acuminate or acute to obtuse or emarginate, the base cuneate or rounded to truncate or cordate, the margin entire, the upper and lower surfaces glabrous or finely pubescent, the petiole 1–8 cm long. Flowers several to many in a terminal or axillary raceme 1–15 cm long, the peduncle 1–5 cm long, the pedicel 2–8 mm long; bract 1–2 mm long, lanceolate to subulate; bracteoles minute; sepals elliptic or oblong to oblanceolate or obovate, 1.5–3.5 mm long, white or green to pink or purplish; stamen filaments 1–2 mm long, the anthers ca. 1 mm long. Fruit subglobose, 3.5–4.5 mm long, red to orange; seed lenticular, 2–3 mm long, covered by a thin "pubescent" membrane derived from the inner layers of the pericarp.

Hammocks. Frequent; peninsula. Florida west to Arizona; West Indies, Mexico, Central

America, and South America; Africa, Asia, and Pacific Islands. Native to North America and tropical America. All year.

Trichostigma A. Rich. 1845.

Woody vines or shrubs. Leaves simple, pinnate-veined, petiolate, estipulate. Flowers in terminal or axillary racemes, bract 1, bracteoles 2; sepals 4, free, subequal, persistent and reflexed in fruit; stamens 8–13, free, the anthers dorsifixed; ovary superior, the style 1, the stigma sessile or subsessile, penicillate. Fruit a 1-seeded berry.

A genus of 3 species; North America, West Indies, Mexico, Central America, and South America. [From the Greek *trichos*, hair, and *stigma*, in reference to the penicillate stigma.]

Trichostigma octandrum (L.) H. Walter [With eight anthers.] HOOPVINE.

Rivina octandra Linnaeus, Cent. Pl. 2: 9. 1756. *Rivina americana* Rafinesque, Fl. Tellur. 3: 56. 1837 ("1836"), nom. illegit. *Trichostigma rivinoides* A. Richard, in Sagra, Hist. Fis. Cuba, Bot. 10: 306. 1845, nom. illegit. *Villamilla octandra* (Linnaeus) Hooker f., in Bentham & Hooker f., Gen. Pl. 3: 81. 1880. *Trichostigma octandrum* (Linnaeus) H. Walter, in Engler, Pflanzenr. 4(Heft 39): 109. 1909.

Suberect shrub or perennial vine, to 10 m; stem glabrous or sparsely pubescent. Leaves with the blade elliptic or elliptic-lanceolate to oblong or ovate, 4–15 cm long, 1.5–6 cm wide, the apex acuminate or acute to obtuse, the base rounded to cuneate, the margin entire, the upper and lower surfaces glabrous to sparsely pubescent on the veins below, the petiole 0.5–3(5) cm long. Flowers many in a raceme 5–10 cm long, the the pedicel 3–10 mm long, the peduncle 1.3–3 cm long; bract lanceolate to subulate, 2–3 mm long; bracteoles minute; sepals ovate to obovate, 3–6 mm long, white or greenish white, pink or purplish in age; stamen filaments 2–3 mm long, the anthers 1.5–2 mm long; stigmas penicillate. Fruit subglobose, 4–6 mm long, black; seed lenticular, reddish brown.

Tropical hammocks and shell middens. Rare; southern peninsula. Florida; West Indies, Mexico, Central America, and South America. All year.

NYCTAGINACEAE Juss. 1789. FOUR-O'CLOCK FAMILY

Annual or perennial herbs, shrubs, or trees, sometimes scandent. Leaves opposite or subopposite, simple, pinnate-veined, petiolate or epetiolate, estipulate. Flowers in terminal or axillary paniculiform, corymbiform, or thyrsiform cymes, rarely solitary, actinomorphic, bisexual or unisexual (plants dioecious), bracteate, the bracts free or connate and involucrate around 1 or more flowers, when containing 1 flower, then simulating a calyx; sepals 5, connate, often petaloid; corolla absent; stamens 2–18, the filaments basally connate, the anthers 2-loculate, dehiscing by longitudinal slits; ovary superior, 1-carpellate and -loculate, the ovule 1, the style 1, the stigma capitate or penicillate. Fruit an achene or utricle enclosed within the persistent, fleshy, leathery, or corky calyx tube base (anthocarp); seed 1.

A family of about 30 genera and about 300 species; nearly cosmopolitan.

Generic classification of the Nyctaginaceae is unresolved. A molecular phylogeny of the family was provided by Douglas and Manos (2007) which showed several large genera to be nonmonophyletic, including *Boerhavia* and *Guapira* of Florida. Further work is needed to provide a stable taxonomy for the family.

Allioniaceae Horan., 1834; *Pisoniaceae* J. Agardh, 1858.

Selected reference: Bogle (1974).

1. Tree, shrub, or woody climber; flowers unisexual (plants dioecious); stigma penicillate.
    2. Fruit oblong-clavate, dry, with rows of glands on the 5 angles.. **Pisonia**
    2. Fruit ellipsoid or ovoid, fleshy, without glands .... **Guapira**
1. Herb (sometimes woody at the base); flowers bisexual; stigma capitate.
    3. Bracts small and inconspicuous.. **Boerhavia**
    3. Bracts evident, involucrate.
        4. Plant erect; involucral bracts 5, connate.. **Mirabilis**
        4. Plant prostrate, rooting at the nodes; involucral bracts 3–4, free .... **Okenia**

Boerhavia L. 1753. SPIDERLING

Annual or perennial herbs. Leaves opposite, simple, in unequal pairs, pinnate-veined, petio-late, estipulate. Flowers in paniculiform cymes, bisexual, the bracts free, 1–3 below each flower, minute; calyx 5-lobed, basally tubular, petaloid; stamens 2–3, the filaments basally connate into a tube; ovary short stipitate, the stigma capitate. Fruit leathery, 5-ribbed; seed 1.

A genus of about 40 species; nearly cosmopolitan. [Commemorates Hermann Boerhaave (1668–1738), Dutch physician and professor of botany at Leiden.]

Selected reference: Spellenberg (2003a).

1. Fruit truncate at the apex; calyx glabrous.. **B. erecta**
1. Fruit rounded at the apex; calyx with short-stalked, glandular-capitate trichomes.
    2. Leaves usually distributed throughout the plant; peduncles pubescent; flowers usually more than 5 per cluster; stamens exserted from the perianth ....**B. coccinea**
    2. Leaves usually concentrated in the proximal ½ of the plant; peduncles glabrate; flowers 2–5 per cluster; stamens included or only barely exserted from the perianth................................**B. diffusa**

Boerhavia coccinea Mill. [Deep red, in reference to the flower.] SCARLET SPIDERLING.

Boerhavia coccinea Miller, Gard. Dict., ed. 8. Boerhavia no. 4. 1768. *Boerhavia hirsuta* Jacquin, Hort Bot. Vindob. 3. 1771, nom. illegit. *Boerhavia diffusa* Linnaeus var. *hirsuta* Kuntze, Revis. Gen. Pl. 2: 533. 1891.

Boerhavia viscosa Lagasca y Segura & Rodríguez, Anales Ci. Nat. 4: 256. 1801. *Boerhavia diffusa* Linnaeus var. *viscosa* (Lagasca y Segura & Rodríguez) Heimerl, Beitr. Syst. Nyctag. 27. 1897. *Boerhavia repens* Linnaeus subsp. *viscosa* (Lagasca y Segura & Rodríguez) Maire, Mém. Soc. Hist. Nat. Afrique N. 3: 88. 1933. *Boerhavia coccinea* Miller var. *viscosa* (Lagasca y Segura & Rodríguez) Moscoso, Cat. Fl. Doming. 180. 1943.

Prostrate to decumbent perennial herb; stem to 1.5 m long, usually branched throughout, mi-nutely pubescent, often glandular, sometimes spreading villous or hirsute proximally, sometimes glabrate or glabrous distally. Leaves usually distributed throughout the plant and into much of the inflorescence, the blade broadly lanceolate to ovate or suborbicular, the apex acute to obtuse or rounded, the base truncate or cordate to rounded, the margin sinuate, ciliolate to glandular-ciliolate, the upper and lower surfaces glabrous or sparsely puberulent to glandular-pubescent, the lower surface paler in color. Flowers in a paniculiform cyme, the branches terminating in pedunculate, subumbellate clusters of usually 5 or more; calyx tube ca. 1 mm long, the limb campanulate, ca. 0.5 mm long, purple or maroon, rarely white; stamens exserted. Fruit narrowly obovate to clavate, 2.5–4 mm long, ca. 1 mm wide, the apex rounded to round-conic, with short-stalked glandular-capitate trichomes on the ribs and in the sulci.

Dry, disturbed sites. Occasional; Manatee and Lee Counties, southern peninsula. Nearly cosmopolitan. Native to the lower central and western North America, tropical America, Africa, Asia, and Australia.

Boerhavia diffusa L. [Spreading, in reference to habit.] RED SPIDERLING; WINE-FLOWER.

> *Boerhavia diffusa* Linnaeus, Sp. Pl. 3. 1753. *Boerhavia diffusa* Linnaeus var. *normalis* Kuntze, Revis. Gen. Pl. 2: 533. 1891, nom. inadmiss. *Boerhavia diffusa* Linnaeus var. *eudiffusa* Heimerl, Beitr. Syst. Nytag. 26. 1897, nom. inadmiss.
> *Boerhavia paniculata* Richard, Actes Soc. Hist. Nat. Paris 1: 105. 1792. *Boerhavia decumbens* Vahl, Enum. Pl. 1: 284. 1804, nom. illegit. *Boerhavia laxa* Persoon, Syn. Pl. 1: 36. 1805, nom. illegit. *Boerhavia coccinea* Miller var. *paniculata* (Richard) Moscoso, Cat. Fl. Doming. 180. 1943.
> *Boerhavia polymorpha* Richard, Actes Soc. Hist. Nat. Paris 1: 185. 1792.

Erect or decumbent perennial herb, to 1 m; stem usually branched throughout, glabrous or minutely pubescent proximally, glabrous or sparsely pubescent distally. Leaves usually concentrated in the proximal ½ of the plant, the blade suborbicular to rhombic-ovate or broadly lanceolate, 2–6 cm long, 1.5–5 cm wide, the apex obtuse or rounded, the base cuneate to cordate, the margin sinuate, ciliolate to glandular-ciliolate, the upper and lower surfaces puberulent to glabrate, the lower surface paler in color, the petiole 1–3 cm long. Flowers in a paniculiform cyme, the branches terminating in pedunculate, compact, subumbellate clusters of 2–5; calyx tube ca. 1 mm long, ca. 0.5 mm wide, the limb campanulate, ca. 0.5 mm long, red to purple, rarely white; stamens 2–3, included or only barely exserted. Fruit obpyramidal, 3.5–4.5 mm long, ca. 1 mm wide, the apex obtuse to obscurely beaked, with short-stalked, glandular-capitate trichomes on the ribs.

Dry disturbed sites. Frequent; nearly throughout. North Carolina south to Florida; West Indies, Mexico, Central America, and South America; Africa, Asia, Australia, and Pacific Islands. Spring–summer.

Boerhavia erecta L. [Erect, in reference to the habit.] ERECT SPIDERLING.

> *Boerhavia erecta* Linnaeus, Sp. Pl. 3. 1753. *Boerhavia elongata* Salisbury, Prodr. Stirp. Chap. Allerton 56. 1796, nom. illegit.
> *Boerhavia atomaria* Rafinesque, Autik. Bot. 40. 1840. TYPE: FLORIDA.

Erect or ascending annual herb, to 1 m; stem glabrous or puberulous. Leaves with the blade ovate-rhombic, 2–9 cm long, 1–4 cm wide, the apex rounded to acute, the base cuneate to subtruncate, the margin undulate or sinuate, glabrous or glabrate, the upper and lower surfaces glabrous, the lower glandular-punctate, the petiole to 3 cm long. Flowers in a pedunculiform cyme, the branches terminating in pedunculate, subumbellate clusters of 2 or 3; calyx tube 1–1.5 mm long, glabrous, the calyx limb campanulate, 0.5–1 mm long, white or pink; stamens 2–3, slightly exserted. Fruit obpyramidal, 3–4 mm long, 1–1.5 mm wide, the apex truncate, glabrous.

Disturbed sites. Occasional; peninsula, central and western panhandle. Maryland, North Carolina south to Florida, west to Arizona; West Indies, Mexico, Central America, and South America. All year.

Guapira Aubl. 1775.

Shrubs or trees. Leaves opposite or subopposite, simple, pinnate-veined, petiolate, estipulate. Flowers in terminal, corymbiform-thyrsoid cymes, unisexual (plant dioecious), the bracts 3 below each flower, free. Staminate flowers campanulate; stamens 5–10, exserted, the filaments connate into a short tube adnate to the pistillode stipe. Carpellate flowers tubular to tubular-campanulate; calyx limb erect or ascending; staminodes about as long as the ovary and with enlarged sterile anthers, the filaments connate into a tube and adnate to the gynoecium stipe; stigma penicillate. Fruit fleshy, 10-ribbed; seed 1.

A genus of about 70 species; North America, West Indies, Mexico, Central America, and South America. [Vernacular name in French Guiana for the type species, derivation unknown.]

Guapira and the genus *Neea*, with about 40 species, form a clade and neither genus appears to be monophyletic (Douglas and Manos, 2007). Further work is needed to clarify the taxonomy.

Torrubia Vell., 1825.

Selected reference: Clement (2003).

1. Leaves chartaceous; petiole slender; buds and inflorescences sparsely rusty-pubescent ...... **G. discolor**
1. Leaves coriaceous; petiole stout; buds and inflorescences densely rusty-pubescent .... **G. obtusata**

Guapira discolor (Spreng.) Little [Two-colored, in reference to the upper and lower leaf surfaces being differently colored.] BEEFTREE; BLOLLY.

Pisonia discolor Sprengel, Syst. Veg. 2: 168. 1825. *Torrubia discolor* (Sprengel) Britton, Bull. Torrey Bot. Club 31: 613. 1904. *Guapira discolor* (Sprengel) Little, Phytologia 17: 368. 1968.

Pisonia discolor Sprengel var. *longifolia* Heimerl, in Urban, Bot. Jahrb. Syst. 21: 627. 1896. *Torrubia longifolia* (Heimerl) Britton, Bull. Torrey Bot. Club 31: 614. 1904. *Pisonia longifolia* (Heimerl) Sargent, Man. Trees N. Amer. 314, f. 251. 1905. *Guapira longifolia* (Heimerl) Little, Phytologia 17: 367. 1968.

Pisonia floridana Britton ex Small, Fl. S.E. U.S. 411, 1330. 1903. *Torrubia floridana* (Britton ex Small) Britton, Bull. Torrey Bot. Club 31: 615. 1904. *Guapira floridana* (Britton ex Small) Lundell, Wrightia 4: 80. 1968. *Pisonia discolor* Sprengel var. *floridana* (Britton ex Small) D. B. Ward, Novon 14: 370. 2004. TYPE: FLORIDA: Monroe Co.: Rock Key, 12 mi. W of Key West, s.d., *Blodgett s.n.* (holotype: NY).

Torrubia bracei Britton, Bull. Torrey Bot. Club 31: 614. 1904. *Guapira bracei* (Britton) Little, Phytologia 17: 367. 1968.

Torrubia globosa Small, Man. S.E. Fl. 490, 1504. 1933. *Guapira globosa* (Small) Little, Phytologia 17:

367. 1968. TYPE: FLORIDA: Miami-Dade Co.: Miami Beach, 22 Jul 1924, *Small et al. 11539* (holotype: NY).

Shrub or tree, to 6 m; stem glabrous. Leaves with the blade narrowly oblong to elliptic or oblanceolate, rarely suborbicular, 2–6(11) cm long, 1–3(4) cm wide, chartaceous, the apex rounded to obtuse, the base cuneate to rounded, the margin entire or undulate, the upper and lower surfaces glabrate, the lower paler in color, the petiole 1–1.5 cm long. Flowers in a terminal, corymbose-thyrsiform cyme 2–3(7) cm long, sparsely rusty-pubescent. Staminate flowers with the calyx widely trumpet-shaped, 3–4 mm long, light green to yellow. Carpellate flowers with the calyx 2–3 mm long, green. Fruit ellipsoid to globose-obovoid or oblong-obovoid, 4–8 mm long, ribbed, red.

Tropical hammocks and pinelands. Occasional; Brevard County southward along the east coast, southern peninsula. Florida; West Indies. Spring–fall.

Guapira obtusata (Jacq.) Little [Obtuse, in reference to the leaf apex.] BROADLEAF BLOLLY.

   Pisonia obtusata Jacquin, Pl. Hort. Schoenbr. 3: 35, pl. 314. 1798. *Torrubia obtusata* (Jacquin) Britton, Bull. Torrey Bot. Club 31: 612. 1904. *Guapira obtusata* (Jacquin) Little, Phytologia 17: 368. 1968.

Shrub or tree, to 8 m; stem glabrate. Leaves with the blade suborbicular to oblong-elliptic, 2–6(10) cm long, 2–4(6) cm wide, coriaceous, the apex rounded to emarginate, the base cuneate to rounded, the margin entire or undulate, the upper and lower surfaces glabrate, the lower paler in color, the petiole 0.5–1 cm long. Flowers in a terminal, corymbose-thyrsiform cyme 4–6 cm long, densely rusty pubescent. Staminate flowers with the calyx funnel-shaped, 4–5 mm long, light green to yellow. Carpellate flowers with the calyx 3–4 mm long, green. Fruit oblong, 8–10 mm long, ribbed, red.

Tropical hammocks. Rare; Monroe County keys. Florida; West Indies. Spring–fall.

Mirabilis L. 1753. FOUR-O'CLOCK

Perennial herbs. Leaves opposite, simple, pinnate-veined, petiolate or subsessile, estipulate. Flowers in compact cymes or solitary, terminal and axillary in the distal leaves, bisexual, the bracts 5, connate and involucrate, subtending 1–3 flowers; calyx tubular, petaloid, 5-lobed; stamens 3–6, basally connate into a tube; ovary superior, 1-carpellate and -loculate, the style 1, the stigma capitate. Fruit leathery, 5-ribbed; seed 1.

A genus of about 60 species; nearly cosmopolitan. [*Mirab*, wonderful, in reference to the large colorful flowers.]

Selected references: Le Duc (1995); Spellenberg (2003c).

Mirabilis jalapa L. [From the Spanish *jalap*, a resin obtained from the tubers, originally named after the town Jalapa or Xalapa in Veracruz, Mexico.] FOUR-O'CLOCK; MARVEL-OF-PERU.

Mirabilis jalapa Linnaeus, Sp. Pl. 177. 1753. *Jalapa congesta* Moench, Methodus 508. 1794. *Nyctago versicolor* Salisbury, Prodr. Stirp. Chap. Allerton 57. 1796, nom. illegit. *Nyctago hortensis* Dumont de Courset, Bot. Cult. 1: 654. 1802, nom. illegit. *Nyctago jalapa* (Linnaeus) de Candolle, Fl. Franc. 3: 426. 1805. *Mirabilis pedunculata* Stokes, Bot. Mat. Med. 1: 311. 1812, nom. illegit. *Mirabilis jalapa* Linnaeus forma *eujalapa* Heimerl, Bot. Jahrb. Syst. 21: 617. 1896, nom. inadmiss. *Admirabilis peruana* Nieuwland, Amer. Midl. Naturalist 3: 280. 1914, nom. illegit.

Erect perennial herb, to 1 m; stem much-branched; root tuberous. Leaves with the blade ovate to ovate-deltoid, 4–14 cm long, 2–8.5 cm wide, the apex acuminate, the base truncate to cordate, the margin entire, the upper and lower surfaces glabrous or rarely pubescent, the petiole of the lowermost leaves about half as long as the blade, that of the uppermost leaves subsessile. Flowers solitary or in a glomerate cyme with reduced leaves, the peduncle to 5 cm long, the involucre campanulate, 7–15 mm long, pubescent to short villous or glabrous, the lobes ovate-lanceolate, two times as long as the tube, the apex acute, usually ciliolate, bristle-tipped; calyx trumpet-shaped, 4–6 cm long, the limb 2–3.5 mm long, shallowly 5-lobed, the lobes rounded, various colored, the outer surface glabrous or sparsely villous; stamens 5, equaling the calyx or slightly exserted. Fruit ovoid to obovoid or oval, 9–10 mm long, dark brown to black, 5-angled, wrinkled-tuberculate, verrucose, or rugose, glabrous or puberulent.

Disturbed sites. Occasional; nearly throughout. Escaped from cultivation. Vermont south to Florida, west to Washington, Oregon, and California; West Indies, Mexico, Central America, and South America; Europe, Africa, Asia, and Australia. Native to Mexico. All year.

Okenia Schltdl. & Cham., 1830.

Annual herbs. Leaves opposite, simple, of unequal pairs, pinnate-veined, petiolate, estipulate. Flowers axillary, solitary at each node, bisexual, chasmogamous or rarely cleistogamous, the bracts involucrate, 3- or 4-lobed, subtending a single flower; calyx tubular, 5-lobed; stamens 14–18, connate basally into a tube; stigma capitate. Fruit corky, 10-ribbed, maturing hypogeously; seed 1.

A genus of 2 species; Florida, Mexico, and Central America. [Commemorates Lorenz Oken (1779–1851), German naturalist, physician, philosopher, and philologist.]

Okenia is closely related to *Boerhavia* and possibly derived within it (Douglas and Manos, 2007). Further work is needed.

Selected reference: Spellenberg (2003b).

Okenia hypogaea Schltdl. & Cham. [From the Greek *hypo*, under, and *gaea*, earth, in reference to the underground fruits.] BURROWING FOUR-O'CLOCK; BEACH PEANUT.

Okenia hypogaea Schlechtendahl & Chamisso, Linnaea 5: 92. 1830.

Prostrate annual herb; stem diffusely branching, to 2 m, radiating from a short caudex and forming a mat, densely and finely glandular-pubescent. Leaves markedly unequal, 1 large and 1 small at each node and alternating on the stem, the small one often not as large as the petiole of the larger one, the blade ovate-deltoid, ovate, or elliptic-ovate, 1–5 cm long, 0.3–4.5 cm wide, the apex acute to rounded, the base rounded, truncate, or subcordate, this sometimes unequal, the margin entire or shallowly sinuate, the upper and lower surfaces densely glandular-pubescent. Flowers on a long peduncle, ascending at anthesis, later deflected and pushing underground, the involucre of 3–4 small, narrow, subulate bracts; calyx tube funnel-shaped, the outer surface densely pubescent, constricted above the ovary, magenta, the early flowers 2.5–3 cm in diameter, the later ones smaller; stamens unequal, the filaments basally connate into a short tube, magenta above, white below. Fruit hypogeous, maturing at a depth of 3 dm below the soil surface, ellipsoid, 9–13 mm long, corky, transversely plicate between the ribs, dark brown to whitish.

Dunes. Rare; St. Lucie County southward along the east coast. Florida; Mexico and Central America. All year.

Okenia hypogaea is listed as endangered in Florida (Florida Administrative Code, Chapter 5B-40).

Pisonia L. 1753. CATCHBIRDTREE

Shrub or small tree, often scandent and armed with stout axillary spines. Leaves opposite, simple, of unequal pairs, pinnate-veined, petiolate, estipulate. Flowers usually terminal on modified shoots, in umbelliform or densely corymbiform-thyrsoid cymes, unisexual (plant dioecious), the bracts 2 or 3 in a tight spiral surrounding a single flower. Staminate flowers with the calyx campanulate; stamens usually 5–8, unequal, the filaments connate into a short tube and adnate to the pistillode stipe. Carpellate flowers with the calyx tubular, the staminodes reduced to a low, occasionally glandular-dentate disk adnate to the pistillode stipe; stigma peltate. Fruit leathery, 5-ribbed; seed 1.

A genus of 35–75 species; North America, West Indies, Mexico, Central America, South America, Africa, Asia, Australia, and Pacific Islands. [Commemorates William Piso (Willem Pies) (1611–1678), Dutch physician and naturalist who traveled in Brazil.]

Selected reference: Clement and Spellenberg (2003).

1. Shrubby vine; stem armed with curved spines; fruit with rows of stalked glands along the entire length..**P. aculeata**
1. Tree or shrub; stem unarmed; fruit with rows of stalked glands only near the apex...........**P. rotundata**

Pisonia aculeata L. [Armed with prickles.] DEVIL'S CLAWS; PULLBACK.

Pisonia aculeata Linnaeus, Sp. Pl. 1026. 1753.

Straggly shrubby vine, to 6 m; usually armed with short, stout, recurved spines, the branchlets densely puberulent or short-villous. Leaves with the blade elliptic-oval to ovate or suborbicular, subcoriaceous, 2–11 cm long, 1–6 cm wide, the apex acute to acuminate, the base cuneate to rounded, the margin entire, the upper surface glabrous or puberulent, the lower surface glabrous or short-villous, the petiole 1–3 cm long. Flowers in a loosely or densely corymbiform-thyrsoid or umbelliform cyme 2–6 cm wide, the peduncle 1–5 cm long, the pedicel short, viscid-puberulent. Staminate flowers with the calyx broadly campanulate, 2–4 mm long, densely puberulent to tomentulose, green or greenish yellow; stamens usually 6, twice as long as the calyx. Carpellate flowers with the calyx tubular, 2–3 mm long, green or greenish yellow; stigma and style exserted. Fruit ellipsoid-clavate, 7–12(17) mm long, 2–4 mm wide, 5-angled, with stipitate glands in 1 or 2 series along the ribs, green, the sides puberulent.

Tropical hammocks. Occasional; central and southern peninsula. Florida and Texas; Mexico, West Indies, Central America, and South America; Africa, Asia, Australia, and Pacific Islands. Native to North America and tropical America. Spring–summer.

Pisonia rotundata Griseb. [Rounded, in reference to the leaf shape.] SMOOTH DEVIL'S CLAWS; COCKSPUR.

> *Pisonia rotundata* Grisebach, Cat. Pl. Cub. 283. 1866. *Pisonia subcordata* Swartz var. *rotundata* (Grisebach) Heimerl, Bot. Jahrb. Syst. 21: 630. 1896. *Torrubia rotundata* (Grisebach) Sudworth, Check List For. Trees U.S., Revis. 119. 1927. *Guapira rotundata* (Grisebach) Lundell, Wrightia 4: 83. 1968.

Shrub or small tree, 5 m; branchlets glabrous or finely brown-pubescent. Leaves with the blades ovate-oval to oblong-elliptic, coriaceous, 2–6(9) cm long, 2–5 cm wide, the apex rounded, sometimes retuse or apiculate, the base obtuse to broadly cuneate, the margin entire, the upper surface glabrous or glabrate, the lower surface finely brown-pubescent, at least on the veins, glabrescent in age, the petiole to ca. 1.5 cm long. Flowers in an umbelliform or corymbiform cyme 2–6 cm wide, densely puberulent to glabrate. Staminate flowers with the calyx tubular-campanulate, 2–3 mm long, green or white, the limb recurved in anthesis; stamens 8–10, at least twice as long as calyx. Carpellate flowers with the calyx campanulate, 3–4 mm long, green or white. Fruit clavate, 5–7 mm long, with 5 rows of uniseriate, short-stalked glands above the middle.

Tropical hammocks and pinelands. Rare; Monroe County keys. Florida; West Indies. Spring–summer.

Pisonia rotundata is listed as endangered in Florida (Florida Administrative Code, Chapter 5B-40).

EXCLUDED TAXON

> *Pisonia macranthocarpa* (Donnell Smith) Donnell Smith—Reported by Bogle (1974, as *P. aculeata* var. *macranthocarpa* Donnell Smith) from the Florida Keys based on a single 1881 A. H. Curtiss collection as possibly referable to this species. However, because of the paucity of the material and the similarity to *A. aculeata*, it is probably best referred to *P. aculeata*.

MOLLUGINACEAE Raf. 1837. CARPETWEED FAMILY

Herbs. Leaves whorled, simple, pinnate-veined, petiolate, estipulate. Flowers in axillary, umbellate clusters, actinomorphic, bisexual, ebracteate; sepals 5, free; petals absent; stamens 3, basally connate, the anthers introrse, versatile, 2-thecous, 4-locular; ovary superior, 3-carpellate and -loculate, the styles 3, free. Fruit a many-seeded capsule, loculicidally dehiscent.

A family of about 9 genera and about 90 species; nearly cosmopolitan.

The Molluginaceae is sometimes included in the Aizoaceae or Phytolaccaceae, from which it differs in having anthocyanin pigments rather than betalains. The anomalous genus *Gisekia*, sometimes included here, is referred to the Gisekiaceae.

Selected references: Boetsch (2002); Vincent (2003b).

Mollugo L. 1753. CARPETWEED

Annual herbs. Leaves whorled, simple, pinnate-veined, petiolate, estipulate. Flowers in axillary, umbellate clusters, pedicellate; sepals 5, free; petals absent; stamens basally connate into a shallow ring, the anthers longitudinally dehiscent; ovary superior, 3-carpellate and -loculate, styles 3, the stigmas linear. Fruit a capsule, longitudinally dehiscent, the partitions separating from the persistent central axis; seeds numerous, somewhat compressed, reniform.

A genus of about 35 species; nearly cosmopolitan. [*Mollis*, pliant or delicate.]

Mollugo verticillata L. [Whorled, in reference to the leaves.] INDIAN CHICKWEED; GREEN CARPETWEED.

Mollugo verticillata Linnaeus, Sp. Pl. 89. 1753. *Pharnaceum verticillatum* (Linnaeus) Sprengel, Syst. Veg. 1: 949. 1824 ("1825"). *Mollugo verticillata* Linnaeus var. *vulgaris* Cambessedes, in A. Saint-Hilaire et al., Fl. Bras. Merid. 2: 170. 1830, nom. inadmiss. *Mollugo verticillata* Linnaeus var. *latifolia* Fenzl, Ann. Wiener Mus. Naturgesch. 1: 376. 1836, nom. inadmiss.

Prostrate or spreading annual herb; stem 7–25 cm long, dichotomously branched, radiating from a short taproot, slender, swollen at the nodes, glabrous. Leaves forming a basal rosette, in whorls of 3–8 on the stem, unequal, the blade narrowly to broadly oblanceolate, 1–3 cm long, 0.5–1 cm wide, the apex acute to obtuse, the base cuneate, tapering to a short petiole, the margin entire, the upper and lower surfaces glabrous, the petiole 0.5–4 mm long. Flowers 2–5 in an axillary umbellate cluster, the pedicel 5–15 mm long; sepals oblong to elliptic, 2–3 mm long, ca. 1 mm wide, the outer surface pale green, the inner surface white, the margin white-hyaline. Fruit ovoid to ellipsoid, 2–3 mm long, slightly longer than the persistent sepals; seeds reniform, ca. 0.5 mm long, dark reddish brown, longitudinally 3- to 7-ridged or smooth, lustrous.

Disturbed sites. Frequent; nearly throughout. Nearly throughout North America; West Indies, Mexico, Central America, and South America; Europe, Africa, Asia, and Australia. Native to North America and tropical America. Spring–fall.

BASELLACEAE Moq. 1840. MADEIRAVINE FAMILY

Vines. Leaves alternate, simple, pinnate-veined, petiolate or epetiolate, estipulate. Flowers in axillary or terminal racemes, actinomorphic, bisexual or functionally staminate, bracteate, bracteolate; sepals 5, basally connate; stamens 5, episepalous, the anthers 4-celled, dorsifixed, dehiscing lengthwise; nectariferous disk present; ovary superior, 3-carpellate, 1-loculate, the styles 3. Fruit a utricle; seed 1.

A family of 4 genera and about 25 species; nearly cosmopolitan.

Some authors (e.g., Vincent, 2003a) consider the bracteoles to be sepals.

Selected references: Bogle (1969); Sperling (1987); Vincent (2003a).

1. Flowers sessile or subsessile; sepals fleshy, hardly opening at anthesis .... **Basella**
1. Flowers pedicellate; sepals membranous, spreading-rotate at anthesis .... **Anredera**

Anredera Juss. 1789. MADEIRAVINE

Perennial, herbaceous vines. Leaves simple, pinnate-veined, petiolate, estipulate. Flowers in axillary or terminal racemes or a panicle of racemes, pedicellate, bisexual or functionally staminate, the bracts 3, 1 at the base of the pedicel and 2 subtending the flower, the 2 bracteoles above these larger than the paired bracts below them and sepaloid; sepals petaloid; stamens basally connate and dilated; styles entire or bifid. Fruit enclosed by the persistent calyx; seed 1.

A genus of 12 species; nearly cosmopolitan. [Commemorates Anreder, about whom nothing else is known.]

Boussingaultia Kunth, 1825.

1. Pair of bracts subtending the flower free, caducous; sepaloid bracteoles winged in fruit; flowers drying white .... **A. vesicaria**
1. Pair of bracts subtending the flower basally connate and forming a persistent cuplet at the pedicel apex; sepaloid bracteoles not winged in fruit; flowers drying brown .... **A. cordifolia**

Anredera cordifolia (Ten.) Steenis [Heartleaved.] HEARTLEAF MADEIRAVINE.

Boussingaultia cordifolia Tenore, Ann. Sci. Nat. Bot., ser. 3. 19: 355. 1853. *Anredera cordifolia* (Tenore) Steenis, Fl. Males. 5: 303. 1957.

Herbaceous perennial vine, to 6 m; stem glabrous, producing small, single or clustered axillary tubers. Leaves with the blade ovate to suborbicular, 2–10 cm long, 1–7 cm wide, the apex acute to obtuse, the base subcordate, the margin entire, the upper and lower surfaces glabrous, the petiole 6–12 mm long. Flowers in a simple or 2- to 4-branched raceme, the pedicel 1.5–2 mm long, the lower subtending bract lanceolate, 1–2 mm long, persistent, the paired bracts below the flower broadly triangular, ca. 1 mm long, hyaline, basally connate, the apex acute, persistent and forming a cuplet, the paired sepaloid bracteoles basally adnate to the sepals, broadly elliptic to suborbicular, 1–2 mm long and wide, cream-white, drying brown, the apex acute; sepals petaloid, ovate-oblong to elliptic, 1.5–3 mm long, 1–2 mm wide, white, drying brown, the apex obtuse; stamens 2–4 mm long, white; styles 1–1.5 mm long, basally connate for ½–⅔ their length, white, the stigmas clavate. Fruit globose, ca. 1 mm long, brown, not winged; seed subglobose, glabrous, smooth.

Disturbed sites. Rare; Leon, Alachua, and Glades Counties. Escaped from cultivation. Florida, Louisiana, Texas, and California; nearly cosmopolitan. Native to South America. Spring–summer.

Rarely producing viable seeds and spreading primarily by stem tubers.

Anredera vesicaria (Lam.) C. F. Gaertn. [*Vesicarius*, bladderlike or inflated, in reference to the keeled upper sepaloid bracteoles in fruit.] TEXAS MADEIRAVINE.

Basella vesicaria Lamarck, Encycl. 1: 382. 1785. *Anredera vesicaria* (Lamarck) C. F. Gaertner, Suppl. Carp. 176. 1807.

Boussingaultia leptostachys Moquin-Tandon, in de Candolle, Prodr. 13(2): 229. 1849. *Anredera leptostachys* (Moquin-Tandon) Steenis, Fl. Males. 5: 302. 1957.

Herbaceous perennial vine, to 8 m; stem glabrous, producing tubers at or below the soil surface or in the lower leaf axils. Leaves with the blade ovate to elliptic, 2–16 cm long, 0.6–9 cm wide, the apex acute to obtuse, the base cuneate, decurrent on the petiole, the margin entire, the upper and lower surfaces glabrous, the petiole 3–18 mm long, reddish. Flowers in a raceme or the leaves sometimes bracteate and forming a panicle of racemes, the pedicel 1–2 mm long, the lower subtending bract oblong to oblanceolate, 1–2 mm long, persistent, the upper paired bracts triangular or deltoid, 0.5–1 mm long, free, the apex acute, deciduous, the paired sepaloid bracteoles basally adnate to the sepals, elliptic, 1–2 mm long, ca. 1 mm wide, keeled (weakly so in staminate plants), basally connate ⅓–½ their length, cream-white; sepals elliptic, 1–2 mm, ca. 1 mm wide, cream-white, the apex obtuse; stamens 2–3 mm long, white; styles 1–1.5 mm long, white, the stigma clavate or bifid. Fruit obovoid, slightly triangular, 1–1.5 mm long, crowned by the persistent style base, tan, winged; seed subglobose, glabrous, smooth.

Disturbed sites. Occasional; central and southern peninsula. Escaped from cultivation. Florida and Texas; nearly cosmopolitan. Native to tropical America. All year.

The Florida plants are functionally staminate and apparently spread primarily by means of the vegetative tubers.

EXCLUDED TAXA

Anredera baselloides (Kunth) Baillon—Reported for Florida by Small (1903), the name misapplied to material of *A. vesicaria*.

Anredera scandens (Linnaeus) Moquin-Tandon—Reported for Florida by Correll and Johnston (1970), the name misapplied to material of *A. vesicaria*.

Basella L. 1753.

Annual herbaceous vine. Leaves alternate, simple, pinnate-veined, petiolate, estipulate. Flowers in axillary spikes, sessile, bisexual, the bracts 3, the bracteoles 2, adnate to the sepals; sepals 5, petaloid; petals absent; stamens 5, episepalous; styles 3. Fruit a utricle, enclosed by the persistent calyx; seed 1.

A genus of 5 species; North America, West Indies, Mexico, Central America, South America, Africa, and Asia. [Malayalam or Malabari dialect name (spoken in India) for the plant.]

Basella alba L. [White, in reference to the flower color.] CEYLON SPINACH; INDIAN SPINACH; MALABAR SPINACH.

Basella alba Linnaeus, Sp. Pl. 272. 1753.

Annual herbaceous vine, to 10 m; stem glabrous. Leaves with the blade ovate to subrotund, 3–9 cm long, 2–8 cm wide, the apex acute, the base cuneate, shallowly cordate, or truncate, the margin entire, the upper and lower surfaces glabrous, the petiole 1–3 cm long. Flowers in an axillary spike 3–15(20) cm long, sessile; bracts 3, minute, caducous; bracteoles 2, adnate to the sepals, oblong, 3–4 mm long, the apex acute, persistent; sepals petaloid, white, sometimes reddish or purplish tinged, 3–4 mm long, the lobes ovate-oblong, 2–3 mm long, cucullate, the apex obtuse; stamens adnate to the sepals, the filaments white, the anthers yellowish episepalous; styles 3. Fruit subglobose, 5–6(8) mm long, fleshy, red to black, enclosed by the persistent calyx; seed 1, globose.

Leon and Escambia Counties. Escaped from cultivation. Florida; West Indies, Central America, and South America; Africa, Asia, and Pacific Islands. Native to Asia. Summer–fall.

PORTULACACEAE Adans. 1763. PURSLANE FAMILY

Annual or perennial herbs. Leaves alternate or subopposite, simple, pinnate-veined, petiolate or epetiolate, stipulate or estipulate. Flowers terminal or axillary, solitary or in racemose to paniculiform cymes, actinomorphic, bisexual, bracteate; sepals 2 or 5, free or basally connate; petals 5–7, free or basally connate; stamens 6–45, the anthers 2-loculate; ovary superior or half-inferior to inferior, 2- to 9-carpellate, 1-loculate, the style 1, the stigmas as many as the carpels. Fruit a loculicidal or circumscissile capsule; seeds numerous.

A family of 20–30 genera and about 500 species; nearly cosmopolitan.

Selected references: Bogle (1969); Packer (2003).

1. Flowers axillary, solitary, or clustered; ovary half-inferior to inferior; fruit a circumscissile capsule..... .... **Portulaca**
1. Flowers in a racemose or paniculiform cyme; ovary superior; fruit a loculicidal capsule.........**Talinum**

Portulaca L. 1753. PURSLANE

Succulent annual herbs. Leaves alternate or subopposite, simple, pinnate-veined, petiolate or epetiolate, stipulate. Flowers solitary and axillary or few in terminal clusters, sessile or subsessile; sepals 2, one larger than the other, basally connate, caducous; petals 5–7, free or basally connate; stamens 6–45; ovary inferior to half-inferior, 2- to 9-carpellate, 1-loculate. Fruit a circumscissile capsule; seeds numerous.

A genus of 100–125 species; nearly cosmopolitan. [*Portula*, a small gate or door, in reference to the lid of the circumscissile capsule.]

Selected reference: Matthews (2003).

1. Leaf axils lacking trichomes or with inconspicuous trichomes to 1 mm long.
 2. Stems and roots tuberous; style lobes 3...**P. minuta**
 2. Stems and roots not tuberous; style lobes 4–6 ....**P. oleracea**
1. Leaf axils with conspicuous trichomes more than 3 mm long.
 3. Leaves flat, oblanceolate to obovate, the largest involucral leaves 3–6(8) mm wide.............**P. amilis**
 3. Leaves terete or nearly so, the largest involucral leaves 1–2(2.5) mm wide.
  4. Flowers 3–5.5 cm wide..**P. grandiflora**
  4. Flowers to 1.5 cm wide.
   5. Stems decumbent, creeping and rooting; petals pink to purple ....**P. pilosa**
   5. Stems erect or ascending; petals yellow or rarely white.......................................**P. rubricaulis**

Portulaca amilis Speg. [*Amabil*, lovely.] PARAGUAYAN PURSLANE.

Portulaca amilis Spegazzini, Anales Soc. Ci. Argentina 92: 104, f. 6. 1921.

Prostrate or ascending annual herb; stem to 2.5 dm long, with tufts of trichomes more than 3 mm long at the nodes, these usually conspicuous. Leaves with the blade oblanceolate, spatulate, or obovate, 5–30 mm long, 2–15 mm wide, flattened, the apex acute, the base cuneate, the margin entire, the upper and lower surfaces glabrous, sessile or the petiole to 2 mm long. Flowers subtended by 6–8(9) involucral leaves; sepals triangular-ovate, 2–3 mm long; petals obovate, 7–10 mm long and wide, pinkish purple; stamens 15–45; stigmas 7–10. Fruit ovoid, 2–5.5 mm long; seeds flattened, ca. 0.5 mm long, black, lustrous, the surface obscurely tuberculate and stellate-puberulent to nearly smooth.

   Disturbed sites. Frequent; northern counties, central peninsula. Virginia south to Florida; South America. Native to South America. Spring–fall.

Portulaca grandiflora Hooker [Large-flowered.] ROSE MOSS.

Portulaca grandiflora Hooker, Bot. Mag. 56: pl. 2885. 1829. *Portulaca megalantha* Steudel, Nomencl. Bot., ed. 2. 2(2): 383. 1840, nom. illegit. *Portulaca pilosa* Linnaeus subsp. *grandiflora* (Hooker) Geesink, Blumea 17(2): 297. 1969.

Prostrate or ascending annual herb; stem to 3 dm long, with tufts of trichomes more than 3 mm long at the nodes, these usually conspicuous. Leaves with the blade linear, 8–25 mm long, 2–5 mm wide, terete, the apex acute, the base cuneate, the margin entire, the upper and lower surfaces glabrous, sessile or the petiole to 2 mm long. Flowers subtended by 6–9 involucral leaves; sepals obovate, 7–10 mm long; petals obovate, 1.5–2.5 cm long and wide, red, pink, purple, salmon, yellow, or white; stamens 40 or more; stigmas 5–9. Fruit ovoid, 4–6 mm long; seeds flattened, ca. 1 mm long, gray-black, sometimes iridescent, the surface tuberculate and stellate-puberulent.

   Disturbed sites. Rare; Polk County. Escaped from cultivation. Vermont south to Florida, west to Manitoba, North Dakota, South Dakota, Kansas, Nebraska, Oklahoma, and Texas, west from Texas to California; West Indies, Mexico, Central America, and South America; Europe. Native to South America. Spring–fall.

Portulaca minuta Correll [Very small, in reference to its size.] TINY PURSLANE.

Portulaca minuta Correll, J. Arnold Arbor. 60: 154, f. 1. 1979.

Prostrate or ascending perennial herb. Stem rarely more than 3 cm long, minutely glandular, usually with a few short trichomes to 1 mm long in the leaf axil. Leaves with the blade obovate to obovate-elliptic or spatulate, ca. 5 mm long, ca. 3 mm wide, flat, the apex rounded, the base cuneate, the margin entire, the upper and lower surfaces glabrous, the petiole ca. 1 mm long. Flowers subtended by 1–4 involucral leaves; sepals obovate to obovate-elliptic, 3–5 mm long and wide; petals 3–8 mm long, 2–3 mm wide, yellow; stamens 8; stigmas 3. Fruit ovoid, ca. 3 mm long; seeds ca. 0.5 mm long, flattened, black, iridescent, the surface pebbly.

Shallow, ephemeral freshwater pools in open limestone flats transitional between pine rockland and tidal mangroves. Rare; Monroe County keys. Florida; West Indies. Spring and fall.

Portulaca oleracea L. [*Oler*, greens or vegetables.] LITTLE HOGWEED.

Portulaca oleracea Linnaeus, Sp. Pl. 445. 1753. *Portulaca oleracea* Linnaeus var. *sylvestris* de Candolle, Prodr. 3: 353. 1828, nom. inadmiss. *Portulaca officinarum* Crantz, Inst. Rei Herb. 2: 428. 1766, nom. illegit. *Portulaca sylvestris* Friche-Joset & Montandon, Syn. Fl. Jura 109. 1856, nom. illegit. *Portulaca oleracea* Linnaeus subsp. *sylvestris* Thellung, Fl. Adv. Montpellier 222. 1912, nom. inadmiss.

Portulaca oleracea Linnaeus var. *granulatostellulata* Poellnitz, Occas. Pap. Bernice Pauahi Bishop Mus. 9: 5. 1936. *Portulaca oleracea* Linnaeus subsp. *granulatostellulata* (Poellnitz) Danin & H. G. Baker, in Danin et al., Israel J. Bot. 27: 189. 1979 ("1978").

Portulaca oleracea Linnaeus subsp. *nicaraguensis* Danin & H. G. Baker, in Danin et al., Israel J. Bot. 27: 186. 1979 ("1978").

Portulaca oleracea Linnaeus subsp. *nitida* Danin & H. G. Baker, in Danin et al., Israel J. Bot. 27: 194. 1979 ("1978").

Portulaca oleracea L. subsp. *papillatostellulata* Danin & H. G. Baker, in Danin et al., Israel J. Bot. 27: 200. 1979 ("1978").

Prostrate or ascending annual herb; stem to 5 dm long, with tufts of trichomes to 1 mm long, few and inconspicuous at the nodes or lacking. Leaves with the blade obovate or spatulate, 5–25 mm long, 2–15 mm wide, flat, the apex round or retuse, the base cuneate, the margin entire, the upper and lower surfaces glabrous, the petiole 2–5 mm long. Flowers subtended by 1–4 involucral leaves; sepals ovate to suborbicular, 3–5 mm long and wide; petals 3–8 mm long, 2–3 mm wide, yellow; stamens 6–10(20); stigmas 3–6. Fruit ovoid, 4–9 mm long; seeds flattened, 0.7–1 mm long, black or dark brown, the surface short-tuberculate, granular, or nearly smooth.

Disturbed sites. Frequent; nearly throughout. Nearly cosmopolitan. Spring–fall.

Danin and H. G. Baker (in Danin et al., 1978) recognized nine subspecies in *Portulaca oleracea* using seed size and seed coat morphology as the main distinguishing features. Danin and Anderson (1986) report five of these subspecies for Florida. We prefer to treat *P. oleracea* as a single highly variable, nearly cosmopolitan taxon, following Matthews (2003), Matthews and Levins (1985), and Matthews et al. (1993).

Portulaca pilosa L. [Pilose.] PINK PURSLANE; KISS-ME-QUICK.

Portulaca pilosa Linnaeus, Sp. Pl. 445. 1753.

Prostrate to suberect annual herb; stem to 1.5 dm long, with tufts of conspicuous trichomes more than 3 mm long at the nodes. Leaves with the blade linear, terete, 5–20 mm long, 1–3 mm wide, the apex acute, the base cuneate, the margin entire, the upper and lower surfaces glabrous, sessile or subsessile. Flowers subtended by 6–10 involucral leaves; sepals triangular-ovate, 2–3 mm long; petals obovate, 3–6 mm long and wide, rose-purple, the apex retuse; stamens 5–12(35); stigmas 3–6. Fruit ovoid, 3–4 mm long; seeds flattened, ca. 0.5 mm long, black or blue-black, the surface short-tuberculate and obscurely stellate-puberulent, but sometimes clearly stellate-puberulent and lacking tubercles.

Dry, often disturbed sites. Frequent; nearly throughout. New Jersey and North Carolina south to Florida, west to Colorado and New Mexico; West Indies, Mexico, Central America, and South America. Spring–fall.

Portulaca rubricaulis Kunth [*Rubr*, red, and *caul*, stem.] REDSTEM PURSLANE.

Portulaca rubricaulis Kunth, in Humboldt et al., Nov. Gen. Sp. 6: 73. 1823.
Portulaca phaeosperma Urban, Symb. Antill. 4: 233. 1905.

Erect or ascending annual herb; stem to 3.5 dm long, with tufts of trichomes more than 3 mm long at the nodes, these usually conspicuous. Leaves with the blade linear-oblong, 5–15 mm long, 1–5 mm wide, flat or subterete, the apex acute, the base cuneate, the margin entire, the upper and lower surfaces glabrous, sessile or subsessile. Flowers subtended by 5–8 involucral leaves; sepals ovate, 3–4 mm long; petals oblong to obovate, 6–8 mm long and wide, yellow, the apex obtuse; stamens 12–30; stigmas 5–7. Fruit subglobose, 3–6 mm long; seeds flattened, ca. 0.6 mm long, black, dark gray, or brown, the surface obscurely tuberculate and stellate-puberulent, sometimes granular.

Beaches and shell middens. Occasional; Pinellas County southward along the west coast, southern peninsula. Florida; West Indies, Mexico, Central America, and South America. All year.

EXCLUDED TAXON

Portulaca halimoides Linnaeus—Reported by Gray (1887), Chapman (1897), and Small (1903, 1913a), all of whom misapplied the name to material of *P. rubricaulis*. Also reported by Legrand (1962), based on the specimen *Nash 187* (S) as this taxon. Duplicates of this collection (F, MICH, NY) are all *P. pilosa*, which implies either a mistake in labeling of the Stockholm specimen, or an error by Legrand in his paper.

Talinum Adans. 1763. FLAME FLOWER

Perennial herbs. Leaves alternate or subopposite, simple, pinnate-veined, petiolate, estipulate. Flowers in lateral and/or terminal racemiform or paniculiform cymes; sepals 5, free, persistent or caducous; petals 5, free; stamens 15–35; ovary superior, 3-carpellate, 1-loculate. Fruit a loculicidal capsule; seeds numerous.

A genus of about 15 species; North America, West Indies, Mexico, Central America, South

America, Africa, and Asia. [Believed to be derived from the aboriginal name of an African species.]

Selected reference: Kiger (2003b).

1. Inflorescence a racemose cyme, the pedicels 3-angled, distinctly thickened distally .... **T. fruticosum**
1. Inflorescence a paniculiform cyme, the pedicels terete, uniformly slender .... **T. paniculatum**

Talinum fruticosum (L.) Juss. [Shrubby.] VERDOLAGA-FRANCESA.

Portulaca fruticosa Linnaeus, Syst. Nat., ed 10. 1045. 1759. *Portulaca triangularis* Jacquin, Enum. Syst. Pl. 22. 1760, nom. illegit. *Portulaca racemosa* Linnaeus, Sp. Pl., ed. 2. 640. 1762, nom. illegit. *Ruelingia triangularis* Ehrhart, Beitr. Naturk. 3: 134. 1788, nom. illegit. *Talinum fruticosum* (Linnaeus) Jussieu, Gen. Pl. 312. 1789. *Talinum triangulare* Willdenow, Sp. Pl. 2: 862. 1799, nom. illegit. *Talinum racemosum* Rohrbach, in Martius, Fl. Bras. 14(2): 297. 1872, nom. illegit. *Claytonia triangularis* Kuntze, Revis. Gen. Pl. 1: 56. 1891, nom. illegit.

Erect herb or sometimes basally suffrutescent, to 1 m; stem glabrous. Leaves with the blade oblanceolate to obovate, 2–6 cm long, 0.5–2.5 cm wide, the apex rounded or acute, the base cuneate, the margin entire, the upper and lower surfaces glabrous, the petiole to ca. 1 cm. Flowers few to many in a racemiform cyme, the pedicel 7–11 mm long, 3-angular, distinctly thicker distally; sepals lanceolate, 5–6 mm long, persistent; petals broadly elliptic to ovate, 7–13 mm long, white, pink, or purple; stamens 20–35; stigmas 3. Fruit subglobose, 4–6 mm long; seeds ca. 1 cm long, dark brown or black, lustrous, minutely striolate.

Dry, disturbed sites. Rare; central and southern peninsula. Escaped from cultivation. Florida; West Indies, Mexico, Central America, and South America; Africa. Native to tropical America and Africa. All year.

Talinum paniculatum (Jacq.) Gaertn. [Paniculate, in reference to the inflorescence.] PINK BABY'S-BREATH; JEWELS-OF-OPAR.

Portulaca paniculata Jacquin, Enum. Syst. Pl. 22. 1760. *Portulaca patens* Linnaeus, Mant. Pl. 242. 1771, nom. illegit. *Ruelingia patens* Ehrhart, Bietr. Naturk. 3: 135. 1788, nom. illegit. *Talinum reflexum* Cavanilles, Icon. 1: 1. 1791, nom. illegit. *Talinum paniculatum* (Jacquin) Gaertner, Fruct. Sem. Pl. 2: 219. 1791. *Talinum dichotomum* Ruiz López & Pavón, Syst. Veg. Fl. Peruv. Chil. 118. 1798, nom. illegit. *Talinum fruticosum* Willdenow, Sp. Pl. 2: 864. 1799, nom. illegit; non (Linnaeus) Jusseau, 1789. *Portulaca reflexa* Haworth, Misc. Nat. 141. 1803, nom. illegit. *Claytonia paniculata* (Jacquin) Kuntze, Revis. Gen. Pl. 1: 57. 1891.

Talium sarmentosum Engelmann, Boston J. Nat. Hist. 6: 153. 1850. *Talinum patens* Willdenow var. *sarmentosum* (Engelmann) A. Gray, Proc. Amer. Acad. Arts 22: 275. 1887. *Claytonia sarmentosa* (Engelmann) Kuntze, Revis. Gen. Pl. 1: 57. 1891. *Talinum reflexum* Cavanilles forma *sarmentosum* (Engelmann) Small, Fl. S.E. U.S. 415, 1330. 1903. *Talinum paniculatum* (Jacquin) Gaertner var. *sarmentosum* (Engelmann) Poellnitz, Ber. Deutsch. Bot. Ges. 51: 123. 1933.

Erect herb or sometimes basally suffruticose, to 1 m; stem glabrous. Leaves with the blade elliptic to obovate, 2.5–12 cm long, 1.5–5 cm wide, the apex obtuse to acute or acuminate, the base rounded or cuneate, the margin entire, the upper and lower surfaces glabrous, the petiole to 1 cm long. Flowers few to many in an elongate paniculiform cyme 7–25 cm, the pedicel 1–2 cm long, uniformly slender; sepals ovate to suborbicular, 3–4 mm long, caducous; petals ovate to suborbicular, 3–6 mm long, red, pink, or yellowish; stamens 15–20. Fruit subglobose, 3–5 mm long; seeds ca. 1 cm long, black, the surface minutely striolate or tuberculate, lustrous.

Dry, disturbed sites. Occasional; peninsula, Jackson County. Escaped from cultivation. Kentucky and North Carolina south to Florida, west to Arizona; West Indies, Mexico, Central America, and South America; Asia and Africa. Native to North America west of the Mississippi River and tropical America. All year.

EXCLUDED GENUS

> *Phemeranthus teretifolius* (Pursh) Rafinesque—This northern species was reported by Small (1903, 1913a, 1933, all as *Talinum teretifolium* Pursh). No Florida specimens known. Kiger (2003a) does not include this species in Florida.

CACTACEAE Juss. 1789. CACTUS FAMILY

Shrubs, trees, or vines; stems segmented or unsegmented; areoles in the leaf axils, with or without spines, woolly trichomes, and glochids. Leaves well developed or vestigial or absent, alternate, simple, petiolate or epetiolate, estipulate. Flowers arising from the areoles or tubercle axils, solitary or in terminal or lateral paniculiform, corymbiform, or aggregate cymes, bisexual or unisexual; perianth segments 10–50+, intergrading from bracteate to sepaloid to petaloid; stamens 12–500+; ovary inferior, 3- to 24-carpellate, 1-loculate, the style 1, the stigma lobes 3–24. Fruit a fleshy berry, indehiscent or dehiscent; seeds 5–500+, arillate or exarillate.

A family of 125–30 genera and about 1,800 species; nearly cosmopolitan. Principally a New World family, but now widely naturalized in the Old World tropics and subtropics.

Opuntiaceae Martinov, 1821.

Selected references: Anderson (2001); Benson (1982); Britton and Rose (1919, 1920, 1922, 1923).

1. Leaves persistent, with broad, flat blades; flowers in paniculiform, corymbiform, or aggregate cymes. ....**Pereskia**
1. Leaves caducous, small and terete, vestigial or absent; flowers solitary in the areoles.
    2. Stems flattened (rarely subcylindric); areoles evenly distributed on the stems, bearing minute, barbed bristles (glochids); inner flower petals brightly colored (red to yellow).
        3. Stems indeterminate, elongate, unsegmented; flower and fruit areoles usually spiny... **Consolea**
        3. Stems determinate, elliptic, ovate, orbicular, to obovate, segmented; flower and fruit areoles not spiny...**Opuntia**
    2. Stems cylindric to flattened; areoles restricted to ribs, angles, or edges of the stem, lacking glochids (areoles evenly distributed on cylindric stems of *Rhipsalis*); inner flower petals whitish.
        4. Branches less than 7 mm wide; flowers rotate, less than 2 cm long; floral tube obsolete; perianth segments less than 12 .... **Rhipsalis**
        4. Branches more than 7 mm wide; flowers funnelform, salverform, or campanulate, more than 4 cm long; floral tube elongate; perianth segments more than 12.
            5. Clambering to erect shrubs or trees; stems without adventitious roots (unless broken or fallen to the ground).

6. Stems with 3–5 ribs; usually at least some flower and fruit areoles with a few spines ....
....**Acanthocereus**
6. Stems with 8–14 ribs; flower and fruit areoles without spines.
   7. Treelike, with stout, erect stems; flowers 4–6 cm long, glabrous, scales rounded at the
      tip; fruit with few or no areoles..**Pilosocereus**
   7. Shrubs, with slender, weakly erect stems; flowers 12–20 cm long, areoles pilose, scales
      acute at the tip; fruit with numerous areoles ....**Harrisia**
  5. Vines or epiphytes; stems with adventitious roots.
   8. Stems flat (basally cylindric), rarely 3-winged..**Epiphyllum**
   8. Stems 3- to 8-ribbed, -angled, or -winged, rarely 2-winged.
      9. Flower and fruit areoles with conspicuous trichomes or spines; bracts of the fruit
         withering-deciduous .... **Selenicereus**
      9. Flower and fruit areoles without conspicuous trichomes; bracts of the fruit persistent
         ....**Hylocereus**

Acanthocereus (Engelm. ex A. Berger) Britton & Rose 1909. TRIANGLE CACTUS

Shrubs; stems unsegmented, ribbed; areoles with spines, glochids and woolly trichomes absent. Leaves vestigial or absent. Flowers solitary in lateral areoles, funnelform, sessile, bisexual, nocturnal; floral tube conspicuous, the perianth segments numerous; stigma lobes 10–15. Fruit indehiscent or irregularly dehiscent; seeds numerous, exarillate.

A genus of 6 species; North America, West Indies, Mexico, Central America, and South America. [From the Greek *akantha*, thorn or spine, and *Cereus*, in reference to this genus of columnar cacti.]

Selected reference: Parfitt and Gibson (2003b).

Acanthocereus tetragonus (L.) Hummelinck [Four-angled, in reference to the juvenile stems, which often have four angles.] TRIANGLE CACTUS; DILDOE CACTUS; BARBED-WIRE CACTUS.

Cactus tetragonus Linnaeus, Sp. Pl. 466. 1753. *Cereus tetragonus* (Linnaeus) Miller, Gard. Dict., ed. 8. 1768. *Acanthocereus tetragonus* (Linnaeus) Hummelinck, Succulenta (Netherlands) 20: 165. 1938.

Cactus pentagonus Linnaeus, Sp. Pl. 467. 1753. *Cereus pentagonus* (Linnaeus) Haworth, Syn. Pl. Succ. 180. 1812. *Acanthocereus pentagonus* (Linnaeus) Britton & Rose, Contr. U.S. Natl. Herb. 12: 432. 1909.

Acanthocereus floridanus Small ex Britton & Rose, Cact. 4: 276. 1923. TYPE: FLORIDA: Monroe Co.: Key Largo, Dec 1917. *Small s.n.* (lectotype: NY; isotype: US). Lectotypified by Benson (1982: 935).

Shrub; stem reclining or clambering, to 10 m long, green, glabrous, 3- to 5-ribbed, the ribs narrowly triangular to winglike, 3–5 cm high, to 1 cm wide; areoles 2–3 cm apart, the spines 4–8 per areole, 2–4 cm long, subulate, straight, brown or reddish gray. Flowers 14–20 cm long; floral tube funnelform, 8–15 cm long, the areoles few, each subtended by a small scale, the outer perianth segments deltoid to triangular-lanceolate or lanceolate to linear, the longer ones 3.5–4 cm long, ca. 1 cm wide, greenish white, the apex acuminate, the margin entire, the inner ones broadly linear, 3.5–4.5 cm long, ca. 1 cm wide, white, the apex acuminate, the margin entire; stamens 4–5 cm long; style 17.5–20 cm long, white, the stigma lobes white. Fruit ovoid to oblong, 8–10 cm long, red, with a few large, tuberculate areoles, the surface glossy, the flesh red; seeds broadly obovoid, 4–5 mm long, black, smooth, glossy.

Dry, disturbed coastal hammocks. Rare; Lee and St. Lucie Counties southward along the coast, southern peninsula. Florida and Texas; West Indies, Mexico, Central America, and South America; Pacific Islands. Native to North America and tropical America. All year.

Acanthocereus tetragonus (as *A. pentagonus*) is listed as threatened in Florida (Florida Administrative Code, Chapter 5B-40).

EXCLUDED TAXON

Acanthocereus baxaniensis (Karwinsky ex Pfeiffer) Borg—Reported by Small (1903, as *Cereus baxaniensis* Karwinsky ex Pfeiffer), the name misapplied to material of *A. tetragonus*.

Consolea Lem.

Trees. Stems unsegmented; areoles with glochids and with woolly trichomes. Leaves conic, sessile, caducous. Flowers solitary in the lateral areoles, salverform, sessile, functional unisexual, diurnal; floral tube conspicuous, the perianth segments numerous; stigma lobes 5. Fruit indehiscent; seeds numerous, covered by a bony aril (funicular envelope) of which the vascular bundle forms a girdle around the edge.

A genus of 9 species; North America and West Indies. [Commemorates Michelangelo Consol (1812–1897), of Palermo Botanic Garden, Italy.]

Selected reference: Pinkava (2003c).

Consolea corallicola Small [Coral dwelling, in reference to its preference for limestone soils.] SEMAPHORE PRICKLYPEAR; SEMAPHORE CACTUS.

Consolea corallicola Small, Addisonia 15: 25, pl. 493. 1930. *Opuntia corallicola* (Small) Werdermann, in Backeberg & Werdermann, Neue Kakteen 66. 1931. *Consolea spinosissima* Small subsp. *corallicola* (Small) Guiggi, Cactology 3: 18. 2012. TYPE: FLORIDA: Monroe Co.: Big Pine Key, May 1919, *Small s.n.* (holotype: NY).

Shrub or tree; stem to 3.5 m, the segments from the top of the unsegmented, nearly cylindrical main axis (trunk) tending to grow in one plane, flattened, elliptic to elongate, asymmetrical, slightly falcate, 10–30 cm long, 5–9 cm wide, light green, glabrous, slightly tuberculate, not readily disarticulating; areoles 1–1.5 cm apart, the woolly trichomes tan or gray, the spines (0)2–3(4) per areole, 7–12 cm long, mostly in the marginal and submarginal areoles, straight, slightly barbed, white or gray, brownish in age, the glochids 1–2 mm long, yellow. Flowers 3–7.5 cm long, 1.2–2.5 cm wide; floral tube funnelform, the outer perianth segments ovate-deltoid, 3–6 mm long, green, the apex acute, the margin entire, the inner ones broadly ovate, 2–2.5 cm long and wide, slightly spreading, red, the apex acute, mucronate, the margin entire; stamens ca. 6 mm long, yellow; style 6–7.5 mm long, light red, conspicuously swollen basally, the stigma lobes 5, light red. Fruit obovate to clavate, usually slightly curving, 5–6 cm long, yellow, the surface glabrous, with prominent areoles and spines, the flesh yellow; seeds irregular in outline, 6–8 mm long, yellowish white, the sides woolly, the girdle cristate.

Hammocks. Rare; Miami-Dade County and Monroe County keys. Endemic. All year.

The flowers of *C. corallicola* are bisexual or functionally unisexual (staminate with closed stigmas and normal anthers; carpellate with open stigmas and the anthers lacking pollen).

Consolea corallicola is listed as endangered in Florida (Florida Administrative Code, Chapter 5B-40) and in the United States (U.S. Fish and Wildlife Service, 50 CFR 23).

EXCLUDED TAXON

Consolea spinosissima (Miller) Lemaire—This Jamaican endemic was reported by Long & Lakela (1971, as *Opuntia spinosissima* Miller) and Wunderlin (1998, as *Opuntia spinosissima* Miller), the name misapplied to material of *C. corallicola*. (See Austin et al., 1998).

Epiphyllum Haw 1812. CLIMBING CACTUS

Shrubs; stems segmented, 2- to 3-ribbed; areoles with woolly trichomes and spines, glochids absent. Leaves vestigial or absent. Flowers solitary in lateral areoles, salverform, sessile, bisexual, nocturnal; floral tube conspicuous, the perianth segments numerous; stigma lobes 12–15. Fruit dehiscent along one side; seeds numerous, exarillate.

A genus of 19 species; West Indies, Mexico, Central America, and South America. From the Greek *epi*, upon, and *phyllos*, leaf, in reference to flowers borne on leaflike stems.]

Selected references: Hawkes (2003d); Kimnach (1964).

Epiphyllum phyllanthus (L.) Haw. var. **hookeri** (Haw.) Kimnach. [From the Greek *phyllos*, leaf, and *anthos*, flower, in reference to the flat leaflike flower-bearing branches; commemorates William Jackson Hooker (1785–1865), professor of botany at Glasgow University and later director of Kew.] ORCHID CACTUS.

Epiphyllum hookeri Haworth, Philos. Mag. 6: 108. 1829. *Phyllocactus hookeri* (Haworth) Salm-Reifferscheid-Dyck, Cact. Hort. Dyck. 1841: 35. 1841. *Epiphyllum phyllanthus* (Linnaeus) Haworth var. *hookeri* (Haworth) Kimnach, Cact. Succ. J. (Los Angeles) 36: 113. 1964.

Shrub, epiphytic; stem erect or pendent, to 2 m long, the segments narrowly elongate to oblanceolate, 40–120 cm long, 6–9(10) cm wide, proximally terete or 3-angled, distally 2-winged, broadly flattened and leaflike, with a prominent midvein-like axis, light green, glabrous, the rib margins obtusely serrate; areoles 3–8 cm apart, only in the sinuses along the rib margins, with woolly trichomes, spines absent or occasionally present in the basal ones and then bristlelike. Flowers 19–23 cm long; floral tube 10–15 cm long, slender, straight or curved, the outer perianth segments linear-lanceolate, ca. 8 cm long, ca. 0.5 cm wide, greenish white, often tinged with red, the apex acute, the margin entire, the inner ones linear-lanceolate, ca. 8 cm long, 0.5 cm wide, white, the apex acute, the margin entire; stamens (1)2–3 cm long, white; style 15–20 cm long, purplish or orangish, the stigma lobes yellow. Fruit oblong, 4–7 cm long, usually 5-angled from the base, smooth between the angles, purplish red, the flesh white; seeds ca. 3 mm long, black, minutely pitted.

Disturbed sites, epiphytic. Rare; Lee County. Escaped from cultivation. Florida; West Indies, Mexico, Central America, and South America. Native to West Indies, Mexico, Central America, and South America. All year.

Epiphyllum phyllanthus var. *hookeri* was first reported from Florida by Lima (1996). The Lima specimens from Lee County are said to be at USF but were never deposited there. A report from Collier County by Lima is based on a plant in cultivation (*Watson 1412a*, FTG) said to have been received from George Avery in 1977 and to have been collected in the Fakahatchee Strand. The provenance of the plant has not been verified.

Harrisia Britton 1908. APPLECACTUS

Shrubs or trees; stems unsegmented, cylindric, 8- to 12 ribbed; areoles with spines, glochids absent. Leaves vestigial or absent. Flowers solitary in lateral areoles, funnelform, sessile, bisexual, nocturnal; floral tube conspicuous, the perianth segments numerous; stamens 100+, concentrated in the lower region of the flower, the filaments light green to white, the anthers beige; style light green to white, the stigma lobes 10–15. Fruit indehiscent or irregularly splitting from the apex, 3.5–8 cm long; seeds numerous, exarillate.

A genus of 20 species; North America, West Indies, and South America. [Commemorates William Harris (1887–1908), superintendent of Public Gardens and Plantations of Jamaica.]

Selected references: Austin (1984); Franck (2012); Parfitt and Gibson (2003a).

1. Spines to 3.5 cm long; scales deltoid-acuminate, 7.0–10.3 mm long at the middle of the tube (hypanthium); ovary wall (pericarp) and immature fruits purple to reddish green; mature fruits orange to red, rarely yellow..**H. fragrans**
1. Spines to 1.5 cm long; scales deltoid-acute, 4.5–8.0 mm long at the middle of the tube (hypanthium); ovary wall (pericarp) and immature fruits green; mature fruits yellow........................**H. aboriginum**

Harrisia aboriginum Small ex Britton & Rose [The type plants were found growing in aboriginal shell middens.] PRICKLY APPLECACTUS; WEST COAST PRICKLY-APPLE.

Harrisia aboriginum Small ex Britton & Rose, Cact. 2: 154. 1920. *Cereus aboriginum* (Small ex Britton & Rose) Little, Amer. Midl. Naturalist 33: 495. 1945. *Cereus gracilis* Miller var. *aboriginum* (Small ex Britton & Rose) L. D. Benson, Cact. Succ. J. (Los Angeles) 41: 126. 1969. *Harrisia gracilis* (Miller) Britton var. *aboriginum* (Small ex Britton & Rose) D. B. Ward, Novon 14: 366. 2004. TYPE: FLORIDA: Manatee Co.: W shore of Terra Ceia Island, 29 Apr 1919, *Small et al. s.n.* (lectotype: NY; isolectotype: US). Lectotypified by Benson (1969: 126).

Shrub; stem erect or reclining, to 6 m long, green, glabrous, 8- to 12-ribbed, the ribs rounded, to 1 cm high, shallowly tuberculate; spines to 1.5 cm long, acicular, straight, gray when mature. Flower buds densely white-pilose; flowers 15–20 cm long, with turgid scales and tufts of white woolly trichomes, the outer perianth segments linear, ca. 6 cm long, ca. 6 mm wide, pale pink, the apex acuminate, the margin entire, the inner ones oblanceolate, ca. 6(7.5) cm long, 1.5–2 cm wide, white, the apex caudate-acuminate, the margin erose-denticulate. Fruit dull yellow, with tufts of white woolly trichomes, the flesh white; seeds ovoid to elliptic, 2.7–3 mm long, black, the surface finely pitted.

Coastal hammocks and shell middens. Rare; Manatee, Sarasota, and Lee Counties. Endemic. All year.

Harrisia aboriginum is listed as endangered in Florida (Florida Administrative Code, Chapter 5B-40) and in the United States (U.S. Fish and Wildlife Service, 50 CFR 23).

Harrisia fragrans Small ex Britton & Rose [Fragrant, in reference to the flower odor.] CARIBBEAN APPLECACTUS; INDIAN RIVER PRICKLY-APPLE; SIMPSON'S APPLECACTUS.

> *Harrisia fragrans* Small ex Britton & Rose, Cact. 2: 149. 1920. *Cereus fragrans* (Small ex Britton & Rose) Little, Amer. Midl. Naturalist 33: 496. 1945. *Cereus eriophorus* Pfeiffer & Otto var. *fragrans* (Small ex Britton & Rose) L. D. Benson, Cact. Succ. J. (Los Angeles) 41: 126. 1969. *Harrisia eriophora* (Pfeiffer & Otto) Britton var. *fragrans* (Small ex Britton & Rose) D. B. Ward, Novon 14: 366. 2004. TYPE: FLORIDA: St. Lucie Co.: 6 mi. S of Fort Pierce, 20 Dec 1917, *Small s.n.* (holotype: NY; isotypes: GH, US).
>
> *Harrisia simpsonii* Small ex Britton & Rose, Cact. 2: 152. 1920. *Cereus gracilis* Miller var. *simpsonii* (Small ex Britton & Rose) L. D. Benson, Cact. Succ. J. (Los Angeles) 41: 126, 1969. *Harrisia gracilis* (Miller) Britton var. *simpsonii* (Small ex Britton & Rose) D. B. Ward, Novon 14: 367. 2004. TYPE: FLORIDA: Monroe Co.: near Flamingo, May 1919, *Small s.n.* (lectotype: NY). Lectotypified by L. D. Benson (1969: 126).

Shrub; stem erect, reclining, or clambering, to 6 m long, green, glabrous, 18- to 12-ribbed, the ribs rounded, to 1 cm high, shallowly tuberculate; spines to 3.5 cm long, acicular, straight, gray when mature. Flower buds densely white-pilose; flowers 15–20 cm long, with turgid scales subtending a tuft of white woolly trichomes, the outer perianth segments linear, 5–6 cm long, 5–6 mm wide, green with a purplish brown midvein, the margin pink to white, the apex acuminate, the margin entire, the inner ones linear-spatulate to narrowly oblanceolate, 5–6(7.5) cm long, 1–2 cm wide, white or pinkish, the apex acute or acuminate to caudate, the margin erose-denticulate. Fruit orange-red to dull red, rarely yellow the flesh white; seeds oblong to elliptic, 2.7–3 mm long, black, the surface finely pitted.

Coastal hammocks and shell middens. Rare; Volusia, Brevard, Indian River, St. Lucie, and Palm Beach Counties, southern peninsula. Endemic. All year.

This species had long been considered restricted to the east-central coast and *H. simpsonii* was applied to populations in the southern peninsula. Morphological and DNA-based studies do not indicate any reliable distinction between the two (Franck, 2012) and here they are united as one species under *H. fragrans*.

Harrisia fragrans is listed as endangered in Florida (Florida Administrative Code, Chapter 5B-40) and in the United States (U.S. Fish and Wildlife Service, 50 CFR 23).

EXCLUDED TAXA

Harrisia brookii Britton—Reported for Florida by Small (1913a, 1913d, 1913e), the name of this Bahamian endemic misapplied to material of *H. fragrans*.

"*Harrisia donae-antoniae* Hooten," nom. nud.—No type was designated by Hooten (1991). This is apparently conspecific with *H. aboriginum*, although Anderson (2001) includes it in synonymy with *H. gracilis*, a species that does not occur in Florida.

Harrisia eriophora (Pfeiffer) Britton—This Cuban endemic is reported for Florida by Anderson (2001). No Florida specimens known.

Harrisia gracilis (Miller) Britton—This Jamaican endemic is reported for Florida by Anderson (2001), the name misapplied to material of *H. aboriginum*.

Hylocereus (A. Berger) Britton & Rose 1909. NIGHTBLOOMING CACTUS

Vines; stems segmented, 3-ribbed; areoles with woolly trichomes and spines, glochids absent. Leaves vestigial or absent. Flowers solitary in lateral areoles, funnelform, sessile, bisexual, nocturnal; floral tube conspicuous, the perianth segments numerous; stigma lobes 15–24. Fruit irregularly dehiscent along one side; seeds numerous, exarillate.

A genus of 18 species; nearly cosmopolitan. [From the Greek *hyle*, forest, and the genus *Cereus* which it resembles.]

Selected reference: Hawkes (2003b).

Hylocereus undatus (Haw.) Britton & Rose [Wavy, in reference to the undulate margins on the stem wings.] NIGHTBLOOMING CACTUS.

Cereus undatus Haworth, Philos. Mag. Ann. Chem. 7: 110. 1830. *Hylocereus undatus* (Haworth) Britton & Rose, in Britton, Fl. Bermuda 256. 1918.

Cereus tricostatus Gosselin, Bull. Soc. Bot. France 54: 664. 1907. *Hylocereus tricostatus* (Gosselin) Britton & Rose, Contr. U.S. Nat. Herb. 12: 429. 1909.

Vine; stem clambering or climbing, to 5 m long, the segments sharply 3-ribbed, 4–7.5 cm wide, green, glabrous, the rib crenately lobed; areoles 4–5 cm, apart, the spines 1–4 per areole, to 3 mm long, conic, brownish gray. Flowers 25–27 cm long; floral tube 12–14 cm long, smooth, with turgid scales the outer perianth segments lanceolate to linear, 10–15 cm long, 1–1.5 cm wide, white with a yellowish green midvein, the apex mucronate, the margin entire, the inner ones oblanceolate, 10–15 cm long, ca. 2.5 cm wide, white, the apex mucronate, the margin entire; stamens 5–7.5 cm long, creamy white; style 17–20 cm long, yellowish white, the stigma lobes yellowish white. Fruit globose to oblong, 5–12 cm long, reddish, smooth, with several to many persistent, foliaceous scales, the flesh white; seeds ovate, ca. 2 mm long, black.

Disturbed sites. Rare; St. Lucie and Martin Counties. Escaped from cultivation. Florida; nearly cosmopolitan. Native to tropical America. All year.

Hylocereus undatus is frequently cultivated in tropical and subtropical areas worldwide as an ornamental and for its edible fruit called "dragonfruit" or "pitaya."

EXCLUDED TAXON

Hylocereus triangularis (Linnaeus) Britton & Rose—Reported by Chapman (1860, 1883, 1897, all as *Cereus triangularis* (Linnaeus) Haworth), and Small (1903, 1913a, both as *Cereus triangularis* (Linnaeus) Haworth), who misapplied the name to material of *H. undatus*.

Opuntia Mill. 1754. PRICKLYPEAR

Shrubs or trees; stems segmented; areoles with glochids and usually also with woolly trichomes and/or acicular spines. Leaves conic or somewhat flattened, sessile, caducous. Flowers solitary in the lateral areoles, funnelform or salverform, sessile, bisexual, diurnal; floral tube conspicuous, the perianth segments numerous; stigma lobes 5–10. Fruit indehiscent; seeds numerous, covered by a bony aril (funicular envelope) of which the vascular bundle forms a girdle around the edge.

A genus of 150–80 species; nearly cosmopolitan. [A name used by Theophrastus for some cactus-like plant, perhaps based on the name Opuntia Locris (present-day Greece), where a cactus-like plant grew].

Nopalea Salm-Dyck, 1850.

Selected references: Pinkava (2003a, 2003b).

1. Petaloid perianth segments red to pinkish red; spines absent; stamens and style much exserted...........
....**O. cochenillifera**
1. Petaloid perianth segments yellow, rarely reddish; spines present or absent; stamens and style included.
  2. Stem segments pubescent ....**O. leucotricha**
  2. Stem segments glabrous.
    3. Distal stem segments narrowly elliptic to linear, 1–2.5 cm wide, subcylindric to slightly flattened, loosely attached and readily disarticulating; spines strongly barbed apically.............. **O. pusilla**
    3. Distal stem segments obovate to elliptic, (2)2.5–14(20) cm wide, flattened, firmly attached or disarticulating; spines barbed apically or not.
      4. Mature stem segments often (2.5)3–6 dm long, often with more than 15 areoles per face (not counting areoles on the margin), often spineless, with inconspicuous glochids to 2 mm long...**O. ficus-indica**
      4. Mature stem segments usually less than 2.5 dm long, with fewer than 15 areoles per face (not counting areoles on the margin), often with spines, usually with conspicuous glochids to 4 mm long.
        5. Spines disposed from areoles in the same plane or porrect; mature spines gray.................
.... **O. humifusa**
        5. Spines disposed from areoles in a stellate pattern or in different planes; mature spines yellow to gray or brown.
          6. Mature spines yellow to brown; stem segments often with a scalloped or undulate margin..**O. stricta**
          6. Mature spines gray (immature spines yellow or brown); stem segments entire to faintly undulate ....**O. ochrocentra**

Opuntia cochenillifera (L.) Mill. [In reference to the bright red dye (cochineal) and the vernacular name of the scale insect from which the dye is derived, and *ferre*, to bear.] COCHINEAL CACTUS.

Cactus cochenillifer Linnaeus, Sp. Pl. 468. 1753. *Opuntia cochenillifera* (Linnaeus) Miller, Gard. Dict., ed. 8. 1768. *Nopalea cochenillifera* (Linnaeus) Salm-Reifferscheid-Dyck, Cact. Hort. Dyck. 1849: 64. 1850. *Nopalea coccifera* Lemaire, Cactees 89. 1868, nom. illegit.

Shrub or tree; stem to 5 m, the segments flattened, linear to narrowly obovate, sometimes falcate, (1)1.5–3.5(5) dm long, 5–15 cm wide, green, glabrous, strongly tuberculate, not readily disarticulating; areoles 2–3 cm apart, the woolly trichomes tan, white in age, the spines (0)1(3), to 2 cm long, straight or curved, slightly barbed, brown, gray in age, the glochids 1–2 mm long, yellow, caducous. Flowers 4–7 cm long; floral tube salverform, the outer perianth segments obovate, to 12 mm long, to 7.5 mm wide, red, the midvein sometimes green, the apex acute, the margin entire or undulate, the inner ones spatulate, to 2.2 cm long, to 1 cm wide, erect, red, the apex rounded, the margin entire; stamens 30–40, to 1.5 cm long, pink, much exserted; style 4–4.5 cm long, much exserted, conspicuously swollen basally, pink, the stigma lobes 6–7. Fruit ellipsoid, 2.5–4 cm long, red, the surface glabrous, smooth, with conspicuous areoles, the flesh red; seeds ellipsoid, 3–5 mm long, tan to gray, slightly pubescent, the girdle protruding 1–2 mm.

Disturbed sites. Occasional; central peninsula. Escaped from cultivation. Florida; West Indies, Mexico, Central America, and South America; Asia, Australia, and Pacific Islands. Native to Mexico. All year.

Opuntia ficus-indica (L.) Mill. [The genus *Ficus* (Moraceae) and of the West Indies, in reference to the shape of the fruit and its place of discovery.] TUNA CACTUS.

Cactus ficus-indica Linnaeus, Sp. Pl. 468. 1753. *Opuntia ficus-indica* (Linnaeus) Miller, Gard. Dict., ed. 8. 1768.
Cactus opuntia Linnaeus, Sp. Pl. 468. 1753. *Opuntia vulgaris* Miller, Gard. Dict., ed. 8. 1768. *Cactus compressus* Salisbury, Prodr. Stirp. Chap. Allerton 348. 1796, nom. illegit. *Cactus opuntia* Linnaeus var. *vulgaris* (Miller) de Candolle, Hist. Pl. Succ. pl. 138. 1804. *Opuntia vulgaris* Miller var. *major* Salm-Reifferscheid-Dyck, Observ. Bot. Horto Dyck. 3: 9. 1822, nom. inadmiss. *Opuntia opuntia* (Linnaeus) H. Karsten, Deut. Fl. 888. 1882, nom. inadmiss. *Opuntia compressa* J. F. Macbride, Contr. Gray Herb. 65: 41. 1922, nom. illegit. *Platyopuntia vulgaris* (Miller) Ritter, Kakt. Suedamer. 1: 35. 1979, nom. illegit.

Tree or shrub; stem to 6 m, the segments flattened, broadly oblong to ovate or narrowly elliptic, (2)3–6 dm long, 2–3 dm wide, green, glabrous, low tuberculate, not readily disarticulating; areoles 2–5 cm apart, the woolly trichomes brown, the spines 2–6 per areole or absent, to 1–2.5(4) cm long, to 5 mm long, straight to slightly curved, basally angular-flattened, slightly barbed, whitish, tan, or brown, 0–2 small, bristlelike, deflexed, the glochids ca. 1 mm long, yellow or brown, caducous. Flowers 5–7(10) cm long; floral tube funnelform, the outer perianth segments broadly cuneate to obdeltoid or ovate, 1–2 cm long, 1.5–2 cm wide, yellow to orange, the apex truncate or acute, mucronate, the margin entire or denticulate, the inner ones obovate to cuneate 2.5–3(5) cm long, spreading, yellow to orange, the apex truncate to rounded, mucronate, the margin entire; stamens 6–8 mm long, yellow; style ca. 1.5 cm long, red, the stigma lobes 8–10, yellow. Fruit ovoid to oblong, 5–10 cm long, yellow to orange or purple, the surface glabrous, areoles 45–60, without spines or with spines 5–10 cm long, the flesh yellow to red; seeds subcircular, 4–5 mm long, pale tan, slightly bent, the girdle protruding to 1 mm.

Disturbed sites. Rare; Volusia, Washington, and Escambia Counties. Escaped from cultivation. North Carolina, Georgia, and Florida, Texas west to California; nearly cosmopolitan. Native to Mexico. Summer.

Opuntia humifusa (Raf.) Raf. [Spreading on the ground.] PRICKLYPEAR.

Cactus humifusus Rafinesque, Ann. Nat. 1: 15. 1820. *Opuntia humifusa* (Rafinesque) Rafinesque, Med. Fl. 2: 247. 1830. *Opuntia rafinesquei* Engelmann, Proc. Amer. Acad. Arts 3: 295. 1856, nom. illegit. *Opuntia vulgaris* Miller var. *rafinesquei* A. Gray, Manual, ed. 2. 136. 1856.

Cactus opuntia Linnaeus var. *nana* de Candolle, Pl. Hist. Succ. pl. 138 (opposite). 1804. *Opuntia nana* (de Candolle) Visiani, Fl. Dalmat. 3: 143. 1850. *Opuntia vulgaris* Miller var. *nana* (de Candolle) K. Schumann, Gesamtbeschr. Kakt. 715. 1898. *Opuntia vulgaris* Miller forma *nana* (de Candolle) Schelle, Handb. kakteenkult. 50. 1907.

Opuntia austrina Small, Fl. S.E. U.S. 816, 1335. 1903. *Opuntia compressa* J. F. Macbride var. *austrina* (Small) L. D. Benson, Cact. Succ. J. (Los Angeles) 41: 125. 1969. *Opuntia humifusa* (Rafinesque) Rafinesque var. *austrina* (Small) Dress, Baileya 19: 165. 1975. TYPE: FLORIDA: Miami-Dade Co.: Miami, 28 Oct–18 Nov 1903, *Small & Carter s.n.* (neotype: NY; isoneotypes: GH, PH). Neotypified by Benson (1982: 924).

Opuntia pollardii Britton & Rose, Smithsonian Misc. Coll. 50: 523. 1908.

Opuntia youngii C. Z. Nelson, Trans. Illinois Acad. Sci. 12: 119. 1919. TYPE: FLORIDA: Hillsborough Co.: 9 mi N of Tampa.

Opuntia ammophila Small, J. New York Bot. Gard. 20: 29. 1919. *Opuntia compressa* J. F. Macbride, var. *ammophila* (Small) L. D. Benson, Cact. Succ. J. (Los Angeles) 41: 124. 1969. *Opuntia humifusa* (Rafinesque) Rafinesque, var. *ammophila* (Small) L. D. Benson, Cact. Succ. J. (Los Angeles) 48: 59. 1976. TYPE: FLORIDA: St. Lucie Co.: St. Lucie Sound 6 mi. S of Fort Pierce, 20 Dec 1917, *Small 8456* (holotype: NY; isotype: NY).

Opuntia lata Small, J. New York Bot. Gard. 20: 26. 1919. *Opuntia mesacantha* Rafinesque subsp. *lata* (Small) Majure, Phytoneuron 2014-106: 1. 2014. TYPE: FLORIDA: Alachua Co.: 12 mi. W of Gainesville, 13 Dec 1917, *Small s.n.* (holotype: NY).

Opuntia abjecta Small ex Britton & Rose, Cact. 4: 257. 1923. TYPE: FLORIDA: Monroe Co.: Big Pine Key, 12 Apr 1921, *Small & Matthews s.n.* (lectotype: NY). Lectotypified by Benson (1982: 922).

Opuntia eburnispina Small ex Britton & Rose, Cact. 4: 260, f. 237. 1923. TYPE: FLORIDA: Collier Co.: Cape Romano, 10 May 1922, *Small s.n.* (holotype: NY).

Opuntia impedata Small ex Britton & Rose, Cact. 4: 257, f. 235. 1923. TYPE: FLORIDA: Duval Co.: Atlantic Beach, E of Jacksonville, Apr 1921, *Small s.n.* (holotype: NY; isotype: US).

Opuntia pisciformis Small ex Britton & Rose, Cact. 4: 258, f. 236. 1923. TYPE: FLORIDA: Duval Co.: Pilot Island, Atlantic Beach, 26 Apr 1921, *Small s.n.* (holotype: NY).

Opuntia turgida Small ex Britton & Rose, Cact. 4: 265. 1923. TYPE: FLORIDA: Volusia Co.: ca. 5 mi. S of Daytona, 31 Nov 1919, *Small et al. s.n.* (holotype: NY).

Opuntia cumulicola Small, Man. S.E. Fl. 907, 1506. 1933. TYPE: FLORIDA: Miami-Dade Co.: Bull Key, opposite Lemon City, 6 Nov 1903, *Small & Carter 970* (holotype: NY; isotype: US).

Opuntia polycarpa Small, Man. S.E. Fl. 906, 1506. 1933. TYPE: FLORIDA: Collier Co.: Caxambas Island, 11 May 1922, *Small s.n.* (holotype: NY; isotype: US).

Opuntia turbinata Small, Man. S.E. Fl. 910, 1506. 1933. TYPE: FLORIDA: Duval Co.: "St. Scorge Island" [Fort George Island], 22 Aug 1922, *Small s.n.* (holotype: NY?).

Shrub; stem prostrate to erect, to 2 m, the segments flattened, orbicular to broadly oblong to obovate, 5–17.5 cm long, 4–12 cm wide, dark or shiny green, glabrous, wrinkling when stressed, tuberculate, not readily disarticulating; areoles 1–2 cm apart, the woolly trichomes tan to brown, the spines 0–2(3) per areole, 2.5–6 cm long, straight, occasionally 1 deflexed, slightly barbed, whitish to brownish, the glochids 3–4 mm long, yellow to red-brown. Flowers 4–6 cm long; floral tube funnelform, the outer perianth segments ovate, 0.5–2.5 cm long, 0.5–2 cm wide, green with yellow margin, the apex short-acuminate or acute, the margin undulate, the inner ones cuneate-obovate, 2.5–4 cm long, 1–2 cm wide, spreading, yellow, the apex rounded, the margin entire; stamens ca. 1 cm long, yellow to orange; style ca. 1 cm long, greenish yellow, the stigma lobes 5–10, white. Fruits elongate to obovate, 3–5 cm long, orange or brownish red, the surface glabrous, with 10–18 areoles, the flesh orange or red; seeds 3–4 mm long, tan, the girdle protruding to 1 mm.

Sandhills, dry, open hammocks, and scrub. Common; nearly throughout. Massachusetts south to Florida, west to Montana, Colorado, and New Mexico. All year.

Some authors recognize *O. pollardii* as the common species in Florida, and treat *O. humifusa* s.s. as a northern species absent from Florida (Ward, 2009; Majure et al., 2012a; Majure et al., 2012b). Other segregates have been recognized as Florida endemics: *O. abjecta* (Florida Keys), *O. ammophila* (having a well-defined trunk and found in the Ocala region), and *O. austrina* (with narrowly obovate stems and found in the central peninsula). Due to the uncertainty of discrete morphological differences and the degree of introgression, *O. humifusa* s.l. is here tentatively treated as a widespread and variable species.

Opuntia leucotricha DC. [From the Greek *leukos*, white, and *trichos*, spine.] ARBORESCENT PRICKLYPEAR.

Opuntia leucotricha de Candolle, Mém. Mus. Hist. Nat. 17: 119. 1828.

Tree or shrub; stem erect, to 3 m, the segments flattened, oblong to orbicular, 1–2.7 dm long, 0.7–1.5(2) dm wide, green or blue-green, puberulent, low tuberculate, not readily disarticulating; areoles 1–2 cm apart, the woolly trichomes white, the spines 1–5(6) per areole, to 7 cm long, filiform, flexible, irregularly curving, slightly compressed, slightly barbed, white or gray, the glochids ca. 1 mm long, yellow or brown. Flowers 4–5 cm long; floral tube funnelform, the outer perianth segments lanceolate, 0.5–2 cm long, 3–4(6) mm wide, greenish, the apex acuminate to mucronate, the margin entire, the inner ones obovate to broadly spatulate, 2–2.5 cm long, 1–1.5 cm wide, spreading, yellow, the apex mucronate, the margin minutely denticulate; stamens ca. 1 cm long, yellow; style ca. 1 cm long, yellow, the stigma lobes 10, yellow. Fruits globose, 4–6 cm long, purple-red, the surface puberulent, with deciduous clusters of spines, the flesh red; seeds discoid, ca. 4 mm long, tan, smooth, the girdle protruding to 1 mm.

Hammocks. Rare; St. Lucie County. Escaped from cultivation. Not recently collected. Florida; Mexico. Native to Mexico. All year.

Opuntia ochrocentra Small ex Britton & Rose [With the center of yellowish brown, in reference to the flower.] BULLSUCKERS.

> *Opuntia ochrocentra* Small ex Britton & Rose, Cact. 4: 262. 1923. TYPE: FLORIDA: Monroe Co.: S end of Big Pine Key, 11 Dec 1921, *Small et al. s.n.* (holotype: NY; isotype: US).

Shrub; stem to 1 m, the segments flattened, elliptic, (4)10–17.5 cm long, (3)5–7 cm wide, dark green, glabrous, moderately tuberculate, not readily disarticulating; areoles 1–2 cm apart, the woolly trichomes pale tan to white, the spines 2–3(5) per areole, 2.5–6 cm long, straight, slightly flattened, spirally twisted, distinctly barbed, yellowish, gray-white in age, the glochids 3–5 mm long, yellow, red to brown in age. Flowers 6–7 cm long; floral tube funnelform, the outer perianth segments ca. 5, cuneate, 1.5–2.5 cm long, 0.5–1.5 cm wide, greenish yellow or purplish, the apex truncate, mucronate, the margin entire, the inner ones narrowly obovate or narrowly obcordate, 2.5–3(4) cm long, 1–2.5 cm wide, spreading, yellow, the apex retuse or cuspidate, the margin entire; stamens to 1.5 cm long, pale yellow; style 1–2 cm long, greenish yellow, the stigma lobes 5, pale cream. Fruit obovoid, 3–4 cm long, reddish or purplish, the surface glabrous, with 17–21 areoles with a few spines to 9 mm long, the flesh red; seeds suborbicular, ca. 3 mm long, tan, the girdle protruding to 0.3 mm.

Tropical hammocks. Rare; Monroe County keys. Florida; West Indies. All year.

This taxon is apparently derived from *O. humifusa* s.l. and *O. stricta* (as *O. abjecta* and *O. dillenii* in Majure et al., 2012b.)

Opuntia pusilla (Haw.) Haw. [Small.] COCKSPUR PRICKLYPEAR.

> *Cactus pusillus* Haworth, Misc. Nat. 188. 1803. *Opuntia pusilla* (Haworth) Haworth, Syn. Pl. Succ. 195. 1812.
>
> *Opuntia drummondii* Graham, Botanist 5: pl. 246. 1846. TYPE: FLORIDA: St. Johns Co.: 5 mi. S of Ponte Vedra, 2 Sep 1954, *Benson & Benson 15388* (neotype: POM). Neotypified by Benson (1982: 923).
>
> *Opuntia pes-corvi* Leconte, in Engelmann, Proc. Amer. Acad. Arts 3: 346. 1856. TYPE: FLORIDA: Franklin Co.: cultivated at St. Louis, 1860, *Engelmann s.n.* (neotype: MO; isotype: F). Neotypified by Benson (1982: 923).
>
> *Opuntia tracyi* Britton, Torreya 11: 152. 1911.

Shrub; stem to 7.5 cm long, the segments flattened or subcylindric, narrowly elliptic to nearly linear, 2.5–5(8) cm long, 1.2–2.5 cm wide, green or purplish red under stress, glabrous, tuberculate, not readily disarticulating; areoles 6–9 mm apart, the woolly trichomes tan to gray, the spines (0)1–2(4) per areole, 2–2.5 cm long, straight, erect or spreading, strongly barbed, yellowish or red-brown, gray in age, the glochids 3–4 mm long, pale yellow, gray or brown in age. Flowers 4–6 cm long; floral tube funnelform, the outer perianth segments ovate, 5–25 mm long, 5–15 mm wide, green, the apex acute, the margin slightly toothed and undulate, yellow, the inner ones obovate, 2–3 cm long, 1.5–2 cm wide, spreading, yellow, the apex rounded or emarginate, the margin slightly toothed; stamens ca. 1 cm long, yellow; style 1.5–2 cm long, white, the stigma lobes 5–10, white. Fruit ellipsoid, 2.5–3 cm long, red-purple, the surface glabrous, with 8–16 areoles, spines absent, the flesh red; seeds suborbicular, 4–6 mm long, tan, the girdle slightly protruding.

Dunes. Occasional; northern counties, Pinellas, Hillsborough, and Sarasota Counties. North Carolina south to Florida, west to Texas. Summer.

Opuntia stricta (Haw.) Haw. [Erect, in reference to the habit.] ERECT PRICKLYPEAR; SHELL-MOUND PRICKLYPEAR.

Cactus strictus Haworth, Misc. Nat. 188. 1803. *Opuntia stricta* (Haworth) Haworth, Syn. Pl. Succ. 191. 1812.

Cactus opuntia Linnaeus var. *inermis* de Candolle, Pl. Hist. Succ. pl. 138. 1804. *Opuntia inermis* (de Candolle) de Candolle, Prodr. 3: 473. 1828.

Cactus dillenii Ker Gawler, Bot. Reg. 3: pl. 255. 1818. *Opuntia dillenii* (Ker Gawler) Haworth, Suppl. Pl. Succ. 79. 1819. *Opuntia stricta* (Haworth) Haworth var. *dillenii* (Ker Gawler) L. D. Benson, Cact. Succ. J. (Los Angeles) 41: 126. 1969.

Opuntia bartramii Rafinesque, Atl. J. 146. 1832. TYPE: FLORIDA: Volusia Co.: New Smyrna region.

Opuntia maritima Rafinesque, Atl. J. 146. 1832. TYPE: "Florida to Carolina."

Opuntia spinalba Rafinesque, Atl. J. 147. 1832. TYPE: FLORIDA: Monroe Co.: "keys of Florida."

Opuntia bentonii Griffiths, Rep. (Annual) Missouri Bot. Gard. 22: 25, pl. 1–2. 1912. TYPE: FLORIDA: Baker Co.: McClenny, cultivated at Washington, DC, 24 Apr. 1910, *Griffiths 8374* (holotype: US; isotype: POM, US).

Opuntia keyensis Britton ex Small, J. New York Bot. Gard. 20: 31, pl. 225. 1919. TYPE: FLORIDA: Monroe Co.: Boot Key, 7–12 Apr 1909, *Britton 537* (holotype: NY; isotype: US).

Opuntia zebrina Small, J. New York Bot. Gard. 20: 35, pl. 226. 1919. TYPE: FLORIDA: Monroe Co.: East Cape Sable, 28 Nov 1916, *Small s.n.* (lectotype: NY). Lectotypified by Benson (1982: 932).

Opuntia atrocapensis Small, Man. S.E. Fl. 905, 1506. 1933. TYPE: FLORIDA: Monroe Co.: Middle Cape Sable, 28 Nov 1916, *Small s.n.* (Neotype: US). Neotypified by Majure (2014: 1).

Opuntia magnifica Small, Man. S.E. Fl. 910, 1506, 1933. TYPE: FLORIDA: Nassau Co.: southern part of Amelia Island, 21 Aug 1922, *Small et al. s.n.* (holotype: NY; isotype: US).

Opuntia nitens Small, Man S.E. Fl. 906, 1506. 1933. TYPE: FLORIDA: Volusia Co.: Green Mound, 5 mi. S of Daytona, 23 Aug 1922, *Small et al. s.n.* (holotype: NY; isotype).

Opuntia tenuiflora Small, Man. S.E. Fl. 908, 1506. 1933. TYPE: Florida: Monroe Co.: Lignum Vitae Key, 30 Mar 1916, *Small s.n.* (holotype: NY).

Shrub; stem erect or sprawling, to 2 m, the segments flattened, narrowly elliptic or obovate, 1–2.5(4) dm long, 7.5–15(25) cm wide, green or blue-green, glabrous, tuberculate, glabrous, not readily disarticulating; areoles 3–5 cm apart, the woolly trichomes dense, tan, the spines 0–11 per areole, 2–4(6) cm long, straight or slightly curving, slightly flattened, slightly barbed, yellow, brown in age, the glochids 4–7 mm long, yellow, brown in age. Flowers 5–6 cm long; floral tube funnelform, the outer perianth segments broadly deltoid-obovate to obovate, 1–2.5 cm long, 0.6–1.2 cm wide, the apex mucronate to short-acuminate, greenish yellow with yellow margins, the margin entire or slightly crisped, the inner ones obovate to cuneate, 2.5–3 cm long, 1.2–2 cm wide, spreading, light yellow, the apex rounded or truncate, emarginate, the margin entire or slightly undulate; stamens 1–1.3 cm long, yellow; style 1.2–2 cm long, yellow, the stigma lobes 5–10, yellowish. Fruit ellipsoidal or obovoid, 4–6 cm long, purplish, the surface glabrous, with 6–10 areoles lacking spines, the flesh red; seeds subcircular, 4–5 mm long, tan, the surface slightly irregular, the girdle protruding to 1 mm.

Shell middens, dunes, and coastal hammocks. Frequent; peninsula, central and western panhandle. Virginia south to Florida, west to Texas; nearly cosmopolitan. Native to North America and tropical America. All year.

EXCLUDED TAXA

Opuntia cubensis Britton & Rose—This taxon is apparently restricted to Cuba and was long misapplied to *O. ochrocentra* in Florida (see Majure et al., 2014).

Opuntia englemannii Salm-Reifferscheidt-Dyck ex Engelmann var. *lindheimeri* (Engelmann) B. D. Parfitt & Pinkava)—Small (1933, as *Opuntia lindheimeri* Engelmann) reported the Texas prickly-pear as an escape from cultivation in Florida. No Florida specimens known; possibly a misidentification of *O. stricta.*

Opuntia monacanthos (Willdenow) Haworth—Reported for Florida by Benson (1982, as *O. vulgaris*), followed by Wunderlin (1998, as *O. vulgaris*) and Wunderlin and Hansen (2003, 2011). The vouchering specimens have turned out to be from plants in cultivation.

Opuntia polyacantha Haworth—Reported by Chapman (1897), the name misapplied to material of *O. stricta.*

Opuntia triacanthos (Willdenow) Sweet—This taxon is perhaps endemic to the West Indies and reports of it for Florida have apparently been misapplied to material of the *O. humifusa* complex (Majure et al., 2012a).

Opuntia tuna (Linnaeus) Miller—Reported by Small (1903), the name misapplied to material of *O. stricta.*

Pereskia Mill. 1754.

Trees, shrubs, or vines; stems unsegmented, lacking ribs and tubercles; areoles in the leaf axils with woolly trichomes and spines, glochids absent. Leaves well-developed, alternate, the blade pinnate-veined, petiolate, estipulate. Flowers in terminal or axillary paniculiform, corymbiform, or aggregate cymes, pedicellate or the terminal one sessile, bisexual, diurnal; floral tube inconspicuous, the perianth segments numerous; stigma lobes 5 or 10. Fruit indehiscent; seeds numerous, exarillate.

A genus of 9 species; North America, West Indies, Mexico, Central America, South America, Africa, Asia, and Australia. [Commemorates Nicolas Claude Fabri de Peiresc (1580–1637), French scholar.]

Recently *Pereskia* has been found to be paraphyletic and the segregate genus *Leuenbergeria* was proposed for the most basal group in the Cactaceae.

Selected reference: Hawkes (2003a).

1. Clambering shrub or vine; spines dimorphic, those of the young branches recurved, those of the older branches straight; flowers many in a paniculiform or corymbiform cyme; filaments white ....
....**P. aculeata**
1. Shrub or tree; spines of one kind, all straight; flowers few in a terminal aggregate cyme; filaments red
....**P. grandifolia**

Pereskia aculeata Mill. [Prickly.] BARBADOS GOOSEBERRY; LEMON VINE.

Cactus pereskia Linnaeus, Sp. Pl. 469. 1753. *Pereskia aculeata* Miller, Gard. Dict., ed. 8. 1768. *Cactus lucidus* Salisbury, Prodr. Strip. Chap. Allerton 349. 1796, nom. illegit. *Pereskia pereskia* (Linnaeus) H. Karsten, Deut. Fl. 888. 1882, nom. inadmiss.

Clambering shrub or vine, to 10 m; stem with the areoles to 12 mm long, the spines yellow to gray, of 2 kinds, the primary ones (produced on young branches)(1)2(3) per areole, 4–8 mm long, recurved and clawlike, somewhat flattened, the secondary ones (produced on older branches) to 25 per areole, 10–35 mm long, straight. Leaves with the blade lanceolate to ovate or oblong, 4.5–11 cm long, 1.5–5 cm wide, succulent, the apex short-acuminate, the base acute to rounded, the margin entire, the upper and lower surfaces glabrous, the lateral veins 4–7, the petiole 0.5–1 cm long. Flowers to 70 in a terminal or lateral panicle or corymb ca. 3 cm long, 2.5–5 cm wide, the pedicel 5–15 mm long; floral tube cupulate, the outer perianth segments oblanceolate to oblong, 2–2.5 cm long, ca. 1 cm wide, pale green, the apex acute to slightly acuminate, the margin entire, the inner ones oblanceolate to oblong, 2–2.5 cm long, ca. 1 cm wide, white to pale yellow or light pink, the apex acute to slightly acuminate, the margin entire; stamens 6–9 mm long, white; style ca. 12 mm long, the stigma lobes 5, white. Fruit elliptical, ca. 4 cm long, yellow to orange, the surface smooth, glabrous, the flesh yellow to orange; seeds lenticular, 4–5 mm long, dark brown to black, glossy.

Disturbed shell middens. Rare; central and southern peninsula. Escaped from cultivation. Florida and Texas; West Indies, Mexico, Central America, and South America; Africa, Asia, and Australia. Native to West Indies, Central America, and South America. Fall.

Pereskia grandifolia Haw. [Large-leaved.] ROSE CACTUS.

> *Pereskia grandifolia* Haworth, Suppl. Pl. Succ. 85. 1819. *Cactus grandifolius* (Haworth) Link, Enum. Pl. Hort. Berol. 2: 25. 1822. *Rhodocactus grandifolius* (Haworth) F. M. Knuth, in Backeberg & F. M. Knuth, Kaktus-ABC 97. 1936.

Shrub or tree, to 5 m; stem with the areoles to 12 mm long, the spines 1–6 per areole, all of one kind, straight, 1–6.5 cm long, dark gray. Leaves with the blade elliptic to obovate to lanceolate, 6–30 cm long, 3–6 cm wide, succulent, the apex short-acuminate, the base cuneate, the margin entire, the upper and lower surfaces glabrous, the lateral veins 7–13, the petiole 0.5–1.2 cm long. Flowers 10–50 in a terminal or lateral cymose-paniculate cluster 1.5–3 cm long, 3–7 cm wide, the pedicel of the lateral flowers 1–3 cm long, the terminal flower sessile; floral tube cupulate, the outer perianth segments 1–2 cm long, ca. 0.5 cm wide, green or reddish with green midvein, the apex mucronate, the margin entire, the inner ones 2–4 cm long, ca. 1 cm wide, pink, the apex mucronate or short-acuminate, the margin entire; stamens 6–9 mm long, yellow; style ca. 1.2 cm long, white, the stigma lobes 10, white. Fruit obovoid, 5–10 cm long, reddish green to yellowish, smooth, glabrous, the flesh reddish or yellowish; seeds obovoid-ellipsoidal, 5–7 mm long, dark brown to black, striate with minute vertical bands of pits, glossy.

Disturbed sites. Rare; Hillsborough and Highlands Counties. Escaped from cultivation. Florida; West Indies, Mexico, Central America, and South America. Native to South America. Fall.

Pilosocereus Byles & G. D. Rowley 1957. TREE CACTUS

Trees or shrubs; stems unsegmented, cylindric, ribbed; areoles with woolly trichomes and spines, glochids absent. Leaves vestigial or absent. Flowers solitary in the lateral or subterminal areoles, sessile, bisexual, nocturnal; floral tube short, the perianth segments numerous; stigma lobes 5. Fruit dehiscent along one side or indehiscent; seeds numerous, exarillate.

A genus of about 40 species; North America, West Indies, Mexico, Central America, and South America. [*Pilosus*, shaggy, apparently in reference to the trichomes at the base of the spines, and the genus *Cereus*, which it resembles.]

Pilocereus Lem., 1839.

Selected reference: Parfitt and Gibson (2003c).

1. Flowering areoles densely woolly, the trichomes more than 8 mm long.............................**P. polygonus**
1. Flowering areoles glabrate or short-pubescent, the trichomes less than 5 mm long................**P. robinii**

Pilosocereus polygonus (Lam.) Byles & G. D. Rowley [From the Greek *poly*, many, and *gonia*, angle, in reference to the many-angled stems.] GREATER ANTILLEAN TREE CACTUS.

Cactus polygonus Lamarck, Encycl. 1: 539. 1783. *Cereus polygonus* (Lamarck) de Candolle, Prodr. 3: 466. 1828. *Cephalocereus polygonus* (Lamarck) Britton & Rose, Contr. U.S. Natl. Herb. 12: 418. 1909. *Pilocereus plumieri* Lemarie, Rev. Hort. 1862: 427. 1862, nom. illegit. *Pilocereus polygonus* (Lamarck) K. Schumann, Gesamtbeschr. Kakt. 196. 1897. *Pilosocereus polygonus* (Lamarck) Byles & G. D. Rowley, Cact. Succ. J. Gr. Brit. 19: 67. 1957.

Tree or shrub, to 10 m; stem erect or ascending, 5–10 cm wide, dull green or blue-green when young, glabrous, 9- to 14-ribbed, the ribs blunt or acute, usually higher than wide; areoles 1–1.5 cm apart, the woolly trichomes 8–20 mm long, the spines 9–20(30) per areole, 1–3.5(5.5) cm long, acicular, radially spreading and ascending, yellowish with a darker base, gray-brown to yellow-brown in age. Flower 5–6 cm long; floral tube cupulate, the outer perianth segments ovate to obovate, 1.2–2.5 cm long, ca. 7.5 mm wide, light green with a brownish midvein, the apex acute to obtuse, the margin entire or minutely denticulate, the inner ones creamy white, ovate to elliptic, the innermost oblanceolate to broadly linear, 1.2–2.5 cm wide, ca. 7.5 mm wide, the apex acuminate, the margin entire; stamens 6–12 mm long, white; style ca. 3 cm long, creamy white, the stigma creamy white. Fruit depressed-globose, 3–4 cm long, red, the surface smooth, sometimes with 1–2 scales, the flesh red; seeds obovate, ca. 2 mm long, black, smooth, glossy.

Rocky tropical hammocks. Rare; Monroe County keys. Florida; West Indies. All year.

Pilosocereus robinii (Lem.) Byles & G. D. Rowley [Commemorates Charles-Philippe Robin (1821–1885), French human anatomist, histologist, and natural historian.] KEY TREE CACTUS.

Pilocereus robinii Lemaire, Ill. Hort. 11 (misc.): 74. 1864. *Cephalocereus robinii* (Lemaire) Britton & Rose, Cact. 2: 39. 1920. *Pilosocereus robinii* (Lemaire) Byles & G. D. Rowley, Cact. Succ. J. Gr. Brit. 19: 67. 1957. *Cereus robinii* (Lemaire) L. D. Benson, Cact. Succ. J. (Los Angeles) 41: 126. 1969.

Cephalocereus bahamensis Britton, Contr. U.S. Natl. Herb. 12: 415. 1909. *Cereus bahamensis* (Britton) Vaupel, Monatsschr. Kakteenk. 23: 23. 1913. *Pilocereus bahamensis* (Britton) F. M. Knuth, in Backeberg & F. M. Knuth, Kaktus-ABC 329. 1935. *Pilosocereus bahamensis* (Britton) Byles & G. D. Rowley, Cact. Succ. J. Gr. Brit. 19: 66. 1957.

Cephalocereus keyensis Britton & Rose, Contr. U.S. Natl. Herb. 12: 416. 1909. *Cereus keyensis* (Britton & Rose) Vaupel, Monatsschr. Kakteenk. 23: 23, 26. 1913. *Pilocereus keyensis* (Britton & Rose) F. M. Knuth, in Backeberg & F. M. Knuth, Kaktus-ABC 331. 1936. *Pilosocereus keyensis* (Britton & Rose) Byles & G. D. Rowley, Cact. Succ. J. Gr. Brit. 19: 67. 1957. TYPE: FLORIDA: Monroe Co.: Key West, 7 Apr 1909, *Britton 518* (holotype: NY).

Cephalocereus deeringii Small, J. New York Bot. Gard. 18: 201. 1917. *Pilocereus deeringii* (Small) F. M. Knuth, in Backeberg & F. M. Knuth, Kaktus-ABC 330. 1935. *Pilosocereus deeringii* (Small) Byles & G. D. Rowley, Cact. Succ. J. Gr. Brit. 19: 66. 1957. *Cereus robinii* (Lemaire) L. D. Benson, var. *deeringii* (Small) L. D. Benson, Cact. Succ. J. (Los Angeles) 41: 126. 1969. *Pilosocereus robinii* (Lemaire) Byles & G. D. Rowley var. *deeringii* (Small) Kartesz & Gandhi, Phytologia 71: 276. 1991. TYPE: FLORIDA: Monroe Co.: Lower Matecumbe Keys, 8 Apr 1916, *Small 7790* (holotype: NY).

Tree or shrub, to 10 m; stem erect or ascending, 5–10 cm wide, dull green or blue-green when young, glabrous, 9- to 14-ribbed, the ribs blunt or acute, usually higher than wide; areoles 1–1.5 cm apart, the trichomes to 3–4 mm long, the spines 9–20(30) per areole, to 1–1.5(2.5) cm long, acicular, radially spreading and ascending, yellowish with a darker base, gray-brown to yellow brown in age. Flower 5–6 cm long; floral tube cupulate, the outer perianth segments ovate to obovate, 1.2–2.5 cm long, ca. 7.5 mm wide, light green with a brownish midvein, the apex acute to obtuse, the margin entire or minutely denticulate, the inner ones creamy white, ovate to elliptic, the innermost oblanceolate to broadly linear, 1.2–2.5 cm long, ca. 7.5 mm wide, the apex acuminate, the margin entire; stamens 6–12 mm long, white; style ca. 3 cm long, creamy white, the stigma creamy white. Fruit depressed-globose, 3–4 cm long, red, the surface smooth, sometimes with 1–2 scales, the flesh red; seeds obovate, ca. 2 mm long, black, smooth, glossy.

Rocky tropical hammocks. Rare; Monroe County keys. Florida; West Indies. All year.

Pilosocereus robinii is listed as endangered in Florida (Florida Administrative Code, Chapter 5B-40) and in the United States (U.S. Fish and Wildlife Service, 50 CFR 23).

EXCLUDED TAXA

Pilosocereus monoclonos (de Candolle) Byles & G. D. Rowley—Reported by Chapman (1860, 1883, 1897, all as *Cereus monoclonos* de Candolle) and Small (1903, as *Cereus monoclonos* de Candolle), who misapplied the name to material of *P. polygonus*.

"*Cereus robinii* (Lemaire) L. D. Benson var. *keyensis* (Britton & Rose) L. D. Benson ex R. W. Long & Lakela," nom. nud.—*Pilosocereus robinii* was reported under this name by Long and Lakela (1971), but the combination was never formally made.

Rhipsalis Gaertn. 1788.

Pendent epiphytic shrubs, rarely ascending to erect; stems segmented, cylindric, ribbed; areoles with spines, woolly trichomes and glochids absent. Leaves vestigial or absent. Flowers solitary in the lateral areoles, rotate, campanulate, or funnelform, sessile, bisexual, diurnal; floral tube conspicuous or inconspicuous, the perianth segments 10–15; stamens ca. 36; stigma lobes 3. Fruit indehiscent; seeds few, exarillate.

A genus of about 40 species; North America, West Indies, Mexico, Central America, South

America, Africa, and Asia. [From the Greek *rhips*, wickerwork, in reference to the slender, flexible stem segments].

1. Stems monomorphic, the segments cylindrical or slightly winged; flowers rotate............. **R. baccifera**
1. Stems heteromorphic, the basal segments cylindrical, the distal ones shorter, cylindrical or clavate; flowers campanulate or funnelform ....**R. cereuscula**

Rhipsalis baccifera (Sol.) Stearn [*Bacca*, berry, and *ferre*, to bear.] MISTLETOE CACTUS.

Cassytha baccifera Solander, in J. S. Miller, Ill. Syst. Sex. Linnaei, Class 9, Ord. 1. t. 29. 1771. *Rhipsalis cassutha* Gaertner, Fruct. Sem. Pl. 1: 137. 1788, nom. illegit. *Hariota cassutha* Lemaire, Cact. Gen. Sp. Nov. 75. 1839, nom. illegit. *Rhipsalis baccifera* (Solander) Stearn, Cact. J. (Croydon) 7: 107. 1939.

Epiphytic shrub; stem to 1 m long, pendent, the segments monomorphic, cylindrical or slightly, 10–20 cm long, 3–6 mm wide, green, glabrous, 8- to 10-ribbed; areoles with 1–9 spines 1–2 mm long, bristlelike, white, deciduous in age. Flowers rotate, ca. 6 mm long, ca. 10 mm wide; floral tube saucerlike, the outer perianth segments deltoid, ca. 2 mm long, ca. 1 mm wide, green, the apex acute, the margin entire, the inner ones elliptic, 2–3 mm long, 1–1.5 mm wide, white, the apex acute, the margin entire; stamens 1–2 mm long, white; style ca. 2 mm long, white, the stigma lobes white. Fruit spherical to ellipsoid, 5–8 mm long, translucent, white, the surface smooth, the flesh white; seeds obovoid or ellipsoid, 1–2 mm long, black, finely reticulate but appearing smooth.

Hammocks; epiphytic. Rare; Miami-Dade and Monroe Counties Florida; West Indies, Mexico, Central America, and South America; Africa and Asia. All year.

Rhipsalis baccifera is listed as endangered in Florida (Florida Administrative Code, Chapter 5B-40).

Rhipsalis cereuscula Haw. [To resemble a small *Cereus*.] CORAL CACTUS.

Rhipsalis cereuscula Haworth, Philos. Mag. Ann. Chem. 7: 112. 1830. *Hariota cereuscula* (Haworth) Kuntze, Revis. Gen. Pl. 1: 262. 1891. *Erythrorhipsalis cereuscula* (Haworth) Volgin, Vestn. Moskovsk. Univ., Ser. 16, Biol. 36(3): 19. 1981.

Epiphytic shrub; stem to 1 m long, pendent, the segments heteromorphic, the basal ones 7–40(50) cm long, 3–4 mm wide, cylindrical, green, glabrous, the distal ones shorter and usually clavate, 4- or 5-ribbed in age; areoles with 2–4 spines 1–2 mm long, bristlelike, whitish, deciduous in age. Flowers ca. 1.5 cm long, ca. 1.2 cm wide, white to creamy white, faintly flushed with pink; floral tube campanulate to funnelform, the outer perianth segments subglobose to obconical, 4–5 mm long, the apex acute, the margin entire, the inner ones 8–10 mm long, ca. 3 mm wide, slightly larger than the outer, narrowly elliptic, the apex acute; stamens 7–10 mm long, white with a slightly reddish base; style 9–10 mm long, white, the stigma lobes whitish. Fruit subglobose, 5–7 mm long, translucent, white, the surface smooth, the flesh white; seeds not seen.

Hammocks; epiphytic. Rare; Seminole County. Escaped from cultivation. Florida; South America. Native to South America. All year.

Selenicereus (A. Berger) Britton & Rose 1909. MOONLIGHT CACTUS

Shrubs or vines; stems segmented or unsegmented, cylindric, ribbed; areoles with woolly trichomes and spines, glochids absent. Leaves vestigial or absent. Flowers solitary in lateral areoles, sessile, bisexual, nocturnal; floral tube elongate, the perianth segments numerous; stamens numerous; stigma lobes 10. Fruit indehiscent; seeds numerous, exarillate.

A genus of about 28 species; North America, West Indies, Mexico, Central America, and South America. [From the Greek *selene*, moon, in reference to its nocturnal blooming, and the genus *Cereus*, from which it was removed.]

Selected reference: Hawkes (2003c).

1. Stem armed with short, conic spines 1–3 mm long...**S. pteranthus**
1. Stem armed with slender spines 4–12(15) mm long..**S. grandiflorus**

Selenicereus grandiflorus (L.) Britton & Rose [Large-flowered.]
QUEEN-OF-THE-NIGHT.

> *Cactus grandiflorus* Linnaeus, Sp. Pl. 467. 1753. *Cereus grandiflorus* (Linnaeus) Miller, Gard. Dict., ed. 8, 1768. *Selenocereus grandiflorus* (Linnaeus) Britton & Rose, Contr. U.S. Natl. Herb. 12: 430. 1909.
> *Cereus nycticallis* Link ex A. Dietrich var. *armatus* K. Schumann, Gesamtbeschr. Kakt. 147. 1898. *Cereus grandiflorus* (Linnaeus) Miller var. *armatus* (K. Schumann) L. D. Benson, Cact. Succ. J. (Los Angeles) 41: 126. 1969.
> *Cereus coniflorus* Weingart, Monatsschr. Kakteenk. 14: 118. 1904. *Selenicereus coniflorus* (Weingart) Britton & Rose, Contr. U.S. Natl. Herb. 12: 430. 1909.

Shrub, terrestrial and hemiephytic or sometimes epiphytic; stem to 5 m long, clambering and vinelike, 1–2.5 cm wide, green or bluish green, sometimes becoming purplish along the ribs, glabrous, 5- to 8-ribbed, the ribs low, rounded; areoles usually 1–2 cm apart along the ribs, the woolly trichomes white, the spines 6–18 per areole, 4–12(15) mm, bristlelike or short-acicular, whitish to brownish, deciduous in age. Flowers 18–30 cm long; floral tube 7–9(11) cm long, with numerous white to tawny, hairlike spines, the outer perianth segments linear, to 10 cm long, 4–5 mm wide, brown, orange, or lemon-yellow, the apex acute, the margin entire, the inner ones oblanceolate to narrowly lanceolate, 7.5–10 cm long, 9–12 mm wide, white, the apex acute or apiculate, the margin entire; stamens 4–5 cm long, white; style 1.5–2 cm long, white, the stigma lobes white. Fruit ovoid, 5–9 cm long, pink to whitish, the surface with numerous hairlike spines, the flesh white; seeds ovoid, black, shiny.

Disturbed hammocks. Rare; St. Lucie, Sarasota, and Broward Counties. Escaped from cultivation. Florida; West Indies and Mexico. Native to West Indies and Mexico. All year.

Selenicereus pteranthus (Link ex A. Dietr.) Britton & Rose [From the
Greek *pteron*, wing, and, *anthos*, flower, in reference to angled stems.]
PRINCESS-OF-THE-NIGHT.

> *Cereus pteranthus* Link ex A. Dietrich, Allg. Gartenzeitung 2: 209. 1834. *Cereus nycticallis* Link ex A. Dietrich, Verh. Vereins Beförd. Gartenbaues Königl. Preuss. Staaten 10: 372. 1834, nom. illegit. *Selenicereus pteranthus* (Link ex A. Dietrich) Britton & Rose, Contr. U.S. Natl. Herb. 12: 431. 1909.

Vine, terrestrial or epiphytic; stem to 30 m long, 2.5–5 cm wide, bluish green to purple, glabrous, 4- to 6-ribbed, the ribs acute; areoles 2–2.5 cm apart along the ribs, the woolly trichomes brownish, the spines 1–5, short-conic, 1–3 mm long, whitish to brownish. Flowers 25–30 cm long; floral tube 10–12 cm long, with scales and numerous white, hairlike spines, the outer perianth segments, linear-lanceolate, 8–12 cm long, 1–2 cm wide, brown or purplish, the apex acute, the margin entire, the inner ones oblanceolate, 8–12 cm long, 3–4 cm wide, white, the apex acute, the margin entire; style ca. 20 cm long, white, the stigma lobes white. Fruit globose, 5–7 cm long, red, the surface with yellowish, bristlelike spines ca. 1 cm long and with long, white trichome-like spines, the flesh white; seeds elongate-obovate, 2–3 mm long, black.

Disturbed hammocks. Occasional; central and southern peninsula. Escaped from cultivation. Florida; Mexico. Native to Mexico. All year.

EXCLUDED GENERA

Brasiliopuntia brasiliensis (Willdenow) A. Berger—Reported by Britton and Rose (1919), Small (1933), and Long and Lakela (1971, as *Opuntia brasiliensis* (Willdenow) Haworth). No Florida specimens known.

Cereus repandus (Linnaeus) Miller—Reported for Florida by Wunderlin and Hansen (2003, 2011), based on a sterile specimen from St. Lucie County that cannot be reliably identified. The specimen may likely be *Cereus hexagonus* (Linnaeus) Miller, which is commonly cultivated throughout peninsular Florida.

CORNACEAE Bercht. ex J. Presl, nom. cons. 1825. DOGWOOD FAMILY

Trees or shrubs. Leaves opposite or alternate, simple, pinnate-veined, petiolate, estipulate. Flowers in terminal or axillary cymes, actinomorphic, bisexual or unisexual (plants androdioecious), bracteate or ebracteate, bracteolate or ebracteolate; sepals 4, basally connate or free; petals 4, free; stamens 4, free, the anthers dorsifixed, longitudinally dehiscent; nectariferous disk present; ovary inferior, 2-carpellate and -loculate. Fruit a drupe; seeds 1(2).

A family of 1 genus and about 58 species; North America, Mexico, Central America, Europe, Africa, and Asia.

Nyssa, sometimes included in the Cornaceae, is here placed in the Nyssaceae based on molecular studies (e.g., Xiang et al., 2002, 2011).

Selected references: Ferguson (1966); Xiang et al. (2006).

Cornus L. 1753. DOGWOOD

Trees or shrubs. Leaves opposite or alternate, simple, pinnate-veined, petiolate, estipulate. Flowers in terminal or axillary cymes; sepals 4, basally connate or free; petals 4, free, inserted on the margin of a nectariferous disk; stamens 4, free; ovary 2-carpellate and carpellate, the style and stigma 1. Fruit a drupe; seeds 1(2).

A genus of about 58 species; North America, Mexico, Central America, South America, Africa, and Asia. [*Cornu*, horn, in reference to the hardness of the wood, in having been used for skewers and daggers.]

Cynoxylon Raf., 1838; *Swida* Opiz, 1838.

Selected references: Murrell (1992, 1993).

1. Leaves alternate or sometimes crowded near the branch tips and appearing whorled ....
....**C. alternifolia**
1. Leaves opposite.
    2. Flowers in compact cymes subtended by 4 white petaloid bracts; fruit red ....**C. florida**
    2. Flowers in open cymes, the bracts absent or minute; fruit blue.
        3. Upper leaf surface glabrous; pith of second year stem brown ....**C. amomum**
        3. Upper leaf surface pubescent; pith of second year stems white or tan.
            4. Leaves with a few short, straight, appressed trichomes on the lower surface ....**C. foemina**
            4. Leaves with numerous long, forked, erect trichomes on the lower surface ....**C. asperifolia**

Cornus alternifolia L. f. [With alternate leaves.] PAGODA DOGWOOD; ALTERNATELEAF DOGWOOD.

Cornus alternifolia Linnaeus f., Suppl. Pl. 125. 1782 ("1781"). *Cornus alternifolia* Linnaeus f. var. *corallina* Aiton, Hort. Kew. 1: 159. 1789, nom. inadmiss. *Swida alternifolia* (Linnaeus f.) Small, Fl. S.E. U.S. 853, 1335. 1903. *Bothriocaryum alternifolium* (Linnaeus f.) Pojarkova, Bot. Mater. Gerb. Bot. Inst. Komarova Akad. Nauk SSSR 12: 170. 1950.

Tree, to 9(12) m; bark relatively smooth, greenish streaked with brown or gray, the branchlets green, becoming brown to maroon, pubescent, the second year pith white. Leaves alternate, or sometimes crowded near the branch tips and appearing whorled, the blade elliptic-ovate, ovate, or obovate, 5–12 cm long, 2.5–5 cm wide, the apex short-acuminate, the base broadly tapered to rounded, the margin entire, the upper surface glabrous, the lower surface sparsely appressed-pubescent, sometimes shaggy-pubescent along the major veins, the petiole 1–6 cm long, glabrous or sparsely pubescent. Flowers in a flat-topped or convex cyme 3–6 cm wide, the axis sparsely shaggy-pubescent; the hypanthium densely appressed-pubescent; sepal lobes minute; petals cream, 3–4 mm long. Fruit subglobose, 4–7 mm long, blue, sparsely appressed pubescent, the pyrene subglabrous, laterally compressed, 5–6 mm long, slightly ribbed.

Bluff forests and creek swamps. Rare; central and western panhandle. Newfoundland and Quebec south to Florida, west to Manitoba, Minnesota, Iowa, Missouri, Arkansas, and Mississippi. Spring.

Cornus alternifolia is listed as endangered in Florida (Florida Administrative Code, Chapter 5B-40).

Cornus amomum Mill. [Beautiful, pleasing.] SILKY DOGWOOD.

Cornus amomum Miller, Gard. Dict., ed. 8. 1768. *Swida amomum* (Miller) Small, Fl. S.E. U.S. 854, 1336. 1903. *Thelycrania amomum* (Miller) Pojarkova, Bot. Mater. Gerb. Bot. Inst. Komarova Akad. Nauk SSSR 12: 165. 1950.

Cornus sericea Linnaeus, Mant. Pl. 199. 1771. *Thelycrania sericea* (Linnaeus) Dandy, Watsonia 4: 47. 1957. *Cornus caerulea* Lamarck, Encycl. 2: 116. 1786, nom. illegit.

Shrub, to 5 m; stem relatively smooth, green, reddish brown, or red, with irregular horizontal flecks of corky outgrowth, these often surrounding lenticels, the branchlets green or maroon, shaggy-pubescent with silvery or rusty-brown trichomes, becoming glabrous, the second year pith tawny to brown. Leaves opposite, the blade ovate to oblong, 3–10 cm long, 2–8 cm wide, the apex short-acuminate, the base truncate, rounded, or cuneate, the margin entire, the upper surface glabrous or sparsely pubescent, the lower surface appressed- and/or shaggy-pubescent with gray or rusty-brown trichomes, the main veins shaggy-pubescent with gray trichomes, the smaller ones appressed pubescent with rust-brown trichomes, the petiole 0.5–2 cm long, shaggy pubescent. Flowers in a flat-topped to hemispheric cyme, ca. ½ as long as the subtending leaves; hypanthium shaggy-pubescent; sepal lobes subulate, 1–2 mm long, shaggy-pubescent; petals cream, 3–5 mm long; styles swollen just below the stigma. Fruit subglobose, 5–9 mm long, blue with areas of cream, the pyrene globose, 4–6 mm long, irregularly longitudinal-ridged.

Floodplain forests and lake margins. Rare; central panhandle, Escambia County. Maine south to Florida, west to Michigan, Iowa, Missouri, Tennessee, and Mississippi. Spring.

Cornus asperifolia Michx. [With rough leaves.] ROUGHLEAF DOGWOOD.

Cornus asperifolia Michaux, Fl. Bor.-Amer. 1: 93. 1803. *Cornus sericea* Linnaeus var. *asperifolia* (Michaux) de Candolle, Prodr. 4: 272. 1830. *Cornus stricta* Lamarck var. *asperifolia* (Michaux) Feay ex A. W. Wood, Class-book Bot., ed. 1861. 392. 1861. *Swida asperifolia* (Michaux) Small, Fl. S.E. U.S. 854, 1336. 1903. *Thelycrania asperifolia* (Michaux) Pojarkova, Bot. Mater. Gerb. Bot. Inst. Komarova Akad. Nauk SSSR 12: 165. 1950.

Cornus microcarpa Nash, Bull. Torrey Bot. Club 23: 103. 1896. *Swida microcarpa* (Nash) Small, Fl. S.E. U.S. 853, 1336. 1903. *Thelycrania microcarpa* (Nash) Pojarkova, Bot. Mater. Gerb. Bot. Inst. Komarova Akad. Nauk SSSR 12: 165. 1950. *Cornus foemina* Miller subsp. *microcarpa* (Nash) J. S. Wilson, Trans. Kansas Acad. Sci. 67: 797. 1965. TYPE: FLORIDA: Gadsden Co.: River Junction, s.d., *Nash 2589* (holotype: NY; isotype: US).

Shrub, to 5 m; bark gray, splitting into small platelets, the branchlets brown becoming gray, short-shaggy-pubescent to appressed pubescent, the second year pith white. Leaves opposite, the blade elliptic, ovate, or lanceolate, 2–8 cm long, 1–4 cm wide, the apex short-acuminate, the base rounded to cuneate, the upper surface moderately pubescent with spreading to appressed trichomes, rough to the touch, the lower surface densely pubescent with spreading to gray or rusty-brown appressed trichomes, the petiole 2–5 mm long, shaggy- or appressed-pubescent. Flowers in a flat-topped or convex cyme 3–4 cm wide, ca. ½ as long as the subtending leaves; hypanthium densely appressed-pubescent; sepal lobes to 1 mm long, sparsely pubescent; petals cream, ca. 2 mm long, pubescent externally; style slightly swollen just below the stigmas. Fruit subglobose, ca. 8 cm long, light blue, sparsely pubescent, the pyrene globose, 3–5 mm long, smooth or slightly grooved.

Wet hammocks. Occasional, northern counties south to Hernando and Sumter Counties. North Carolina south to Florida, west to Mississippi. Spring.

Cornus florida L. [Flowering, in reference to the conspicuous petaloid floral bracts.] FLOWERING DOGWOOD.

Cornus florida Linnaeus, Sp. Pl. 117. 1753. *Benthamidia florida* (Linnaeus) Spach, Hist. Vég. 8: 107. 1839. *Cynoxylon floridum* (Linnaeus) Rafinesque ex Small, Fl. S.E. U.S. 854. 1903. *Benthamia florida* (Linnaeus) Nakai, Bot. Mag. (Tokyo) 23: 41. 1909.

Tree, to 12(20) m; bark gray to nearly black, irregularly broken into scaly blocks, the branchlets green or red, eventually brown, sparsely appressed-pubescent, becoming glabrous, the second year pith white. Leaves ovate, broadly elliptic, oblong-elliptic, subrotund, or lanceolate, 3–10 cm long, 2–7 cm wide, somewhat decurrent on the petiole, the apex acute, the base cuneate to rounded, sometimes inequilateral and oblique, the margin entire, the upper surface gray appressed-pubescent, the lower gray appressed-pubescent, often with gray shaggy pubescence along the major veins, the petiole 0.5–1.5 cm long, sparsely short-pubescent. Flowers in a cyme compacted into heads, surrounded by 4 petaloid white or pale cream-white, spreading, obovate bracts, the apex notched; hypanthium densely appressed-pubescent; sepal lobes to 1 mm long, appressed-pubescent; petals yellow or greenish yellow, 3–4 mm long, appressed-pubescent exteriorly. Fruit oblong-elliptic, 8–14 mm long, red, the pyrene ellipsoid, 10–12 mm long, smooth.

Mesic hammocks. Frequent; northern counties south to Manatee County. Maine south to Florida, west to Ontario, Michigan, Illinois, Kansas, Oklahoma, and Texas; Mexico. Spring.

Cornus foemina Mill. [Female.] SWAMP DOGWOOD; STIFF DOGWOOD.

Cornus foemina Miller, Gard. Dict., ed. 8. 1768. *Swida foemina* (Miller) Rydberg, Brittonia 1: 94. 1931. *Cornus stricta* Lamarck, Encycl. 2: 116. 1786. *Swida stricta* (Lamarck) Small, Fl. S.E. U.S. 853, 1335. 1903. *Thelycrania stricta* (Lamarck) Pojarkova, Bot. Mater. Gerb. Bot. Inst. Komarova Akad. Nauk SSSR 12: 165. 1950.

Shrub or small tree, to 8 m; bark gray to brownish gray, somewhat plated, the branchlets reddish brown to maroon or tan, glabrous, the second year pith white. Leaves with the blade lanceolate, elliptic, or ovate, 2–10 cm long, 1–4 cm wide, the apex acute or acuminate, the base cuneate or rounded, the margin entire, the upper and lower surfaces glabrous or with sparse, short, appressed trichomes, smooth to the touch, the petiole to ca. 1 cm long, grayish appressed-pubescent. Flowers in a convex cyme 3–7 cm wide, the axis glabrous or sparsely pubescent with short, rusty-brown trichomes; hypanthium densely grayish appressed-pubescent; sepal lobes triangular, ca. 1 mm long, sparsely pubescent; petals cream, 3–4 mm long, with a few minute, appressed trichomes exteriorly; style somewhat swollen just below the stigmas. Fruit subglobose to ellipsoid, 4–6 mm long, blue, the pyrene globose to oblong, 3–4 mm long, slightly ribbed.

Wet hammocks and floodplain forests. Frequent; nearly throughout. New Jersey south to Florida, west to Illinois, Missouri, Oklahoma, and Texas. Spring.

EXCLUDED TAXA

Cornus drummondii C. A. Meyers—Reported for Florida by Correll & Johnston (1970), who misapplied the name to material of *C. asperifolia*.

Cornus racemosa Lamarck—Reported for Florida by Correll and Johnston (1970). No Florida specimens known.

NYSSACEAE Juss. ex Dumort., nom. cons. 1829. TUPELO FAMILY

Trees. Leaves alternate, simple, pinnate-veined, petiolate, estipulate. Flowers axillary, solitary or in ball-like clusters, racemes, or spikes, actinomorphic, bisexual or unisexual (plants monoecious, polygamomonoecious, or polygamodioecious), bracteate or ebracteate; sepals 5 or absent; petals 5 or absent; stamens 5–10, or absent, the anthers dorsifixed, longitudinally dehiscent; ovary inferior, 1-carpellate and -loculate. Fruit a drupe.

A family of 2 genera and about 15 species; North America, Mexico, Central America, and Asia.

The Nyssaceae is segregated from the Cornaceae based on molecular studies (e.g., Xiang et al., 2002, 2011).

Selected reference: Eyde (1966).

Nyssa L. 1753. TUPELO

Trees. Leaves alternate, simple, pinnate-veined, petiolate, estipulate. Flowers axillary, solitary or in ball-like clusters, racemes, or spikes; floral tube closed with a nectariferous disk; sepals 5 or absent; petals 5 or absent; stamens 5–10, free; ovary 1-carpellate and -loculate. Fruit a drupe; seed 1.

A genus of about 12 species; North America, Mexico, Central America, and Asia. [*Nyssa*, the Greek water nymph; the genus so called because the original species grows in water.]

Nyssa, sometimes included in Cornaceae, is placed here based on molecular studies (e.g., Xiang et al. 2002, 2011).

Selected reference: Burckhalter (1992).

1. Petioles (2)3–6 cm long..**N. aquatica**
1. Petioles 0.5–2(2.5) cm long.
    2. Larger leaves 10–30 cm long; staminate flowers sessile, in a ball-like cluster; carpellate flowers 1(2); fruit 2.5–4 cm long, red..**N. ogeche**
    2. Larger leaves 2–10 cm long; staminate flowers pedicellate, in a short raceme; carpellate flowers usually (2)3–5(8), rarely solitary; fruit 1–1.5 cm long, blue-black..........................**N. sylvatica**

Nyssa aquatica L. [Growing in water.] WATER TUPELO.

Nyssa aquatica Linnaeus, Sp. Pl. 1058. 1753.
Nyssa uniflora Wangenheim, Beytr. Teut. Forstwiss. 83, pl. 27(57). 1787.

Nyssa denticulata Aiton, Hort. Kew. 3: 446. 1789. *Nyssa palustris* Salisbury, Prodr. Stirp. Chap. Allerton 175. 1796, nom. illegit.

Nyssa tomentosa Michaux, Fl. Bor.-Amer. 2: 259. 1803; non Poiret, 1798. TYPE: "ad amnem S. Mary
et in Florida."

Bumelia denticulata Rafinesque, New Fl. 3: 29. 1838 ("1836"). *Streblina denticulata* (Rafinesque) Rafinesque, Autik. Bot. 75. 1840. TYPE: FLORIDA.

Tree, to 30 m; branchlets pubescent or glabrous. Leaves with the blade ovate or ovate-oblong,
6–30 cm long, 2–15 cm wide, the apex acute to acuminate, the base rounded to subcordate or
broadly cuneate, the margin entire or with 1–several coarse dentate teeth or lobes on a side,
the upper surface pubescent on the veins, becoming glabrous, the lower surface densely pubescent, becoming sparsely pubescent only along the veins, much lighter than the upper, the
petiole 3–6 cm long, sparsely pubescent. Flowers bisexual or unisexual (plants monoecious or
polygamomonoecious); staminate flowers 1(2–5), sessile in a ball-like cluster; carpellate and
bisexual flowers 1–2, subtended by 2–several unequal bracts at the end of a peduncle 1.5–2 cm
long. Fruit oblong or obovate, 2–4 cm long, red or green with a reddish blush, the pyrene with
papery wings.

Swamps, floodplain forests, and pond and lake margins. Frequent; northern counties, Levy
County. Virginia south to Florida, west to Missouri, Arkansas, and Texas. Spring.

Nyssa ogeche W. Bartram ex Marshall [In reference to the Ogeechee River of Georgia
where first collected.] OGEECHEE TUPELO.

Nyssa ogeche W. Bartram ex Marshall, Arbust. Amer. 97. 1785.
Nyssa capitata Walter, Fl. Carol. 253. 1788.

Tree, to 20 m; branchlets pubescent. Leaves with the blades elliptic, oblanceolate, or oblong-
ovate, to 1.5 cm long, to 8 cm wide, the apex acuminate, acute, or rounded, the base cuneate,
rounded, or subcordate, the margin entire or with 1–several coarsely dentate teeth on a side,
the upper surface sparsely pubescent, becoming glabrous, the lower surface densely pubescent,
becoming sparsely pubescent, much lighter than the upper, the petiole 1–2(3) cm long, pubescent. Flowers bisexual or unisexual (plants polygamodioecious); staminate flowers (1)2–5,
sessile, in a ball-like cluster; carpellate or bisexual flowers 1–2, subtended by 2–several unequal
bracts at the end of a peduncle 1.5–2 cm long. Fruit red or green with a red blush, the pyrene
with papery wings.

Floodplain forests, swamps, sloughs, and pond and lake margins. Frequent; northern counties, introduced and naturalized in Hillsborough County. South Carolina, Georgia, Alabama,
and Florida. Spring.

Nyssa sylvatica Marshall [Growing in woods.]

Tree, to 30 m. Leaves with the blades elliptic, lance-elliptic, lanceolate, ovate, oblanceolate,
obovate to elliptic-oblong or suborbicular, 3–10(15) cm long, 2–9 cm wide, the apex acuminate,
acute, or obtuse, the base cuneate to rounded, the margin entire or with 1–several dentate teeth,
the upper surface pubescent, becoming glabrous, the lower surface pubescent or glabrous.

Flowers bisexual or unisexual (plants monoecious or polygamomonoecious); staminate flowers (1)2–5(8) in a short, loose raceme; carpellate or bisexual flowers (1)2–4(8) in a bracted pedunculate cluster or 5–6 in an interrupted spike. Drupe ellipsoid, oblong-ellipsoid or subglobose, ca. 1 cm long, blue-black, the pyrene ribbed.

1. Carpellate or bisexual flowers or the fruits (2)3–5(8) in a bracteate, pedunculate cluster or an interrupted spike; plant of mesic habitats..var. **sylvatica**
1. Carpellate or bisexual flowers or the fruits (1)2(3) in a bracteate pedunculate cluster; plant of hydric habitats..var. **biflora**

Nyssa sylvatica var. **sylvatica** BLACKGUM.

> *Nyssa sylvatica* Marshall, Arbust. Amer. 97. 1785. *Nyssa multiflora* Wangenheim var. *sylvatica* (Marshall) S. Watson, Bibl. Index N. Amer. Bot. 442. 1878. *Nyssa sylvatica* Marshall var. *typica* Fernald, Rhodora 37: 343. 1935, nom. inadmiss.
> *Nyssa multiflora* Wangenheim, Beytr. Teut. Forstwiss. 46, pl. 16(39). 1787.
> *Nyssa caroliniana* Poiret, in Lamarck, Encycl. 4: 507. 1798. *Nyssa sylvatica* Marshall var. *caroliniana* (Poiret) Fernald, Rhodora 37: 436. 1935.
> *Nyssa sylvatica* Marshall var. *dilatata* Fernald, Rhodora 37: 436. 1935.

Leaves with the blade membranaceous, obovate to elliptic-oblong or suborbicular, 3–10(15) cm long, the apex short-acuminate, the base cuneate or rounded. Carpellate or bisexual flowers (2)3–5(8) in a bracteate pedunculate cluster or 5–6 in an interrupted spike. Fruit with the pyrene only slightly ribbed.

Mesic hammocks. Frequent; northern counties south to Manatee County. Maine south to Florida, west to Ontario, Wisconsin, Illinois, Missouri, Kansas, Oklahoma, and Texas. Spring.

Nyssa sylvatica var. **biflora** (Walter) Sarg. [Two-flowered.] SWAMP TUPELO.

> *Nyssa biflora* Walter, Fl. Carol. 253. 1788. *Nyssa sylvatica* Marshall var. *biflora* (Walter) Sargent, Sylva 5: 76. 1893. *Nyssa sylvatica* Marshall subsp. *biflora* (Walter) E. Murray, Kalmia 13: 10. 1983.
> *Nyssa ursina* Small, Torreya 27: 92. 1927. *Nyssa sylvatica* Marshall var. *ursina* (Small) Wen & Stuessy, Syst. Bot. 18: 79. 1993. *Nyssa biflora* Walter var. *ursina* (Small) D. B. Ward, Novon 11: 362. 2001. TYPE: FLORIDA: Gulf Co.: swamp N of Port St. Joe, 27 Nov 1923, *Small et al. 10995* (lectotype: NY; isolectotypes: NY) Lectotypified by Wen and Stuessy (1993: 79).

Leaves with the blade thick, elliptic to lance-elliptic, lanceolate, or oblanceolate, 3–8(10) cm long, the apex acute, obtuse, or rounded, the base cuneate. Carpellate or bisexual flowers (1)2(3) in a bracteate pedunculate cluster. Fruit with the pyrene conspicuously ribbed.

Swamps, pond margins, and bogs. Frequent; northern counties, central peninsula. New Jersey south to Florida, west to Illinois, Missouri, Arkansas, and Texas. Spring.

Shrublike or small trees in pineland swamps and bogs in the panhandle, with narrow leaves and subglobose (versus ellipsoid or ovate) drupes, have been treated as *N. ursina* or *N. biflora* var. *ursina* by various workers (for a brief discussion, see Ward, 2001). These are here treated as morphological variation within *N. sylvatica* var. *biflora*.

HYDRANGEACEAE Dumort., nom. cons. 1829. HYDRANGEA FAMILY

Woody vines or shrubs. Leaves opposite, simple, pinnate- or pinnipalmate-veined, petiolate, estipulate. Flowers in cymes, actinomorphic, bisexual or sterile; hypanthium present; sepals 4—12, basally connate; petals 4–12, basally connate and forming a floral tube; stamens 8–90, adnate to the hypanthium; ovary inferior or nearly so, 2- to 12-carpellate and -loculate, the styles 1–4. Fruit a capsule.

A family of 17 genera and about 190 species; North America, Mexico, Central America, South America, Europe, Asia, and Pacific Islands.

The genera included here have sometimes been placed in the Saxifragaceae by various authors.

Selected reference: Spongberg (1972).

1. Woody vine; petals 7–12..**Decumaria**
1. Shrub; petals 4 or 5.
    2. Some flowers larger than the others and sterile..**Hydrangea**
    2. Flowers all the same size and fertile.
        3. Stamens twice as many as the petals; trichomes stellate.....................................**Deutzia**
        3. Stamens more than twice as many as the petals; trichomes simple.........................**Philadelphus**

Decumaria L. 1763.

Woody vines. Leaves opposite, simple, pinnate-veined, petiolate, estipulate. Flowers in terminal corymbiform cymes, actinomorphic, bisexual; sepals 7–12; petals 7–12; stamens 20–30, inserted near the top of the hypanthium, free; ovary 6- to 12-carpellate and -loculate, the style 1, the stigma capitate. Fruit a capsule, dehiscence intercostal (lateral walls separating from the ribs and leaving a cage-like remnant).

A genus of 2 species; North America and Asia. [Related to tenths, for the sometimes 10-merous flowers.]

Decumaria barbara L. [*Barbatus*, bearded, with tufts of long, weak trichomes.] COWITCH VINE; CLIMBING HYDRANGEA; WOODVAMP.

Decumaria barbara Linnaeus, Sp. Pl., ed. 2. 1663. 1763. *Decumaria radicans* Moench, Methodus 107. 1794, nom. illegit. *Decumaria forsythia* Michaux, Fl. Bor.-Amer. 1: 282. 1803, nom. illegit.

Vine, to 1 m; stem with adventitious roots, glabrous. Leaves with the blade 10–12 cm long, 6–8 cm wide, the apex acute, obtuse, rounded, or mucronate, the base cuneate, truncate, cordate, or rounded, the margin entire, the upper surface glabrous, the lower surface finely pubescent along the veins, the petiole 10–30 mm long, slightly winged proximally, glabrous or short-pubescent. Flowers in a terminal, corymbiform cyme 3–8 cm long, 4–10 cm wide, the peduncle 2–6 cm long, glabrous, the pedicel 1–6 mm long, glabrous; hypanthium ca. 2 mm long; sepals lanceolate, ca. 1 mm long, glabrous; petals linear-lanceolate, ca. 3 mm long, white; style stout, the stigma capitate with 7–12 radiating stigmatic lines. Fruit 3–5 mm long, dehiscence intercostal (the lateral walls separating from the ribs and leaving a cage-like remnant); seeds numerous, fusiform, 2–3 mm long, yellow.

Moist to wet hammocks and swamps. Frequent; northern counties, central peninsula. New York south to Florida, west to Arkansas and Louisiana. Spring.

Deutzia Thunb. 1781. PRIDE-OF-ROCHESTER

Shrubs. Leaves opposite, simple, pinnate-veined, petiolate, estipulate. Flowers in terminal racemiform or paniculiform cymes, actinomorphic, bisexual; sepals 5; petals 5; stamens 10, inserted near the top of the hypanthium; ovary 3- to 5-carpellate and -loculate, the styles 3(4), free. Fruit a capsule, dehiscence septicidal.

A genus of about 60 species; North America, Central America, Europe, Asia, and Pacific Islands. [Commemorates Johann van der Deutz (1743–1784), Dutch merchant, amateur botanist, and patron of Carl Peter Thunberg.]

Deutzia scabra Thunb. [Roughened.] FUZZY PRIDE-OF-ROCHESTER.

> *Deutzia scabra* Thunberg, Nov. Gen. Pl. 19. 1781. *Deutzia scabra* Thunberg var. *typica* C. K. Schneider, Ill. Handb. Laubholzk. 1: 379. 1905, nom. inadmiss.

Shrub, to 30 dm; branches stellate-pubescent. Leaves with the blade ovate-lanceolate to ovate, 3–8 cm long, 1.5–5 cm wide, the apex acute to acuminate, the base rounded to cuneate, the margin crenate-dentate, the upper surface stellate-pubescent, the lower surface densely stellate-pubescent, the petiole 1–3 mm long, sparsely to densely stellate-pubescent. Flowers in a racemiform or paniculiform cyme, the pedicel 2–10 mm long, sparsely stellate-pubescent; hypanthium campanulate, 3–5 mm long, densely stellate-pubescent; sepal lobes triangular to ovate, ca. 2 mm long, stellate-pubescent; petals elliptic to oblong, 7–15 mm long, white or pinkish, the outer surface stellate pubescent, the inner surface glabrous or stellate-pubescent; stamens with the filaments 2-toothed at the apex, the outer ones 7–9 mm long, the inner ones 5–6 mm long. Fruit hemispheric, 4–5 mm long, dehiscence septicidal, acropetally along the septum at the base of the fruit and apically; seeds numerous, ellipsoidal, dark brown.

Disturbed hammocks. Rare; Franklin County. Escaped from cultivation. Vermont south to Florida, west to Illinois and Kentucky, also Florida; Asia. Native to Asia. Spring–summer.

Hydrangea L. 1753.

Shrubs. Leaves opposite, simple, pinnate-veined, petiolate, estipulate. Flowers in terminal paniculiform cymes, actinomorphic, bisexual or sterile; sepals 5; petals 5; stamens 8–10, inserted near the top of the hypanthium; ovary 2- to 4-carpellate and -loculate, the styles 2, free. Fruit a capsule, dehiscence interstylar, creating a pore at the base of the styles.

A genus of about 30 species; North America, Central America, South America, and Asia. [From the Greek *hydro*, water and *angeion*, the diminutive of *angos*, a vessel or container, in reference to the shape of dehisced fruit.]

1. Leaves unlobed, pubescent only on the principal veins of the lower surface; flowers in a corymbiform cyme; sterile flowers on the periphery of the inflorescence or absent..............................**H. arborescens**
1. Leaves deeply lobed, conspicuously pubescent on the lower surface; flowers in a paniculiform cyme; sterile flowers interspersed in the inflorescence with the perfect ones..............................**H. quercifolia**

Hydrangea arborescens L. [Becoming treelike.] WILD HYDRANGEA; MOUNTAIN HYDRANGEA.

Hydrangea arborescens Linnaeus, Sp. Pl. 397. 1753. *Hydrangea vulgaris* Michaux, Fl. Bor.-Amer. 1: 268. 1803, nom. illegit. *Hydrangea arborescens* Linnaeus var. *vulgaris* Seringe, in de Candolle, Prodr. 4: 14. 1830, nom. inadmiss. *Hydrangea arborescens* Linnaeus forma *typica* C. K. Schneider, Ill. Handb. Laubholzk. 1: 387, nom. inadmiss. *Hydrangea arborescens* Linnaeus forma *vulgaris* (Seringe) C. K. Schneider, Ill. Handb. Laubholzk. 1: 387. 1905, nom. inadmiss.

Shrubs, to 2 m; bark brown, exfoliating in irregular papery patches, the branchlets pubescent. Leaves with the blade ovate to elliptic-ovate, (3)6–17(20) cm long, (1.5)2.5–12(15) cm wide, the apex acute to acuminate, the base cordate, truncate, or cuneate, the margin dentate to serrate, the upper surface glabrous or sparsely pubescent, the lower surface glabrous or sparsely pubescent along the midribs and sometimes also the lateral veins, the petiole 1.5–8.5(11) cm long, glabrous or sparsely tomentose. Flowers 100–500 in a convex to flat-topped, paniculiform cyme 4–12 cm long, 4–15 cm wide, glabrous or sparsely pubescent. Sterile flowers present or absent, terminating the lateral branches on the periphery of the inflorescence; hypanthium 6–16 mm long, glabrous, with 3–4(5) petaloid sepal lobes, these oval or obovate, 4–15 mm long, white, greenish white, or yellowish white. Fertile flowers with the hypanthium ca. 1 mm long, adnate to the ovary to near the apex, 8–10(11) ribbed, glabrous, the sepal lobes deltate to triangular, ca. 0.5 mm long; petals elliptic to narrowly ovate, 1–2 mm long, white or yellowish white, caducous; styles 2, each with a broad, short stigma. Fruit hemispheric, 1–2 mm long, brown, 8- to 10-ribbed, opening by a pore between the persistent styles; seeds fusiform, ca. 1 mm long, longitudinally ribbed, brown.

Mesic hammocks. Rare; Liberty and Walton Counties. New Brunswick south to Florida, west to Kansas and Oklahoma. Summer.

Hydrangea quercifolia W. Bartram [With leaves like oak (*Quercus*, Fagaceae).] OAKLEAF HYDRANGEA; GRAYBEARD.

Hydrangea quercifolia W. Bartram, Travels Carolina 382, t. 6. 1791.

Shrubs, to 3 m; bark gray, peeling into papery patches and producing a varicolored mottling, the branchlets with a rust-colored loose pubescence. Leaves with the blade suborbicular to ovate, (3)7–26 cm long and wide, 3- to 9- lobed or unlobed, the apex acute to acuminate, the base truncate to cuneate, the margin coarsely serrate, the upper surface glabrous or sparsely pubescent, the lower surface gray-tomentose, the petiole (1)1.5–8(12) cm long, densely tomentose. Flowers 500–1,000 in an ovoid or conic paniculiform cyme, 9–32 cm long, (6)8–14 cm wide, sparsely pubescent or glabrous. Sterile flowers always present, interspersed in the inflorescence with the perfect ones; hypanthium 1–3 cm long; petaloid sepal lobes 4, obovate to ovate or suborbicular, 6–20 mm long, white, turning pink or purplish. Fertile flowers with the hypanthium ca. 1 mm long, adnate to the ovary to near the apex, (7)8- to 10-ribbed, glabrous; petals elliptic, oblong, or spatulate, 2–3 mm long, white or yellowish white, caducous; styles 2, each with a broad, short stigma. Fruit hemispheric or campanulate, 2–3 mm long, 8- to 10-ribbed, brown, dehiscent by a pore between the persistent styles; seeds fusiform-falcate, ca. 1 mm long, longitudinally ribbed, brown.

Bluff forests and wooded ravines, often along stream banks. Occasional; Marion County, central and western panhandle. North Carolina south to Florida, west to Tennessee, Mississippi, and Louisiana. Spring.

Philadelphus L. 1753. MOCKORANGE

Shrubs. Leaves opposite, simple, pinnipalmate-veined, petiolate, estipulate. Flowers in terminal or subterminal cymes, actinomorphic, bisexual; sepals 4; petals 4; stamens 60–90, inserted near the top of the hypanthium; ovary 4-carpellate and -loculate, the style 1, the stigma 4-lobed. Fruit a capsule, dehiscence loculicidal.

A genus of about 25 species; North America, Mexico, Central America, Europe, and Asia. [From the Greek *phila*, loving, and a*delphos*, brother, said to be named for Ptolemy Philadelphus (309–246 BC), King of Ptolemaic Egypt (283–247 BC).]

Philadelphus inodorus L. [Scentless, in reference to the flowers.] SCENTLESS MOCKORANGE.

> *Philadelphus inodorus* Linnaeus, Sp. Pl. 470. 1753. *Syringa inodora* (Linnaeus) Moench, Methodus 678. 1753.
> *Philadelphus grandiflorus* Willdenow, Enum. Pl. 511. 1809. *Philadelphus inodorus* Linnaeus var. *grandiflorus* (Willdenow) A. Gray, Manual, ed. 2. 146. 1856.

Shrub, to 4 m; bark brown, gray, or stramineous, the branchlets glabrous to sparsely strigose, the bark reddish, flaking off. Leaves with the blade lanceolate to ovate or elliptic, (3.5)5–12(14) cm long, 2–5(7) cm wide, the base cuneate to rounded, the margin entire or irregularly serrate, crenate, or dentate, the upper surface glabrous or sparsely strigose, the apex acute to acuminate, the lower surface glabrous or strigose, the petiole 1–8 mm long, strigose. Flowers 1–3 in a terminal or subterminal cyme, the pedicel 3–8 mm long, glabrous or sparsely strigose; hypanthium adnate to the ovary to near the apex, obconic or turbinate, 4–6 mm long, glabrous or sparsely strigose; sepal lobes ovate to ovate-lanceolate, 7–14 mm long, glabrous or strigose on the outer surface, densely villous on the inner; petals oblong, obovate, or suborbicular, 1.5–2.5(3) cm long, white; stamens 60–90, 5–11 mm long; style 1–1.5 cm long, divided at the apex into 4 lobes 4–8 mm long. Fruit obconic to obovoid, 1–1.3 cm long, longitudinally ribbed; seeds linear-fusiform, 2–3 mm long, rusty-brown.

Calcareous hammocks. Rare; Gadsden, Liberty, Jackson, and Okaloosa Counties. Massachusetts south to Florida, west to Ontario, Wisconsin, Illinois, Missouri, Arkansas, and Louisiana. Spring.

LOASACEAE Juss., nom. cons. 1804. LOASA FAMILY

Herbs. Leaves alternate, simple, pinnate-veined, petiolate, estipulate. Flowers solitary, axillary, actinomorphic, bisexual; sepals 5, basally connate; petals 6, free, but adnate with the stamens to form a ring; stamens numerous, the anthers basifixed, dehiscent by longitudinal slits; ovary inferior, 3-carpellate, 1-loculate, the style and stigma 1. Fruit a dehiscent capsule.

A family of 20 genera and about 350 species; North America, West Indies, Mexico, Central America, South America, Africa, Asia, and Pacific Islands.

Selected reference: Ernst and Thompson (1963).

Mentzelia L. 1753. BLAZINGSTAR

Perennial herbs. Leaves alternate, simple, pinnate-veined, petiolate, estipulate. Flowers solitary, axillary, actinomorphic, bisexual; sepals 5, basally connate; petals 6, free, but adnate with the stamens to form a ring; stamens 20–35; ovary inferior, 3-carpellate, 1-loculate, the style and stigma 1. Fruit a capsule, Dehiscence loculicidal by apical valves; seeds 4–8, exarillate.

A genus of about 95 species; North America, West Indies, Mexico, Central America, and South America. [Commemorates Christian Mentzel (1622–1701), German physician and botanist.]

Selected reference: Darlington (1934).

Mentzelia floridana Nutt. ex Torr. & A. Gray [Of Florida.] POORMAN'S PATCH; STICKLEAF.

Mentzelia floridana Nuttall ex Torrey & A. Gray, Fl. N. Amer. 1: 533. 1840. TYPE: FLORIDA: s.d., *Peale s.n.* (lectotype: PH). Lectotypified by Darlington (1934: 132).

Ascending perennial herb, to 6 dm; stem rough-pubescent with barbed trichomes. Leaves with the blade ovate to triangular-ovate, sometimes 3-lobed, 2–9 cm long, 2–6 cm wide, brittle, the apex acute to acuminate, the base subtruncate to broadly cuneate, the margin dentate, the upper and lower surfaces scabrous-pubescent with barbed trichomes. Flowers solitary in the axil of the upper leaves, sessile; sepal lobes lanceolate to linear-lanceolate, 5–6 mm long; petals ovate, 0.6–1.3 cm long, 4–7 mm wide, creamy yellow to orange, adnate to the stamens and forming a ring; stamens 6–11 mm long, the filaments narrowly spatulate; ovary densely hispid, the style 8–10 mm long. Fruit ellipsoid-clavate, 1–1.8 cm long, 4–6 mm wide, hispid; seeds (4)6–8, oblong or pyriform.

Hammocks, shell middens, and dunes. Common; peninsula. Florida; West Indies. All year.

BALSAMINACEAE A. Rich., nom. cons. 1822. JEWELWEED FAMILY

Herbs. Leaves alternate, simple, pinnate-veined, petiolate or epetiolate, estipulate. Flowers axillary, solitary, paired, or in racemiform cymes, bracteate, ebracteolate, zygomorphic, bisexual;

sepals 3, free; petals 5, free or connate; stamens 5, partly connate into a ring around the ovary; ovary superior, 5-carpellate and -loculate, the styles 5, connate. Fruit a dehiscent capsule.

A family of 2 genera and about 1,000 species; North America, Europe, Africa, and Asia.

Impatiens L. 1753. TOUCH-ME-NOT

Annual or perennial herbs. Leaves alternate, pinnate-veined, petiolate, estipulate. Flowers axillary, solitary, paired, or in racemiform cymes, resupinate; sepals with the lower one large, petaloid, saccate and constricted into a nectariferous spur, the 2 lateral ones small; petals with the lateral ones connate into 2 pairs, the upper petal variously modified and differentiated from the 4 lateral ones; stamens wholly or partly connate into a ring around the ovary; styles connate, the stigmas partly or wholly connate. Fruit 5-valvate, explosively loculicidally dehiscent, the valves coiling when opening; seeds numerous.

A genus of about 1,000 species; nearly cosmopolitan. [Impatient, in reference to the sudden bursting of the ripe capsules.]

1. Sepal spur straight or only slightly recurved..I. walleriana
1. Sepal spur strongly recurved.
    2. Flower on a short, stout, erect to spreading pedicel; capsule pubescent .... I. balsamina
    2. Flower on a long, slender, drooping pedicel; capsule glabrous...I. capensis

Impatiens balsamina L. [Balsam-scented.] GARDEN BALSAM; ROSE BALSAM.

> *Impatiens balsamina* Linnaeus, Sp. Pl. 938. 1753. *Balsamina lacca* Medikus, Malvenfam. 71. 1787. *Balsamina foeminea* Gaertner, Fruct. Sem. Pl. 2: 151. 1790, nom. illegit. *Balsamina hortensis* Desportes, in Cuvier, Dict. Sci. Nat. 3: 485. 1817, nom. illegit. *Impatiens balsamina* Linnaeus var. *vulgaris* Wight & Arnott, Prodr. Fl. Ind. Orient. 136. 1834, nom. inadmiss.

Erect annual herb, to 5 dm; stem pubescent. Leaves with the blade ovate-lanceolate, elliptic, or oblanceolate, 3–15 cm long, 1–4 cm wide, the apex acute to acuminate, mucronate, the base cuneate, often decurrent onto the petiole, the margin serrate or dentate, the teeth mucronate, usually glandular toward the leaf base, the upper and lower surfaces glabrous or pubescent, sessile or the petiole to 1 cm long, glabrous or pubescent, usually glandular toward the apex. Flowers solitary or in a racemiform cyme of 2–3(4) in the axil of the upper leaves, the peduncle 0.5–2 cm long; bracts linear, 1–2 mm long, the margin entire; lower sepal abruptly constricted into a recurved spur 1–2.5 cm long, purple-red or pink, the lateral sepals ovate, 2–3 mm long, colored as the lower one; upper petal orbicular to obovate, 6–10 mm long, purple, red, or pink, the apex retuse, the lateral pairs 2–3.5 cm long, colored as the lower one. Fruit ovoid, ellipsoid, or fusiform, 1–2 cm long, pubescent; seeds numerous, subglobose, 2–3 mm long, tuberculate, brown.

Disturbed sites. Rare; Leon, Bay, and Miami-Dade Counties. Escaped from cultivation. New York south to Florida, west to Wisconsin, Illinois, Missouri, and Louisiana; nearly cosmopolitan. Native to Asia. Spring–fall.

Impatiens capensis Meerb. [Believed to be native to the Cape of Good Hope, South Africa.] TOUCH-ME-NOT; JEWELWEED.

Impatiens capensis Meerburgh, Afb. Zeldz. Gew. pl. 10. 1775.
Impatiens biflora Walter, Fl. Carol. 219. 1788. *Impatiens fulva* Nuttall, Gen. N. Amer. Pl. 1: 146. 1818, nom. illegit. *Chrysaea biflora* (Walter) Nieuwland & Lunell, Amer. Midl. Naturalist 4: 473. 1916. *Impatiens noli-tangere* Linnaeus subsp. *biflora* (Walter) Hultén, Ark. Bot., ser. 2. 7(1): 84. 1968.

Erect annual herb, to 2 m; stem sometimes with red or purple streaks or spots, glabrous, sometimes glaucous, the nodes swollen. Leaves with the blade ovate to elliptic, 2–10 cm long, 3–6 cm wide, the apex acute to acuminate, mucronate, the base cuneate to rounded, the margin crenate to dentate, the teeth mucronate, sometimes glandular toward the base, the upper and lower surfaces glabrous, the petiole 1–5 cm long, glabrous, sometimes glandular toward the apex. Flowers solitary or in a racemiform cyme of (2)3–6, in the axil of the upper leaves, the peduncle 1–4 cm long, glabrous; bracts linear-subulate to lanceolate, 3–5 mm long, the margin entire; lower sepal saccate, conic, or helmet-shaped, 1.5–3 cm long, orange, yellow, pink, or white, usually with reddish brown spots, usually abruptly constricted to a recurved spur (0.5)1–2(2.5) cm long, the lateral sepals ovate to obovate, 4–10 mm long, light green or purplish; petals orange, yellow, pink, or white, usually with reddish spots, the upper petal orbicular to obovate, 0.5–2 cm long, the apex retuse, the lateral pairs 1.2–2 cm long, colored as the adaxial one. Fruit clavate, 1.5–2.5 cm long, glabrous; seeds numerous, elliptic, 2–3 mm long, with 4 longitudinal ridges, tuberculate, brown.

Floodplain forests. Rare; Volusia, Liberty, and Jackson Counties. Nearly throughout North America; Europe and Asia. Native to North America. Spring–fall.

Impatiens walleriana Hook. f. [Commemorates Horace Waller (1833–1896), botanist who traveled with David Livingston up the Zambezi River.] GARDEN IMPATIENS; BIZZY-LIZZIE.

Impatiens walleriana Hooker f., in Oliver, Fl. Trop. Afr. 1: 302. 1868.

Erect perennial herb, to 6 dm; stem tinged with purple, sometimes glabrous or nearly so. Leaves with the blade ovate to elliptic, 4–8 cm long, 3–5 cm wide, the apex acute to acuminate or cuspidate, sometimes mucronate, the base cuneate to rounded, sometimes decurrent onto the petiole, the margin crenate to serrate, the teeth mucronate, often glandular toward the base, the upper and lower surfaces glabrous, the petiole 1–4 cm long, glabrous, usually glandular near the apex. Flowers solitary or in a racemiform cyme of 2–3(5) in the axil of the upper leaves, the peduncle (0.5)1–3(5) cm long, glabrate; bracts linear to lanceolate, 3–4(6) mm long, the margin entire; lower sepal 1–1.5 cm long, naviculate, greenish white, abruptly constricted to a straight or slightly distally curved spur 2.5–4.5(5) cm long, the lateral sepals ovate to lanceolate, 3–6 mm long; upper petal cupped, obovate, (1)1.5–2(2.5) cm long, red, pink, or white, the apex retuse, the lateral pairs 1.2–2.5 mm long, colored as adaxial one, the apex retuse or entire. Fruit fusiform, 1.5–2 cm long, glabrous or nearly so; seeds numerous; pyriform, 2–2.5 mm long, with short, thick trichomes on a papillate base, brown.

Wet, disturbed sites. Rare; Hillsborough, Polk, and Leon Counties. Escaped from cultivation. Florida; West Indies, Mexico, Central America, and South America; Africa, Asia, Australia, and Pacific Islands. Native to Africa. Spring–fall.

POLEMONIACEAE Juss., nom. cons. 1789. PHLOX FAMILY

Herbs. Leaves alternate or opposite, simple, pinnate-veined, petiolate or epetiolate, estipulate. Flowers axillary or in terminal cymes or panicles, bracteate, actinomorphic or zygomorphic, bisexual; sepals 5, basally connate; petals 5, basally connate; stamens 4, epipetalous, the anthers dorsifixed, longitudinally dehiscent; ovary superior, 3-carpellate and -loculate, the styles 3, basally connate. Fruit a loculicidally dehiscent capsule.

A family of about 18 genera and about 380 species; North America, Mexico, Central America, South America, Europe, and Asia.

Selected reference: Wilson (1960).

1. Leaves pinnatifid ....**Ipomopsis**
1. Leaves entire ....**Phlox**

Ipomopsis Michx. 1803.

Biennial herbs. Leaves alternate, simple, pinnately dissected, pinnate-veined, petiolate, estipulate. Flowers axillary, solitary or few in panicles, bracteate; sepals 5, basally connate with an interconnecting membrane; petals 5, basally connate, zygomorphic; stamens 5, epipetalous; ovary superior, 3-carpellate and -loculate, the style 1, the stigmas 3. Fruit a loculicidally dehiscent capsule.

A genus of about 25 species; North America and Mexico. [*Ipomoea*, and the Greek *opsis*, aspect, in reference to the similarity of the corolla to *Quamoclit* (*Ipomoea* sect. *Quamoclit*, Convolvulaceae).]

Ipomopsis rubra (L.) Wherry [Red, in reference to the flower color.] STANDING CYPRESS; SPANISH LARKSPUR.

Polemonium rubrum Linnaeus, Sp. Pl. 162. 1753. *Ipomoea rubra* (Linnaeus) Linnaeus, Syst. Veg., ed. 13. 171. 1774. *Cantua coronopifolia* Willdenow, Sp. Pl. 1: 879. 1798, nom. illegit. *Cantua rubra* (Linnaeus) Dumont de Courset, Bot. Cult. 2: 185. 1802. *Ipomopsis elegans* Michaux, Fl. Bor.-Amer. 1: 142. 1803, nom. illegit. *Gilia coronopifolia* Persoon, Syn. Pl. 1: 187. 1805, nom. illegit. *Ipomeria coronopifolia* Nuttall, Gen. N. Amer. Pl. 1: 124. 1818, nom. illegit. *Navarretia rubra* (Linnaeus) Kuntze, Revis. Gen. Pl. 2: 433. 1891. *Gilia rubra* (Linnaeus) A. Heller, Bot. Explor. S. Texas 81. 1895. *Ipomopsis rubra* (Linnaeus) Wherry, Bartonia 18: 56. 1936.
Ipomoea erecta Michaux, J. Hist. Nat. 1: 410. 1792. TYPE: FLORIDA.
Cantua floridana Nuttall, J. Acad. Nat. Sci. Philadelphia 7: 110. 1834. TYPE: FLORIDA.

Erect biennial herb, to 1 m; stem simple, with a prominent basal rosette, sparsely pubescent. Leaves with the blade elliptic-ovate, 4–8 cm long, 1–4 cm wide, deeply pinnately dissected, the segments filiform, the upper and lower surfaces sparsely pubescent along the veins, the petiole

1–4 cm long. Flowers in a slender panicle of few-flowered cymes, the pedicel 2–6 mm long; bracts filiform; sepals 8–9 mm long, connate to below the middle, the lobes subulate, the apex with a short awn; corolla salverform, lacking a definite expanded throat, slightly zygomorphic, bright red, the tube 2–2.5 cm long, the lobes ca. 9 mm long, ca. 5 mm wide, the apex obtuse to subacute; stamens included or a few slightly exserted; style exserted well beyond the anthers. Fruit oblong, 8–10 mm long, angulate; seeds 10–12 per locule.

Sandhills and coastal dunes. Occasional; northern counties south to Indian River County. Massachusetts and New York south to Florida, west to Ontario, Wisconsin, Iowa, Kansas, Oklahoma, and Texas. Summer.

Phlox L. 1753.

Annual or perennial herbs. Leaves all opposite or the upper alternate and the lower ones opposite, simple, pinnate-veined, petiolate or epetiolate, estipulate. Flowers in terminal or axillary, simple or compound cymes, bracteate; sepals 5, basally connate, with interconnecting membranes; petals 5, basally connate; stamens 5, the filaments unequal, epipetalous; ovary superior, 3-carpellate and -loculate, the styles 3, basally connate. Fruit a loculicidally dehiscent capsule.

A genus of about 70 species; North America, Mexico, Central America, and South America. [From the Greek *phlox*, flame, in reference to the sometimes brilliant flower color.]

Selected reference: Wherry (1955).

1. Upper leaves alternate, the blade elliptic to elliptic-lanceolate .... **P. drummondii**
1. Upper leaves opposite, the blade linear to broadly lanceolate.
    2. Principal leaves less than 2.5 cm long; stem woody, trailing or decumbent and mat-forming with short, erect, flowering stems .... **P. nivalis**
    2. Principal leaves more than 2.5 cm long; stem herbaceous, erect or ascending, the sterile stems sometimes decumbent.
        3. Apex of the uppermost anther level with the corolla throat or protruding from it; sepal lobes lanceolate, glabrous or inconspicuously puberulent .... **P. glaberrima**
        3. Apex of the uppermost anther below the corolla throat; sepal lobes linear, densely stipitate-glandular.
            4. Leaves with long, spreading trichomes, the margin flat, the midrib scarcely raised on the lower surface .... **P. divaricata**
            4. Leaves glabrous or with short trichomes, the margin revolute (at least when dry), the midrib conspicuously raised on the lower surface.
                5. Calyx with few or no glandular trichomes .... **P. amoena**
                5. Calyx with numerous glandular trichomes.
                    6. Upper leaves spreading, distinctly longer than the bracts .... **P. pilosa**
                    6. Upper leaves ascending, grading into the bracts .... **P. floridana**

Phlox amoena Sims [*Amoene*, beautiful, pleasing.] HAIRY PHLOX.

Phlox amoena Sims, Bot. Mag. 32: pl. 1308. 1810. *Phlox pilosa* Linnaeus var. *amoena* (Sims) Pursh, Fl. Amer. Sept. 150. 1814. *Armeria amoena* (Sims) Kuntze, Revis. Gen. Pl. 2: 432. 1891.

Phlox pilosa Linnaeus var. *walteri* A. Gray, Manual, ed. 2. 331. 1856. *Phlox walteri* (A. Gray) Chapman,
Fl. South. U.S. 339. 1860. *Phlox amoena* Sims var. *walteri* (A. Gray) Wherry, Bartonia 12: 53. 1931.
Phlox lighthipei Small, Fl. S.E. U.S. 978, 1337. 1903. *Phlox amoena* Sims var. *lighthipei* (Small) Brand,
in Engler, Pflanzenr. 4(Heft 27): 70. 1907. *Phlox amoena* subsp. *lighthipei* (Small) Wherry, Genus
Phlox 55. 1955. TYPE: FLORIDA: Duval Co.: Jacksonville, 2 Apr 1896, *Lighthipe s.n.* (holotype:
NY).

Erect perennial herb, to 4.5 dm; stem pilose, the sterile ones decumbent. Leaves opposite, the
blade linear or oblong-elliptic to lanceolate, 3.5–4.5(5) cm long, 3–8(12) mm wide, the apex
obtuse, acute, or acuminate, the base cuneate, the margin entire, ciliate, the upper and lower
surfaces pilose, short-petiolate or sessile. Flowers (3)6–12(30) in a terminal or axillary, compact, simple or compound cyme, the pedicel 1–6 mm long, pubescent; bracts narrowly elliptic,
forming a lax involucre; sepals 7–12 mm long, connate ¼–½ their length, the lobes linear-subulate, the apex aristate, the interconnecting membranes flat to subplicate; corolla salverform,
purple, pink, or rarely white, the tube 12–16(19) mm long, glabrous, the lobes obovate, ca. 10
mm long, ca. 7 mm wide, the apex obtuse or apiculate; styles 1–3 mm long, connate ¼–½ their
length. Fruit ovoid, 4–6 mm long, glabrous; seeds 3–9, ellipsoid.

Hammocks and sandhills. Rare; northern peninsula, Jefferson and Escambia Counties.
North Carolina south to Florida, west to Kentucky, Tennessee, and Mississippi. Spring.

Phlox divaricata L. [Spreading at a wide angle, in reference to the leaves.] WILD
BLUE PHLOX.

Phlox divaricata Linnaeus, Sp. Pl. 512. 1753. *Phlox vernalis* Salisbury, Prodr. Stirp. Chap. Allerton 123.
1796, nom. illegit. *Armeria divaricata* (Linnaeus) Kuntze, Revis. Gen. Pl. 2: 432. 1891.
Phlox divaricata Linnaeus var. *laphamii* A. W. Wood, Class-Book Bot., ed. 2. 439. 1847. *Phlox laphamii*
(A. W. Wood) Clute, Amer. Bot. 25: 101. 1918. *Phlox divaricata* Linnaeus subsp. *laphamii* (A. W.
Wood) Wherry, Genus Phlox 41. 1955.

Erect perennial herb, to 3.5(4.5) dm; stem glabrous, the sterile ones decumbent. Leaves opposite, the blade broadly elliptic to ovate or lanceolate, 2–5 cm long, 0.8–2.5 cm wide, the apex
obtuse, acute, or acuminate, the base cuneate, the margin entire, the lower leaves glabrescent,
the upper ones with the margin ciliate, the upper and lower surfaces sparsely glandular-pubescent, short-petiolate or sessile. Flowers 9–24(30) in a terminal or axillary, simple or compound
cyme, the pedicel 4–18 mm long, pubescent with gland-tipped trichomes; bracts linear; sepals
6–11 mm long, united ⅜–½ their length, the lobes linear-subulate, the apex cuspidate, the interconnecting membranes flat to subplicate; corolla salverform, blue, purple, or rarely white, the
tube 1–1.7(2) mm long, glabrous, the lobes obovate to oblanceolate, 1–1.5(2) cm long, 6–9(13)
mm wide, the apex entire or erose, obtuse, apiculate or the apex notched to 4 mm; styles 2–3
mm long, united to ⅝ their length. Fruit ovoid, 4–6 mm long, glabrous; seeds 3–9, ellipsoid.

Calcareous hammocks. Rare; central panhandle. Quebec south to Florida, west to Ontario,
Minnesota, South Dakota, Nebraska, Kansas, and New Mexico. Spring.

Phlox drummondii Hook. [Commemorates Thomas Drummond (1793–1835), Scottish botanist who collected in Texas, Canadian Rockies, and the Gulf coastal states.] ANNUAL PHLOX.

Phlox drummondii Hooker, Bot. Mag. 62: pl. 3441. 1835. *Armeria drummondii* (Hooker) Kuntze, Revis. Gen. Pl. 2: 432. 1891. *Polemonium drummondii* (Hooker) Kuntze, Revis. Gen. Pl. 3(2): 203. 1898. *Phlox drummondii* Hooker subsp. *eudrummondii* Brand, in Engler, Pflanzenr. 4(Heft 27): 70. 1907, nom. inadmiss. *Phlox drummondii* Hooker var. *typica* Whitehouse, Amer. Midl. Naturalist 34: 390. 1945, nom. inadmiss.

Phlox drummondii Hooker var. *peregrina* Shinners, Field & Lab. 19: 127. 1951.

Phlox drummondii Hooker forma *albiflora* Moldenke, Phytologia 26: 224. 1973. TYPE: FLORIDA: Putnam Co.: Bostwick, 26 Mar 1973, *Moldenke & Moldenke 26471* (holotype: LL).

Erect annual herb, to 5 dm; stem pubescent with viscid and sometimes gland-tipped trichomes. Leaves opposite below, alternate above, the lower ones with the blade oblanceolate, the base tapering to petiolate, the upper ones with the blade oblong, sessile-subclasping, 3–7.5 cm long, 1–2 cm wide, the apex acute or short acuminate, subaristate, the base cuneate, the margin entire, the upper and lower surfaces pubescent, with viscid and sometimes gland-tipped trichomes, short-petiolate or sessile. Flowers in a terminal or axillary, helicoid cyme or aggregated into 3–4 cymules, the pedicel 5–10 mm long, glandular-pubescent; bracts linear; sepals 8–18 mm long, connate ca. ¼ their length, the lobes linear-subulate, the apex aristate, the interconnecting membranes flat; corolla salverform, red, purple, or blue, often with a deeper-colored or white eye, the tube 13–17 mm long, glandular-pubescent, the lobes broadly obovate, ca. 12 mm long, ca. 9 mm wide, the apex obtuse, often apiculate; styles 2–3 mm long, connate to ½ their length. Fruit ovoid, 4–6 mm long, glabrous; seeds 3–9, ellipsoid.

Disturbed sites. Frequent; nearly throughout. Escaped from cultivation. New Brunswick and Ontario, scattered in the northern states from Vermont to Minnesota, and from Virginia south to Florida, west to Oklahoma and Texas. Native to Texas. Spring.

Phlox floridana Benth. [Of Florida.] FLORIDA PHLOX.

Phlox floridana Bentham, in de Candolle, Prodr. 9: 304. 1845. *Phlox pilosa* Linnaeus var. *floridana* (Bentham) A. W. Wood, Class-Book Bot., ed. 1861. 568. 1861. *Armeria floridana* (Bentham) Kuntze, Revis. Gen. Pl. 2: 432. 1891. *Phlox floridana* Bentham subsp. *typica* Wherry, Bartonia 22: 1. 1943, nom. inadmiss. TYPE: FLORIDA: s.d., *Chapman s.n.* (holotype: K).

Phlox floridana Bentham subsp. *bella* Wherry, Bartonia 22: 1. 1943. TYPE: FLORIDA: Okaloosa Co.: Valparaiso, 26 Jun 1939, *Henry 1630* (holotype: PH).

Erect or ascending perennial herb, to 5(8) dm; stem glabrous below the inflorescence. Leaves opposite, the blade linear to oblong or lanceolate, 4–8(9) cm long, 3–6(7) mm wide, the apex acute, the base cuneate, the margin entire, the upper and lower surfaces glabrous, the upper ones sometimes glandular-pubescent, short-petiolate or sessile. Flowers (6)12–24(36) in a terminal or axillary, simple or compound, compact cyme, the pedicel (3)4–8(10) mm long; bracts linear to narrowly lanceolate; sepals 7–11 mm long, connate ⅜–½ their length, the interconnecting membranes usually plicate, the lobes subulate, the apex cuspidate to subaristate; corolla salverform, purple to pink, the tube 1.5–2 cm long, glabrous, the lobes obovate, 11–18 mm long, the apex obtuse; styles 2–3 mm long, connate ca. ¼ their length. Fruit ovoid, 4–6 mm long glabrous; seeds 3–9, ellipsoid.

Sandhills, flatwoods, and hammocks. Occasional; northern counties south to Hernando County. Georgia, Alabama, and Florida. Spring.

Phlox glaberrima L. [Most glabrous, in reference to the lack of trichomes on the leaves and stems]. SMOOTH PHLOX.

Phlox glaberrima Linnaeus, Sp. Pl. 152. 1753. *Phlox melampyrifolia* Salisbury, Prodr. Stirp. Chap. Allerton 123. 1796, nom. illegit. *Armeria glaberrima* (Linnaeus) Kuntze, Revis. Gen. Pl. 2: 432. 1891. *Phlox glaberrima* Linnaeus var. *melampyrifolia* Wherry, Bartonia 14: 19. 1932, nom. inadmiss.

Phlox carolina Linnaeus, Sp. Pl., ed. 2. 216. 1762. *Phlox altissima* Moench, Methodus 454. 1794, nom. illegit. *Phlox ovata* Linnaeus forma *carolina* (Linnaeus) Voss, Vilm. Blumengärtn., ed. 3. 1: 679. 1894. *Phlox ovata* Linnaeus var. *carolina* (Linnaeus) Wherry ex Schaffner, Ohio J. Sci. 31: 304. 1931. *Phlox carolina* Linnaeus var. *altissima* Wherry, Bartonia 13: 36. 1932, nom. inadmiss. *Phlox carolina* Linnaeus subsp. *typica* Wherry, Bartonia 23: 2. 1945, nom. inadmiss.

Phlox suffruticosa Ventenat, Jard. Malmaison 2: pl. 107(2). 1804. *Phlox glaberrima* Linnaeus var. *suffruticosa* (Ventenat) A. Gray, Syn. Fl. N. Amer. 2(1): 130. 1878.

Phlox nitida Pursh, Fl. Amer. Sept. 730. 1814. *Phlox carolina* Linnaeus var. *nitida* (Pursh) Bentham, in A. de Candolle, Prodr. 9: 304. 1845. *Phlox maculata* Linnaeus var. *nitida* (Pursh) Chapman, Fl. South. U.S. 338. 1860. *Phlox carolina* Linnaeus var. *angusta* (Wherry) D. B. Ward, Phytologia 94: 477. 2012.

Phlox glaberrima Linnaeus subvar. *angustissima* Brand, in Engler, Pflanzenr. 4(Heft 27): 65. 1907. *Phlox carolina* Linnaeus subsp. *angusta* Wherry, Genus Phlox 108. 1955.

Phlox carolina Linnaeus var. *heterophylla* Wherry, Bartonia 13: 36. 1932. *Phlox carolina* Linnaeus subsp. *heterophylla* (Wherry) Wherry, Bartonia 23: 6. 1945.

Erect perennial herb, to 4(10) dm; stem minutely rough-pubescent, pilose, or glabrous. Leaves opposite, the blade lanceolate, 5–10 cm long, 0.5–3 cm wide, the apex acute, the base rounded or cuneate, the margin entire, the upper and lower surfaces glabrous or rarely finely pubescent, short-petiolate or sessile. Flowers (15)30–60(100) in a terminal or axillary, compound cyme, the pedicel 3–6(10) mm long, glabrous or fine-pubescent; bracts linear; sepals (6)7–9(11) mm long, connate to ca. ½ their length, the interconnecting membranes plicate, the lobes narrowly triangular or broadly subulate, the apex acuminate, usually with a short awn; corolla salverform, reddish purple to violet or white, the tube 1.5–2.5 cm long, the lobes 7–12 mm long, 5–10 mm wide; styles 1.5–2 cm long, connate nearly their length. Fruit ovoid, 5–7 mm long, glabrous; seeds 3–9, ellipsoid.

Bluff and floodplain forests. Occasional; central and western panhandle. Virginia south to Florida, west to Wisconsin, Illinois, Missouri, Oklahoma, and Louisiana. Spring.

Phlox nivalis Lodd. ex Sweet [Snowy, in reference to the low growing, snow-like appearance of the white-flowered form.] TRAILING PHLOX.

Phlox nivalis Loddiges ex Sweet, Brit. Fl. Gard. 2: pl. 185. 1827. *Phlox subulata* Linnaeus forma *nivalis* (Loddiges ex Sweet) Voss, Vilm. Blumengärtn., ed. 3. 1: 680. 1894. *Phlox subulata* Linnaeus subsp. *nivalis* (Loddiges ex Sweet) Brand, in Engler, Pflanzenr. 4(Heft 27): 78. 1907.

Phlox hentzii Nuttall, J. Acad. Nat. Sci. Philadelphia 7: 110. 1834. *Phlox subulata* Linnaeus forma *hentzii*

(Nuttall) Voss, Vilm. Blumengärtn., ed. 3. 1: 680. 1894. *Phlox subulata* Linnaeus subvar. *hentzii* (Nuttall) Brand, in Engler, Pflanzenr. 4(Heft 27): 78. 1907. *Phlox nivalis* Loddiges ex Sweet subsp. *hentzii* (Nuttall) Wherry, Genus Phlox 25. 1955. *Phlox nivalis* Loddiges ex Sweet var. *hentzii* (Nuttall) D. B. Ward, Phytologia 94: 478. 2012.

Phlox nivalis Loddiges ex Sweet forma *roseiflora* Fernald, Rhodora 51: 78. 1949.

Phlox nivalis Loddiges ex Sweet forma *rubella* Moldenke, Phytologia 26: 225. 1973. TYPE: FLORIDA: Bay Co.: Youngstown, 1 Apr 1973, *Moldenke & Moldenke 26696* (holotype: LL).

Prostrate perennial herb; stem pubescent, the flowering stem erect, to 3 dm. Leaves opposite, the blade linear-subulate, 8–20 mm long, 1–5 mm wide, the apex acute, the base cuneate, the margin entire, ciliate, the upper and lower surfaces glabrous or pilose, sessile or subsessile. Flowers 3–6(9), in a terminal or axillary, simple or compound cyme, the pedicel (3)8–30(35) mm long; bracts linear; sepals 6–9 mm long, united ca. ½ their length, the lobes linear-subulate, the interconnecting membrane flat to subplicate, the apex cuspidate; corolla salverform, purple, pink, or white, the tube 10–13(17) mm long, the lobes usually obovate, ca. 11 mm long, ca. 7 mm wide, the apex entire, erose, or notched to 4 mm; styles 1–4 mm long, connate nearly to the apex. Fruit ovoid 4–6 mm long, glabrous; seeds 3–9, ellipsoid.

Sandhills. Occasional; northern counties, central peninsula. Virginia south to Florida, also Michigan, Texas, and Utah. Spring.

Phlox pilosa L. [With long trichomes.] DOWNY PHLOX.

Phlox pilosa Linnaeus, Sp. Pl. 152. 1753. *Armeria pilosa* (Linnaeus) Kuntze, Revis. Gen. Pl. 2: 432. 1891.

Phlox aristata Michaux, Fl. Bor.-Amer. 1: 144. 1803. *Phlox aristata* Michaux var. *virens* Michaux, Fl. Bor.-Amer. 1: 144. 1803, nom. inadmiss. *Phlox pilosa* Linnaeus var. *virens* Wherry, Bartonia 12: 47. 1931.

Phlox pilosa Linnaeus var. *detonsa* A. Gray, Proc. Amer. Acad. Arts 8: 251. 1870. *Phlox detonsa* (A. Gray) Small, Fl. S.E. U.S. 978, 1337. 1903. *Phlox pilosa* Linnaeus subsp. *detonsa* (A. Gray) Wherry, Genus Phlox 51. 1955.

Erect perennial herb, to 5(6.5) dm; stem pilose. Leaves opposite, the blade linear to lanceolate or ovate, 4–8(10) cm long, 3–9(12) mm wide, the apex acute to acuminate, the base cuneate, the margin entire, the upper and lower surfaces of the lower ones glabrescent, the upper and lower surfaces of the medial ones sparsely ciliate, pilose on the midrib, the upper and lower surfaces of the upper ones pubescent, or all the leaves glabrous, sessile or subsessile. Flowers (12)24–48(100), in a terminal, compact to open paniculiform cyme, the pedicel 4–8(12) mm long, densely glandular-pubescent to glabrate or glabrous; bracts linear; sepals 6–12 mm long, connate ⅜–½ their length, the lobes subulate, the interconnecting membranes flat to moderately plicate, the apex aristate; corolla salverform, purple to pink or white, the tube 8–12(15) mm long, pilose, eglandular or sparsely glandular, sometimes glabrous, the lobes oblanceolate or obovate, ca. 11 mm long, ca. 7 mm wide, the apex obtuse, rounded, or apiculate or sometimes erose-emarginate; styles 2–3(4) mm long, connate ca. ½ their length. Fruit ovoid, 4–6 mm long, glabrous; seeds 3–9, ellipsoid.

Dry hammocks. Frequent; northern counties, central peninsula. New York south to Florida, west to Manitoba and Texas. Spring.

EXCLUDED TAXA

Phlox maculata Linnaeus—Reported for Florida by Chapman (1860) and Small (1903, 1913a). Wherry (1955) excluded the species from our state. No Florida specimens known.

Phlox paniculata Linnaeus—Reported by Small (1903, 1913a). Wherry (1955) excluded the species from our state. No Florida specimens known.

Phlox subulata Linnaeus—This northern species was reported for Florida by Chapman (1860), the name misapplied to material of *P. nivalis*.

PENTAPHYLACACEAE Engl., nom. cons. 1897.

Shrubs. Leaves alternate, simple, pinnate-veined, petiolate, estipulate. Flowers axillary, solitary or several clustered on leafless branchlets, actinomorphic, bisexual or unisexual (plants andro-dioecious), bracteolate; sepals 5, basally connate; petals 5, basally connate; stamens numerous, epipetalous; ovary superior, 2-carpellate and -loculate, the style 1. Fruit a berry.

A family of about 12 genera and about 340 species; North America, West Indies, Mexico, Central America, South America, Africa, Asia, Australia, and Pacific Islands.

Ternstroemiaceae Mirb. ex DC. (1816).

Ternstroemia Mutis ex L. f., nom. cons. 1782.

Shrubs. Leaves alternate, simple, pinnate-veined, petiolate, estipulate. Flowers axillary, solitary or several clustered on leafless branchlets, actinomorphic, bisexual or unisexual (plants andro-dioecious), bracteolate; sepals 5, basally connate; petals 5, basally connate; stamens numerous, in 3 series, epipetalous, the anthers basifixed, longitudinally dehiscent, introrse; ovary superior, 2-carpellate and -loculate, the style 1, the stigmas 2. Fruit a berry.

A genus of about 100 species; North America, West Indies, Mexico, Central America, South America, Africa, Asia, Australia, and Pacific Islands. [Commemorates the Swedish naturalist Christopher Ternstroem (1703–1745), student of Linnaeus and traveler in China.]

Ternstroemia is sometimes placed in the Theaceae.

Ternstroemia gymnanthera (Wight & Arn.) Bedd. [From the Greek *gymnos*, naked, and *anthera*, anther, in reference to the exposed stamens.] JAPANESE CLEYERA.

Cleyera gymnanthera Wight & Arnold, Prodr. Fl. Ind. Orient. 1: 87. 1834. *Ternstroemia gymnanthera* (Wight & Arnold) Beddome, Fl. Sylv. S. India 19, pl. 91. 1871.

Shrub, to 3 m; bark grayish brown, smooth, the branchlets purplish red to reddish brown, becoming grayish brown, glabrous. Leaves with the blade obovate, oblong-obovate, or broadly elliptic, (3)4–12 cm long, 1.5–5.5 cm wide, coriaceous, evergreen, the apex acute to short-acuminate, the base cuneate, the margin entire, the upper and lower surfaces glabrous, the petiole ca. 1 cm long, glabrous. Flowers axillary or several clustered on a leafless branchlet, the pedicel 1–1.5 cm long, slightly recurved; bracteoles 2, alternate, close to the sepals, triangular to triangular-ovate, ca. 2 mm long, the margin glandular-dentate; sepals ovate to elongate-ovate,

4–7 mm long, glabrous, the margin glandular-dentate, the apex rounded; petals obovate, 6–9 mm long, pale yellow, the apex rounded and retuse; stamens ca. 30, 4–5 mm long; style 1, the stigmas 2; staminate flowers similar to the bisexual ones, but the ovary reduced to a pistillode. Fruit globose, 1–1.5 cm long, purplish red; seeds reniform, ca. 6 mm long, with a fleshy, red outer layer.

Disturbed sites. Rare; Leon County. Escaped from cultivation. South Carolina, Alabama, and Florida; Asia. Native to Asia. Summer.

SAPOTACEAE Juss., nom. cons. 1789. SAPODILLA FAMILY

Trees or shrubs. Leaves alternate, simple, pinnate-veined, petiolate, stipulate or estipulate. Flowers axillary, solitary or fasciculate, actinomorphic, bisexual; sepals 4–8, free; petals 4–8, basally connate; stamens 4–8, the anthers dorsifixed, versatile, dehiscent by longitudinal slits; ovary superior, 3- to 12-carpellate and -locular, the style 1, terminal, the stigma 1. Fruit a berry; seeds 1–10.

A family of about 50 genera and about 1,100 species; North America, West Indies, Mexico, Central America, South America, Africa, Asia, Australia, and Pacific Islands.

Selected reference: Wood and Channell (1960).

1. Sepals 6 or 8 in 2 whorls, the outer whorl valvate.
    2. Sepals 8 in 2 whorls of 4; seed hilum circular...**Mimusops**
    2. Sepals 6 in 2 whorls of 3; seed hilum linear..**Manilkara**
1. Sepals 4–6 in 1 whorl, imbricate.
    3. Corolla lobes divided...**Sideroxylon**
    3. Corolla lobes undivided.
        4. Leaves densely coppery or silvery brown-tomentose on the lower surface; staminodes absent ..
        ....**Chrysophyllum**
        4. Leaves glabrous or glabrate on the lower surface; staminodes present, petaloid.
            5. Pedicels glabrous; sepals 1.5–2 mm long; fruit yellow to orange ....**Sideroxylon**
            5. Pedicels densely pubescent; sepals 2.5–11 mm long; fruits brown.............................**Pouteria**

Chrysophyllum L. 1753.

Trees. Leaves alternate, simple, pinnate-veined, petiolate, estipulate. Flowers axillary, solitary or in fascicles; sepals 4–5, free; petals (4)5(6), basally connate; stamens 4–5, free; ovary 3- to 6-carpellate and -loculate. Fruit a berry; seed 1.

A genus of about 70 species; North America, Mexico, Central America, South America, Africa, Asia, and Australia. [From the Greek *chrysos*, gold, and *phyllon*, leaf, in reference to the gold-colored lower leaf surface.]

Selected reference: Wunderlin and Whetstone (2009b).

Chrysophyllum oliviforme L. [Olive-shaped, in reference to the fruit.] SATINLEAF.

Chrysophyllum oliviforme Linnaeus, Syst. Nat., ed. 10. 937. 1759. *Guersentia oliviformis* (Linnaeus) Rafinesque, Sylva Tellur. 153. 1838. *Chrysophyllum oliviforme* Linnaeus var. *typicum* Cronquist,

Bull. Torrey Bot. Club 72: 200. 1945, nom. inadmiss. *Cynodendron oliviforme* (Linnaeus) Baehni, Arch. Sci. 17: 78. 1964.

Chrysophyllum monopyrenum Swartz, Prodr. 49. 1788. *Chrysophyllum oliviforme* Linnaeus var. *monopyrenum* (Swartz) Grisebach, Fl. Brit. W.I. 398. 1861.

Tree, to 10 m; branchlets rusty sericeous-tomentose, becoming gray in age. Leaves with the blade broadly elliptic to ovate, (1.3)3–12 cm long, (0.6)1.5–4(6.5) cm wide, the apex acute to acuminate, the base cuneate to rounded, the margin entire, the marginal vein distinct, the upper surface glabrous, glossy green, the lower surface coppery sericeous-tomentose, becoming grayish in age, the petiole 5–10 mm long. Flowers axillary, solitary or 2–10 in a fascicle, the pedicel 3–8 mm long, sericeous-tomentose; sepals broadly ovate, ca. 2 mm long, the outer surface rusty sericeous-tomentose, the margin ciliate, the apex ciliate; petals ca. 4 mm long, connate to ca. ½ their length, greenish white or greenish yellow, the lobes triangular, the outer surface sericeous; ovary pubescent. Fruit ellipsoid to ovoid, 1.5–3 cm long, purple to black, glabrous; seed 1, elongate, 13–15 cm long, brown, laterally compressed, the hilum narrowly ovate to obovate, ca. ⅓ the seed length.

Hammocks, scrub, and coastal thickets, frequently on coquina, coral, sandy, and shelly soils. Occasional; central and southern peninsula. Florida; West Indies. All year.

Chrysophyllum oliviforme is listed as threatened in Florida (Florida Administrative Code, Chapter 5B-40).

Manilkara Adans., nom. cons. 1763.

Trees or shrubs. Leaves alternate, simple, pinnate-veined, petiolate, estipulate. Flowers axillary, solitary or in fascicles; sepals 6 in two whorls of 3, free, the outer whorl valvate; petals 6, basally connate; stamens 6, free; staminodes 6, alternate with the stamens, petaloid; ovary 5- to 12-carpellate and -loculate. Fruit a berry; seeds 2–10.

A genus of about 65 species; North America, West Indies, Mexico, Central America, South America, Africa, Asia, and Pacific Islands. [From the Malabar *Manil*, this in turn from the Portuguese, *Manilhas Insulas* (Manila, Philippines), and *kara*, edible fruit.]

Selected reference: Wunderlin and Whetstone (2009a).

1. Leaves with the apex usually emarginate; corolla tube shorter than the lobes; young fruit with a conspicuous, long, spikelike style .... **M. jaimiqui**
1. Leaves with the apex acute to acuminate or obtuse, rarely emarginate; corolla tube equaling or exceeding the lobes; young fruit with a short style .... **M. zapota**

Manilkara jaimiqui (C. Wright & Griseb.) Dubbard subsp. **emarginata** (L.) Cronquist [Taino Indian name meaning "water crab spirit"; lowly notched, in reference to the leaf apex.] WILD DILLY.

Sloanea emarginata Linnaeus, Sp. Pl. 512. 1753. *Achras sapota* Linnaeus, Sp. Pl., ed. 2. 470. 1762, nom. illegit. *Sapota achras* Miller var. *depressa* A. de Candolle, in de Candolle, Prodr. 8: 174. 1844. *Mimusops depressa* (A. de Candolle) Pierre, Not. Bot. 37. 1891, nom. illegit. *Mimusops emarginata* (Linnaeus) Britton, Torreya 11: 129. 1911. *Manilkara emarginata* (Linnaeus) Britton & P. Wilson, Bot.

Porto Rico 6: 366. 1926; non H. J. Lam, 1925. *Manilkara emarginata* (Linnaeus) Britton & P. Wilson var. *typica* Cronquist, Bull. Torrey Bot. Club 72: 557. 1945, nom. inadmiss. *Manilkara jaimiqui* (C. wright ex Grisebach) Dubard subsp. *emarginata* (Linnaeus) Cronquist, Bull. Torrey Bot. Club 73: 467. 1946. *Achras emarginata* (Linnaeus) Little, Rhodora 49: 292. 1947.

Achras zapotilla (Jacquin) Nuttall var. *parvifolia* Nuttall, N. Amer. Sylv. 3: 28, t. 90. 1849. *Mimusops parvifolia* (Nuttall) Radlkofer, Sitzungsber. Math.-Phys. Cl. Königl. Bayer. Akad. wiss. München 12: 344. 1882; non R. Brown, 1810. *Manilkara parvifolia* (Nuttall) Dubard, Ann. Inst. Bot.-Geol. Colon. Marseille, ser. 3. 3: 16. 1915. TYPE: FLORIDA: s.d. (holotype: Nuttall, N. Amer. Sylv. 3: t. 90. 1849).

Achras bahamensis Baker, Hooker's Icon. Pl. 18: pl. 1795. 1888. *Manilkara bahamensis* (Baker) H. J. Lam & A. Meeuse, Blumea 4: 354. 1941.

Mimusops floridana Engler, Bot. Jahrb. Syst. 12: 524. 1890. TYPE: FLORIDA: Monroe Co.: Boca Chica Key, s.d., *Curtis 1766* (holotype: B (destroyed); isotypes: BM, K, M, NY, US).

Trees or shrubs, to 10 m; branches glabrous, the latex milky. Leaves with the blade oblong or oblong-elliptic, 5–10.5 cm long, 3–5 cm wide, the apex emarginate, the base rounded or truncate, the margin entire, sinuate and somewhat revolute, the upper surface glabrous, the lower surface glabrous or sparsely tomentose, the petiole 1.2–2.5 cm long, glabrous. Flowers axillary, solitary or 2–5 in a fascicle, the pedicel 1–2.5 cm long, puberulous or glabrous; sepals elliptic-lanceolate, 4–8 mm long, the apex acute to obtuse, tomentose to subglabrate; petals 4–9 mm long, yellow, the tube 1–2 mm long, shorter than the lobes, glabrous, the lobes divided to the base into a larger median and 2 lateral segments, the lobes 6–7 mm long, the median segment boat-shaped, clawed, the lateral segments lanceolate to falcate; staminodes triangular-lanceolate; ovary appressed-puberulous. Fruit with a conspicuous, long, spikelike style when young, at maturity depressed-globose, 2.5–4 cm long, the surface roughened, scaly; seeds 3–5, 14–18 mm long, laterally compressed, brown, the hilum linear.

Hammocks, mangroves, and coastal thickets. Occasional; Collier and Miami-Dade Counties, Monroe County keys. Florida; West Indies. Winter–spring.

Manilkara jaimiqui subsp. *emarginata* (as *M. jaimiqui*) is listed as threatened in Florida (Florida Administrative Code, Chapter 5B-40).

Manilkara zapota (L.) P. Royen [Mexican vernacular name for the plant, in turn derived from the Nahuatl *tzapotl.*] SAPODILLA

Achras zapota Linnaeus, Sp. Pl. 1190. 1753. *Achras mammosa* Linnaeus, Sp. Pl., ed. 2. 469. 1762, nom. illegit. *Lucuma mammosa* C. F. Gaertner, Suppl. Carp. 129. 1805, nom. illegit. *Achras zapota* Linnaeus var. *globosa* Stokes, Bot. Mat. Med. 2: 292. 1812, nom. inadmiss. *Vitellaria mammosa* Radlkofer, Sitzungsber. Math.-Phys. Cl. Köengl. Bayer. Akad. Wiss. München 12: 326. 1882, nom. illegit. *Calospermum mammosum* Pierre, Bot. Not. 11. 1890, nom. illegit. *Calocarpum mammosum* Pierre ex Pierre & Urban, in Urban, Symb. Antill. 5: 98. 1904. *Acradelpha mammosa* O. F. Cook, J. Wash. Acad. Sci. 3: 160. 1913, nom. illegit. *Manilkara zapota* (Linnaeus) P. Royen, Blumea 7: 410. 1953.

Sapota achras Miller, Gard. Dict., ed. 8. 1768. *Manilkara achras* (Miller) Fosberg, Taxon 13: 255. 1964. *Nispero achras* (Miller) Aubreville, Adansonia, ser. 2. 5: 19. 1965.

Tree, to 18 m; branches glabrous, the latex milky. Leaves with the blade elliptic to oblong-elliptic, 6–14 cm long, 2–5 cm wide, the apex acute to acuminate or obtuse, rarely emarginate, the base cuneate, the margin entire, sinuate, the upper surface glabrous, the lower surface glabrous or brown-tomentose along the midrib, the petiole 1–3 cm long. Flowers axillary, solitary, the pedicel 1–2 cm long, rufous-tomentose; sepals ovate to lanceolate, 7–10 mm long, the apex acute to obtuse, rufous tomentose; petals 7–11 mm long, white, the tube 4–6(7), equaling or exceeding the lobes, the lobes entire or irregularly 2- to 3-lobed, glabrous or sparsely pubescent; staminodes petaloid; ovary pubescent. Fruit with a short style when young, at maturity ellipsoid or depressed globose to subglobose, 3.5–8 cm long, surface roughed, scaly; seeds 2–10, 1.5–2.5 cm long, laterally compressed, brown, the hilum linear.

Hammocks and disturbed sites. Occasional; Lee and Palm Beach Counties, southern peninsula. Escaped from cultivation. Florida; West Indies, Mexico, and Central America. Native to Mexico and Central America. All year.

Manilkara zapota is listed as a Category I invasive species in Florida by the Florida Exotic Plant Pest Council (FLEPPC, 2017).

The milky latex once was the primary source of chicle used in the manufacture of chewing gum.

EXCLUDED TAXON

Manilkara bidentata (A. de Candolle) A. Chevalier—Reported for Florida by Chapman (1860, 1883, 1897, all as *Mimusops sieberi* A. de Candolle) and Small (1903, 1913a, both as *Mimusops sieberi* A. de Candolle), the name misapplied to material of *M. jaimiqui* subsp. *emarginata*.

Mimusops L. 1753.

Trees. Leaves alternate, simple, pinnate-veined, petiolate, stipulate. Flowers axillary, solitary or in fascicles; sepals 8 in 2 whorls of 4, free; petals (7)8, basally connate, the lobes divided into a large median and 2 smaller lateral segments; stamens (7)4, free; staminodes (7)8, alternating with the stamens, petaloid; ovary (7)8-carpellate and -loculate. Fruit a berry; seed 1.

A genus of about 30 species; North America, Africa, Asia, and Pacific Islands. [From the Greek *mimo*, ape, and *ops*, face, in reference to the appearance of the flowers.]

Selected reference: Elisens (2009a).

1. Leaf apex rounded or emarginate ....**M. coriacea**
1. Leaf apex acuminate or abruptly blunt-tipped..**M. elengi**

Mimusops coriacea (A. DC.) Miq. [Leathery.] MONKEY'S APPLE.

Imbricaria coriacea A. de Candolle, in de Candolle, Prodr. 8: 200. 1844. *Mimusops coriacea* (A. de Candolle) Miquel, in Martius, Fl. Bras. 7: 44. 1863.

Tree, to 20 m; branchlets reddish brown appressed-puberulent to glabrate, the latex milky. Leaves with the blade elliptic to ovate or obovate, 8–12 cm long, 4–8 cm wide, the apex rounded to emarginate, the base broadly cuneate to rounded, the margin entire, undulate, the upper and lower surfaces glabrous, the petiole ca. 1 cm long, glabrous; stipules caducous. Flowers axillary, solitary, the pedicel 3–4 cm long; sepals narrowly ovate, 4–6(10) mm long, the outer surface densely tawny to reddish brown appressed-pubescent; corolla 1–1.5 cm long, white to yellow or pink, the lobes 6–9 mm long, the lateral segments lanceolate, erose; stamens and staminodes 4–6 mm long; ovary pubescent. Fruit subglobose, 2–3 cm long, yellow to orange, glabrate; seed 1.5–2 cm long, brown to black, laterally compressed, the hilum circular.

Disturbed sites. Rare; Martin, Broward, and Palm Beach Counties. Escaped from cultivation. Florida; Africa, Australia, and Pacific Islands. Native to Africa. Spring–summer.

Minusops elengi L. [Malayalam vernacular name.] SPANISH CHERRY; KABIKI; BAKUL.

> *Mimusops elengi* Linnaeus, Sp. Pl. 349. 1753. *Kaukenia elengi* (Linnaeus) Kuntze, Revis. Gen. Pl. 2: 406. 1891.

Tree, to 20 m; branchlets reddish brown appressed-puberulent to glabrate, the latex milky. Leaves with the blade elliptic to ovate or obovate, 4–14.5 cm long, 2–7 cm wide, the apex acute to cuspidate or obtuse, the base cuneate, the margin entire, undulate, the upper surface glabrous, the lower surface glabrate or glabrous, the petiole 1–3 cm long, glabrescent; stipules caducous. Flowers axillary, solitary or 2–5 in a fascicle, the pedicel 1.5–2 cm long; sepals narrowly ovate, 4–6(10) mm long, the outer surface densely tawny to reddish brown appressed-pubescent; corolla 1–1.5 cm long, white to yellow or pink, the lobes 6–9 mm long, the lateral segments lanceolate, erose; stamens and staminodes 4–6 mm long; ovary pubescent. Fruit subglobose, 2–3 cm long, yellow to orange, glabrate; seed 1.5–2 cm long, brown to black, laterally compressed, the hilum circular.

Disturbed sites. Rare; Broward County. Escaped from cultivation. Florida; Asia and Australia. Native to Asia and Australia. Spring–summer.

Pouteria Aubl. 1775.

Trees. Leaves alternate, simple, pinnate-veined, petiolate, estipulate. Flowers axillary, solitary or in fascicles; sepals 4–6, free; petals 4–6, basally connate; stamens 5–7; staminodes 5–7, alternate with the stamens, petaloid; ovary 5- to 7-carpellate and -loculate. Fruit a berry.

A genus of about 325 species; North America, West Indies, Mexico, Central America, South America, Africa, Asia, Australia, and Pacific Islands. [From the Galibi (Suriname) name *Pourama-pouteri*.]

> *Lucuma* Molina, 1782.

Selected reference: Wunderlin and Whetstone (2009c).

Pouteria campechiana (Kunth) Baehni [Of Campeche, Mexico.] EGG FRUIT; CANISTEL.

> *Lucuma campechiana* Kunth, in Humboldt et al., Nov. Gen. Sp. 3: 240. 1819. *Richardella campechiana* (Kunth) Pierre, Not. Bot. 20. 1890. *Vitellaria campechiana* (Kunth) Engler, Bot. Jahrb. Syst. 12: 513. 1890. *Pouteria campechiana* (Kunth) Baehni, Candollea 9: 398. 1942.
>
> *Lucuma nervosa* A. de Candolle, in de Candolle, Prodr. 8: 169. 1844. *Lucuma rivicoa* C. F. Gaertner var.

angustifolia Miquel, in Martius, Fl. Bras. 7: 71. 1863. *Vitellaria nervosa* (A. de Candolle) Radlkofer, Sitzungsber. Math.-Phys. Cl. Königl. Bayer. Akad. Wiss. München 12: 326. 1882. *Richardella nervosa* (A. de Candolle) Pierre, Not. Bot. 20. 1890. *Pouteria campechiana* (Kunth) Baehni var. *nervosa* (A. de Candolle) Baehni, Candollea 9: 401. 1942.

Tree, to 8(25) m; bark irregularly and deeply furrowed, the branchlets rusty appressed-puberulent, soon glabrous. Leaves with the blade elliptic to oblanceolate or obovate, 8–25(33) cm long, 3–8(15) cm wide, the apex acute to broadly obtuse, the base cuneate, the margin entire, revolute, the upper and lower surfaces glabrous, the petiole 1–2.5(4.5) cm long, puberulent. Flowers axillary, solitary or 2–9 in a fascicle, the pedicel 6–12 mm long, puberulent; sepals ovate to suborbicular, 5–11 mm long, sericeous to puberulent; petals 8–12 mm long, white, the tube 5–6(8) mm long, the lobes shorter than the tube, sparsely puberulent; staminodes linear-lanceolate, 2–4 mm long, petaloid; ovary densely rufous-pubescent, the style 5–12 mm long, the stigma capitate, 5-lobed. Fruit ellipsoid to subglobose or pyriform, 2.5–7 cm long and wide, the apex short apiculate, the surface brown to orange or yellow, smooth; seeds 1–4, 2–4 cm long, laterally flattened, brown, glossy, the hilum linear to ovate.

Hammocks. Rare; Miami-Dade County, Monroe County keys. Escaped from cultivation. Florida; West Indies, Mexico, and Central America. Native to Mexico and Central America. All year.

EXCLUDED TAXON

Pouteria domingensis (C. F. Gaertner) Baehni—Reported by Wood and Channell (1960) and Long and Lakela (1971). No Florida specimens known.

Sideroxylon L. 1753.

Shrubs or trees. Leaves alternate, simple, pinnate-veined, petiolate, estipulate. Flowers in axillary fascicles; sepals 4–6, free; petals 4–6, basally connate, the lobes divided into 1 large medial and 2 smaller lateral segments, the lateral segments sometimes vestigial or absent; stamens 4–6, free; staminodes 4–6, alternating with the stamens, petaloid; ovary 4- to 8-carpellate and -loculate. Fruit a berry; seeds 1–2.

A genus of about 70 species; North America, West Indies, Mexico, Central America, South America, and Africa.

Bumelia Sw., nom. cons., 1788; *Dipholis* A. DC., nom. cons., 1844; *Mastichodendron* (Engler) H. J. Lam, 1939.

Selected reference: Elisens and Jones (2009).

1. Corolla lobes lacking lateral segments or these vestigial; fruit yellow to orange at maturity ....**S. foetidissimum**
1. Corolla lobes with a large segment and 2 smaller lateral segments; fruit purple to black at maturity.
   2. Ovary glabrous ....**S. salicifolium**
   2. Ovary pubescent.

3. Leaf blade not conspicuously reticulate...S. celastrinum
3. Leaf blade conspicuously reticulate.
    4. Stem of the current season glabrous or nearly so, any pubescence of a blond or whitish color.
        5. Twigs gray-white...S. alachuense
        5. Twigs dark brown.
            6. Leaves with the upper surface (with magnification) relatively coarsely reticulate-veined, the veins somewhat impressed, not bony-cartilaginous in appearance............ S. thornei
            6. Leaves with the upper surface (with magnification) finely reticulate-veined, the veins usually raised and bony-cartilaginous in appearance.
                7. Leaf blade 1–5(7) cm long, broadest distally...S. reclinatum
                7. Leaf blade 8–12(14) cm long, broadest at or near the middle .... S. lycioides
    4. Stem of the current season with dense to moderately reddish brown, tawny, or dark brown pubescence.
        8. Pubescence on the lower leaf surface sericeous ....S. tenax
        8. Pubescence on the lower leaf surface woolly and lusterless.
            9. Larger leaves more than 3 cm long, the lower surface densely pubescent; sepals densely pubescent throughout..S. lanuginosum
            9. Larger leaves less than 3(3.5) cm long, the lower surface sparsely pubescent; sepals glabrate apically, pubescent basally...S. rufohirtum

Sideroxylon alachuense L. C. Anderson [Of Alachua County, Florida.] SILVER BUCKTHORN; CLARK'S BUCKTHORN; SILVER BULLY.

Bumelia lanuginosa (Michaux) Persoon var. *anomala* Sargent, J. Arnold Arbor. 2: 168. 1921. *Bumelia anomala* (Sargent) R. B. Clark, Ann. Missouri Bot. Gard. 29: 169. 1942. *Bumelia tenax* (Linnaeus) Willdenow forma *anomala* (Sargent) Cronquist, J. Arnold Arbor. 26: 256. 1945. *Sideroxylon alachuense* L. C. Anderson, Sida 17: 565. 1997. TYPE: FLORIDA: Alachua Co.: Gainesville, 17 Jun 1917, *Harbison 47* (holotype: A).

Tree, to 10 m; branchlets gray-white, armed with thorns, glabrous or glabrate. Leaves deciduous, the blade elliptic-ovate, 3.5–6 cm, 1.5–3 cm wide, the apex rounded to obtuse, the base cuneate, the margin entire, flat, the upper surface glabrous, the midrib flat, the marginal vein absent, the lower surface densely silver-sericeous, the petiole 3–7 mm long, sparsely sericeous to glabrate. Flowers 6–20 per fascicle, the pedicel 4–5 mm long, densely white- to tawny-sericeous; sepals 5(6), 2–3 mm long silvery- to tawny-sericeous; petals 4(6), ca. 2 mm long, white, the median segment elliptic to ovate, the lateral segments lanceolate, slightly shorter than the median; stamens 5(6), ca. 2 mm long; staminodes deltate, ca. 2 mm long, minutely erose; ovary 5(6)-carpellate and -loculate, pubescent. Fruit oblong to ovoid, 10–13 mm long, glossy black, glabrate; seeds 9–10 mm long, buff to light brown.

Calcareous hammocks. Rare; northern peninsula south to Orange County. Endemic. Summer.

Sideroxylon alachuense is listed as endangered in Florida (Florida Administrative Code, Chapter 5B-40).

Sideroxylon celastrinum (Kunth) T. D. Penn. [An evergreen tree.] SAFFRON PLUM.

Bumelia celastrina Kunth, in Humboldt et al., Nov. Gen. Sp. 7: 212. 1825. *Sideroxylon celastrinum*
(Kunth) T. D. Pennington, Fl. Neotrop. 52: 123. 1990.
Bumelia angustifolia Nuttall, N. Amer. Sylv. 3: 38, pl. 93. 1849. *Lyciodes angustifolia* (Nuttall) Kun-
tze, Revis. Gen. Pl. 2: 406. 1891. *Bumelia celastrina* Kunth var. *angustifolia* (Nuttall) R. W. Long,
Rhodora 72: 26. 1970. TYPE: FLORIDA: Monroe Co.: Key West, s.d. *Blodgett s.n.* (holotype: PH?;
isotypes: GH, K, NY).

Shrub or tree, to 10 m; branchlets gray or brown, armed with thorns, glabrate. Leaves deciduous, the blade broadly elliptic, obovate, oblanceolate, or spatulate, 0.6–3.5 cm long, 0.3–2.3
mm wide, the apex rounded to obtuse, the base cuneate, the margin entire, flat, the upper and
lower surfaces glabrous, the tertiary and smaller veins inconspicuous, the midrib flat on the
upper surface, the marginal vein present, the petiole 1–6 mm long, glabrous. Flowers 4–12 per
fascicle, the pedicel 3–6 mm long, glabrous; sepals 5, ovate-lanceolate, 2–3 mm long, glabrous;
petals 5, 3–4 mm long, white to yellowish, the median segment elliptic, 2–3 mm long, the lateral segments lanceolate, 1–2 mm long; stamens 5, 2–3 mm long; staminodes 5, lanceolate, ca. 2
mm long; ovary 5-carpellate and -loculate, hirsute to strigose basally. Fruit ellipsoid, 8–12 mm
long, purplish to purplish-black, glabrous; seeds 6–11 mm long, buff to light brown.

Coastal hammocks. Frequent; central and southern peninsula. Florida and Texas; West Indies, Mexico, Central America, and South America. Spring–fall.

Sideroxylon foetidissimum Jacq. [With a very foetid odor.] FALSE MASTIC.

Sideroxylon foetidissimum Jacquin, Enum. Syst. Pl. 15. 1760. *Bumelia foetidissima* (Jacquin) Willde-
now, Sp. Pl. 1: 1086. 1798. *Mastichodendron foetidissimum* (Jacquin) H. J. Lam, Meded, Bot. Mus.
Herb. Rijks. Univ. Utrecht 76: 521. 1939. *Mastichodendron foetidissimum* (Jacquin) H. J. Lam var.
typicum Cronquist, Lloydia 9: 247. 1946, nom. inadmiss.
Sideroxylon mastichodendron Jacquin, Collectanea 2: 253, t. 17(5). 1789. *Bumelia mastichodendron*
(Jacquin) Roemer & Schultes, Syst. Veg. 4: 493. 1819.
Bumelia pallida Swartz, Prodr. 49. 1788. *Achras pallida* (Swartz) Poiret, in Lamarck, Encycl. 6: 533.
1805. *Sideroxylon pallidum* (Swartz) Sprengel, Syst. Veg. 1: 666. 1924 ("1825"). *Sideroxylon mas-
tichodendron* Jacquin var. *pallidum* (Swartz) M. Gómez de la Maza y Jiménez, Anales Soc. Esp.
Hist. Nat. 19: 253. 1890.
Bumelia undulata Rafinesque, New Fl. 3: 28. 1838 ("1836"). TYPE: FLORIDA.

Tree, to 25 m; bark irregularly, deeply furrowed, branches lacking thorns. Leaves deciduous,
the blade elliptic to oblanceolate, 5–11 cm long, 2–6 cm wide, the apex rounded to obtuse, the
base rounded to broadly cuneate, the margin entire, slightly undulate, the upper and lower surfaces glabrous, the midrib sunken on the upper surface, the marginal vein present, the petiole
1.5–5 cm long, glabrous. Flowers 6–12 per fascicle, the pedicel 4–10 mm long, glabrous; sepals
5, suborbicular, ca. 2 mm long, glabrous; petals 5, greenish yellow, the median segment elliptic
to ovate, ca. 2 mm long, the lateral segments vestigial or absent; stamens 5–6, 3–4 mm long;
staminodes lanceolate, 1–2 mm long, erose or toothed; ovary (4)5(6)-carpellate and -loculate,
glabrous. Fruit ellipsoid, ovoid, or pyriform, 1–3 cm long, glabrous; seeds 1.5–2 cm long, buff
or brown.

Coastal hammocks. Occasional; central and southern peninsula. Florida; West Indies and Central America. Summer.

Sideroxylon lanuginosum Michx. [Woolly.] GUM BULLY.

Sideroxylon lanuginosum Michaux, Fl. Bor.-Amer. 1: 122. 1803. *Bumelia lanuginosa* (Michaux) Persoon, Syn. Pl. 1: 237. 1805. *Lyciodes lanuginosa* (Michaux) Kuntze, Revis. Gen. Pl. 2: 406. 1891. *Bumelia lanuginosa* (Michaux) Persoon subsp. *typica* Cronquist, J. Arnold Arbor. 26: 453. 1945, nom. inadmiss.

Bumelia rufa Rafinesque, New Fl. 3: 29. 1838 ("1836"). TYPE: FLORIDA: s.d., *Ware s.n.* (holotype: PH?).

Shrub or tree, to 15 m; branchlets villous when young, glabrescent, with or without thorns. Leaves deciduous, the blade oblong or oblanceolate to spatulate, 5–9 cm long, 1–4 cm wide, the apex rounded to obtuse or acute, the base cuneate, the margin entire, flat, the upper surface glabrate, the midrib flat, the marginal vein absent, the lower surface villous, the trichomes tawny, the petiole 2–14 mm long, villous. Flowers 7–17 per fascicle, the pedicel 2–7, villous; sepals (4)5, ovate, 2–3 mm long, villous; petals (4)5(6), white, the median segment oblong to ovate, 1–2 mm long, the lateral segments lanceolate or falcate, 1–2 mm long, slightly shorter than the median; stamens 5(6), ca. 3 mm long; staminodes lanceolate, shorter than the stamens, entire or erose; ovary 5(8)-carpellate and -loculate, strigose. Fruit broadly ellipsoid to obovoid, 7–12 mm long, purplish black, glabrate; seeds 6–12 mm long, brown.

Sandhills and hammocks. Occasional; northern counties south to Pinellas County. South Carolina south to Florida, west to Arkansas and Louisiana. North America; Mexico. Summer.

Sideroxylon lycioides L. [Resembling *Lycium* (Solanaceae).] BUCKTHORN BULLY; GOPHERWOOD BUCKTHORN.

Sideroxylon spinosum Duhamel du Monceau, Traite Arbr. Arbust. 2: 260., t. 68. 1755; non Linnaeus, 1753. *Sideroxylon lycioides* Linnaeus, Sp. Pl., ed. 2. 279. 1762. *Bumelia lycioides* (Linnaeus) Persoon, Syn. Pl. 1: 237. 1805. *Decateles lycioides* (Linnaeus) Rafinesque, Sylva Tellur. 36. 1838. *Lyciodes spinosa* Kuntze, Revis. Gen. Pl. 2: 406. 1891.

Shrubs, to 1 m; branchlets glabrous, glabrate, or strigose, with or without thorns. Leaves deciduous, the blade elliptic to ovate or oblanceolate, 3–12 cm long, 1.5–5 cm wide, the base attenuate, the apex acute to acuminate, the margin entire, flat, the upper surface glabrous or glabrate, finely reticulate-veined, the veins usually raised and bony cartilaginous in appearance, the marginal vein absent, the lower surface glabrous or glabrate, with the midrib villous, the petiole 3–14 mm long, glabrous. Flowers 7–40 per fascicle, the peduncle 2–10 mm long, glabrous; sepals 5(6), ovate, 1.5–2.5 mm long, glabrous; petals 5(6), white, the median segment elliptic to ovate, ca. 2 mm long, the lateral segments falcate, ca. 1.5 mm long, shorter than the median; stamens 5(6), ca. 3 mm long; staminodes lanceolate, 1.5–2 mm long, entire; ovary (4)5(6)-carpellate and -loculate, glabrous or pilose to hirsute. Fruit ellipsoid to subglobose, 1–1.5 cm long, glabrous or glabrate; seeds 7–9 mm long, buff to light brown.

Hammocks and floodplain forests; occasional; northern counties south to Lake and

Orange Counties. Virginia south to Florida, west to Illinois, Missouri, Arkansas, and Texas. Spring–summer.

Sideroxylon lycioides is listed as endangered in Florida (Florida Administrative Code, Chapter 5B-40).

Sideroxylon reclinatum Michx. [Turned or bent downward, in reference to the branches.] FLORIDA BULLY.

Shrub or tree, to 5 m; branchlets glabrous, glabrate or villous, armed with thorns. Leaves deciduous or persistent, the blade broadly elliptic, oblanceolate, or spatulate, 1–5 cm long, 0.5–2.5 cm wide, the apex rounded to obtuse, sometimes retuse, the base cuneate, the margin entire, flat, the upper surface glabrous or sparsely villous along the midrib, finely reticulate-veined, the veins usually raised and bony cartilaginous in appearance, the midrib flat, the marginal vein absent, the lower surface villous or sparsely villous only along the midrib or glabrous, the petiole 2–6 mm long, glabrous or pubescent at the base. Flowers 4–20 per fascicle, the pedicel 2–16 mm long, glabrous or strigose; sepals 5(6), ovate, 1.5–2 mm long, glabrous, rarely strigose; petals 5, white, the median segment ovate, ca. 1 mm long, the lateral segments lanceolate to falcate, equaling the median; stamens 5, 2–3 mm long; staminodes broadly lanceolate, ca. 1 mm long; ovary (4)5-carpellate and -loculate, sparsely to densely strigose. Fruit ellipsoid to subglobose, 4–9 mm long, purplish black, glabrous or glabrate; seeds 3–7 mm long, brown.

1. Lower surface of the leaf blade persistently pubescent; sepals strigose; ovary densely strigose...............
.... subsp. **austrofloridense**
1. Lower surface of the leaf blade glabrous or pubescent only along the midvein; sepals glabrous; ovary
   glabrate... subsp. **reclinatum**

Sideroxylon reclinatum subsp. **reclinatum**

> *Sideroxylon reclinatum* Michaux, Fl. Bor.-Amer. 1: 122. 1803. *Bumelia reclinata* (Michaux) Ventenat,
> Choix Pl. pl. 22. 1803–1804. *Bumelia lycioides* (Linnaeus) Persoon var. *reclinata* (Michaux) A. Gray,
> Syn. Fl. N. Amer. 2(1): 68. 1878.
> *Bumelia microcarpa* Small, Bull. New York Bot. Gard. 1: 440. 1900. TYPE: FLORIDA: Alachua Co.:
> Gainesville, Mar–Jun 1876, *Garber s.n.* (holotype: NY).

Shrub or tree, to 5 m; branchlets glabrous or glabrate. Leaves deciduous, the lower surface of the blade glabrous or sparsely villous along the midrib; sepals glabrous; ovary sparsely strigose.

Floodplain forests. Common; nearly throughout. South Carolina south to Florida, west to Louisiana. Spring.

Sideroxylon reclinatum subsp. **austrofloridense** (Whetstone) Kartesz & Gandhi [Of south Florida.]

> *Bumelia reclinata* (Michaux) Ventenat var. *austrofloridensis* Whetstone, Ann. Missouri Bot. Gard. 72:
> 545. 1985. *Sideroxylon reclinatum* Michaux subsp. *austrofloridense* (Whetstone) Kartesz & Gandhi,
> Phytologia 68: 425. 1990. TYPE: FLORIDA: Miami-Dade Co.: Long Pine Key, Everglades National
> Park, 7 Jul 1984, *Whetstone 14459* (holotype: JSU; isotypes: A, FLAS, FSU, MO, NCU, USF, VDB).

Shrub, to 2 m; branchlets strigose. Leaves persistent, the lower surface of the blade persistently villous; sepals strigose; ovary densely strigose.

Calcareous glades. Rare; Miami-Dade and Monroe Counties. Endemic. Spring.

Sideroxylum rufohirtum Herring & Judd [Reddish hairy.] RUFOUS BULLY.

Bumelia rufotomentosa Small, Bull. New York Bot. Gard. 1: 440. 1900. *Bumelia reclinata* (Michaux) Ventenat var. *rufotomentosa* (Small) Cronquist, J. Arnold Arbor. 26: 256. 1945. *Sideroxylon reclinatum* subsp. *rufotomentosum* (Small) Kartesz & Gandhi, Phytologia 68: 425. 1990. *Sideroxylon rufohirtum* Herring & Judd, Castanea 60: 358. 1995. TYPE: FLORIDA: Hillsborough Co.: Tampa, May 1876, *Garber s.n.* (holotype: NY).

Shrub, to 1 m; branchlets villous, armed with thorns. Leaves deciduous, the blade elliptic to obovate, 1.5–3 cm long, 0.7–2 cm wide, the apex rounded to obtuse, sometimes retuse, the base cuneate, the margin entire, flat, the upper surface glabrous, the midrib flat, the marginal vein absent, the lower surface reddish brown- to brown-villous, venation visible, the petiole 2–5 mm long, reddish brown- to brown-villous. Flowers 4–24 per fascicle, the pedicel 2–5 mm long, reddish brown- to brown-villous; sepals 5, ovate, ca. 2 mm long, reddish brown- to brown-villous; petals 5, white, the median segment broadly ovate, 1–2 mm long, the lateral segments lanceolate, equaling the median; stamens 5, ca. 2 mm long; staminodes lanceolate, shorter than the stamens, erose; ovary 5-carpellate and -loculate, villous to glabrate. Fruit ellipsoid to subglobose, 8–13 mm long, glabrous or glabrate; seeds 8–11 mm long, brown.

Hammocks. Occasional; northern and central peninsula, Dixie County. Endemic. Spring.

Sideroxylon salicifolium (L.) Lam. [With leaves like *Salix* (Salicaceae).] WILLOW BUSTIC; WHITE BULLY.

Achras salicifolia Linnaeus, Sp. Pl., ed. 2. 470. 1762. *Bumelia salicifolia* (Linnaeus) Swartz, Prodr. 50. 1788. *Sideroxylon salicifolium* (Linnaeus) Lamarck, Tabl. Encycl. 2: 42. 1794. *Dipholis salicifolia* (Linnaeus) A. de Candolle, in de Candolle, Prodr. 8: 188. 1844. *Spondogona salicifolia* (Linnaeus) House, Amer. Midl. Naturalist 7: 131. 1921.

Shrub or tree, to 25 m; branchlets glabrous, without thorns. Leaves deciduous, the blade elliptic, 2.5–11 cm long, 1.2–4 cm wide, the apex acute to acuminate, the base narrowly cuneate, the margin entire, flat, the upper surface glabrous, the midrib slightly raised, the marginal vein present, the lower surface glabrous or glabrate with reddish brown trichomes scattered along the midrib, the venation visible, the petiole 6–14 mm long, glabrous or glabrate. Flowers 5–14 per fascicle, the pedicel 1–5 mm long, reddish brown-sericeous; sepals 5, ovate, ca. 2 mm long, white- to tawny-sericous; petals 5, white to cream, the median segment ovate to suborbicular, ca. 2 mm long, the lateral segments lanceolate, slightly shorter than the median; stamens 5, ca. 3 mm long; staminodes lanceolate, ca. 2 mm long, erose; ovary 5-carpellate and -loculate, glabrous. Fruit ellipsoid to subglobose, 6–10 mm long, glabrous; seeds 4–6 mm long, brown.

Hammocks and pinelands. Occasional; Martin and Palm Beach Counties, southern peninsula. Florida; West Indies, Mexico, and Central America. All year.

Sideroxylon tenax L. [Holding fast, tough.] TOUGH BULLY.

Sideroxylon tenax Linnaeus, Mant. Pl. 48. 1767. *Bumelia tenax* (Linnaeus) Willdenow, Sp. Pl. 1: 1085. 1798. *Sideroxylon chrysophylloides* Michaux, Fl. Bor.-Amer. 1: 123. 1803, nom. illegit. *Bumelia chrysophylloides* Pursh, Fl. Amer. Sept. 155. 1814, nom. illegit. *Sclerocladus tenax* (Linnaeus) Rafinesque, Sylva Tellur. 35. 1838. *Sclerozus tenax* (Linnaeus) Rafinesque, Autik. Bot. 73. 1840. *Lyciodes tenax* (Linnaeus) Kuntze, Revis. Gen. Pl. 2: 406. 1891.

Bumelia megococca Small, Bull. New York Bot. Gard. 1: 441. 1900. TYPE: FLORIDA: Hillsborough Co.: Tampa, Oct 1877, *Garber s.n.* (holotype: NY).

Bumelia lacuum Small, Man. S.E. Fl. 1034, 1507. 1933. TYPE: FLORIDA: Highlands Co.

Shrub or tree, to 5 m; branchlets sericeous to glabrescent, armed with thorns. Leaves deciduous or persistent, the blade oblanceolate to spatulate, 2.5–6 cm long, 0.6–2 cm wide, the apex rounded to obtuse or retuse, the base cuneate, the margin entire, flat, the upper surface glabrous or glabrate, the midrib flat or slightly sunken, the marginal vein absent, the lower surface densely tawny- or reddish brown- to brown-sericeous, the petiole 3–10 mm long, sparsely to densely tawny- to reddish brown- to brown-villous. Flowers 8–40 per fascicle, the pedicel 4–12 mm long, tawny- or reddish brown- to brown-sericeous; sepals 5, ovate, 2–3 mm long, tawny- or reddish brown- to brown-sericeous; petals 5, white, the median segment elliptic to obovate, ca. 2 mm long, the lateral segments lanceolate, subequaling the media; stamens 5, ca. 2 mm long; staminodes broadly lanceolate, slightly shorter than the stamens, entire; ovary 5-carpellate and -loculate, glabrous or strigose distally. Fruit ellipsoid to obovoid 8–13 mm long, purplish black, glabrous; seeds 8–12 mm long, brown.

Sandhills, scrub, and coastal hammocks. Frequent; peninsula. North Carolina south to Florida. Spring.

Sideroxylon thornei (Cronquist) T. D. Penn. [Commemorates Robert Folger Thorne (1920–2015), known for his classification system, biogeographic studies and floristics of western North America.] GEORGIA BULLY; THORNE'S BUCKTHORN.

Bumelia thornei Cronquist, Castanea 14: 103. 1949. *Sideroxylon thornei* (Cronquist) T. D. Pennington, Fl. Neotrop. 52: 170. 1990.

Shrub or tree, to 6 m; branchlets glabrate to villous, glabrescent, armed with thorns. Leaves deciduous, the blade elliptic to oblanceolate or ovate, 1.5–7.5 cm long, 1–5.5 mm wide, the apex rounded to obtuse, sometimes retuse, the base cuneate, the margin entire, flat, the upper surface glabrous or glabrate, the midrib flat, the marginal vein absent, the lower surface tawny- to reddish-brown-villous, especially along the midrib, the venation visible, the petiole 3–7 mm long, tawny-villous. Flowers 3–8 per fascicle, the pedicel 5–8 mm long, glabrous or glabrate; sepals 5, ovate, ca. 2 mm long, glabrous or reddish brown-villous; petals 5, white, the median segment ovate to obovate, ca. 2 mm long, the lateral segments lanceolate, subequaling the median; stamens 5, ca. 1 mm long; staminodes deltate to ovate, ca. 2 mm long; ovary 5-carpellate and -loculate, villous. Fruit subglobose, 8–11 mm long, glabrous or glabrate; seeds 7–8 mm long, brown.

Creek hammocks or margin of cypress ponds and swamps. Rare; central and western panhandle. Georgia, Alabama, and Florida. Spring.

Sideroxylon thornei is listed as endangered in Florida (Florida Administrative Code, Chapter 5B-40).

EBENACEAE Gürke 1891. EBONY FAMILY

Trees or shrubs. Leaves alternate, simple, pinnate-veined, petiolate, estipulate. Flowers in axillary cymes or solitary, actinomorphic, unisexual (plants functionally dioecious); sepals 4, basally connate; petals 4, basally connate. Staminate flowers with the stamens 16, connate in pairs, the anther 2-locular, longitudinally dehiscent; pistillodes present. Carpellate flowers with staminodes present; ovary 1, superior, 4- to 8-carpellate and -loculate. Fruit a berry.

A family of 2-6 genera and about 500 species; nearly cosmopolitan.

Selected references: Eckenwalder (2009); Wood and Channell (1960).

Diospyros L. 1753.

Trees or shrubs. Leaves alternate, simple, pinnate-veined, petiolate, estipulate. Staminate flowers in axillary cymes or solitary; sepals 4, basally connate; petals 4, basally connate; stamens with 16 stamens in 2 whorls of 4 pairs; pistillode present. Carpellate flowers solitary; sepals 4, basally connate; petals 4, basally connate; staminodes 8 in 2 whorls of 4 pairs; ovary 1, 4- to 8-carpellate and -loculate, the styles 4, basally connate. Fruit a berry; seeds 1–2 per locule or fewer by abortion.

A genus of about 450 species; nearly cosmopolitan. [From the Greek *Dios*, of Zeus, and *pyr*, grain, used by Theophrastus for an unknown fruit.]

1. Lower surface of the leaf blade with a pair of basal glands ....**D. maritima**
1. Lower surface of the leaf blade lacking basal glands.
   2. Leaf blade broadly ovate to elliptic, the apex acute to acuminate; fruit yellow to orange or dark red, rarely purple ....**D. virginiana**
   2. Leaf blade elliptic-lanceolate, rarely ovate, the apex rounded to obtuse or short acuminate; fruit black...**D. ebenum**

Diospyros ebenum J. König ex Retz. [Pre-Linnaean name for the species.] EBONY.

Diospyros ebenum J. König ex Retzius, Physiogr. Saelsk. Handl. 1: 176. 1781. *Diospyros glaberrima* Rottbøll, Nye Saml. Kongel. Danske Vidensk. Selsk. Skr. 2: 540. 1783, nom. illegit. *Diospyros ebenum* König ex Retzius var. *glaberrima* Bakhuizen, Bull. Jard. Bot. Buitenzorg, ser. 3. 15: 216, nom. inadmiss.

Tree, to 20(30) m; bark, gray to black, shallowly furrowed, flaking in rectangular plates. Leaves persistent, the blade elliptic-lanceolate, rarely ovate, 6–10(15) cm long, (2)3–5 cm wide, the apex rounded to obtuse or short-acute, the base broadly cuneate, the margin entire, the upper and lower surfaces glabrous, the petiole 4–6(8) mm long, glabrous. Flowers 3–15 in a cyme or solitary; sepals 4, ovate; petals 4, 12–18 cm long. Staminate flowers with the 16 stamens in 2 whorls of 4 pairs. Carpellate flowers usually with 8 staminodes; ovary 4- to 8-carpellate, glabrous, the styles 4, connate nearly to the stigmas. Fruit depressed-globose, (3)4–5 cm long, black, glabrous; seeds triangular-ellipsoid, 1–1.5 cm long, black.

Disturbed sites. Rare; Palm Beach and Broward Counties. Escaped from cultivation. Florida; Asia. Native to Asia. Spring.

Diospyros maritima Blume [Of the sea.] MALAYSIAN PERSIMMON.

Diospyros maritima Blume, Bijdr. 669. 1826. *Cargillia maritima* (Blume) Hasskarl, Cat. Hort. Bot. Bogor. 159. 1844.

Shrub or tree, to 12 m; bark dark brown to black, smooth and flaking. Leaves persistent, the blade elliptic-lanceolate to elliptic-oblong or slightly obovate, 7–15(17) cm long, 3–6(8) cm, wide, the apex obtuse, the base rounded, the margin entire, the lower surface glabrous or sparsely pubescent when young, soon glabrous, with a pair of basal glands, the petiole 0.5–1 cm long. Flowers 2–3 in a cyme or solitary; sepals 4, ovate; petals 4, 1.5–1.8 mm long. Staminate flowers with 16 stamens in 2 whorls of 4 pairs. Carpellate flowers usually with 8 staminodes, the ovary 4- to 8-carpellate, pubescent, the styles connate for most of their length. Fruit depressed-globose, (1.5)2–3 cm long, yellow to orange or black, glabrous, except at the apex; seeds ellipsoid, ca. 1.5 cm long, brown.

Disturbed sites. Rare; Miami-Dade County. Escaped from cultivation. Florida; Asia, Australia, and Pacific Islands. Native to Asia, Australia, and Pacific Islands. Spring.

Diospyros virginiana L. [Of Virginia.] COMMON PERSIMMON.

Diospyros virginiana Linnaeus, Sp. Pl. 1057. 1753. *Diospyros concolor* Moench, Methodus 471. 1794, nom. illegit.
Diospyros ciliata Rafinesque, New. Fl. 3: 25. 1838 ("1836"). TYPE: FLORIDA.
Diospyros mosieri Small, J. New York Bot. Gard. 22: 33. 1921. *Diospyros virginiana* Linnaeus var. *mosieri* (Small) Sargent, J. Arnold Arbor. 2: 170. 1921. *Diospyros virginiana* Linnaeus subsp. *mosieri* (Small) E. Murray, Kalmia 12: 20. 1982. TYPE: FLORIDA: Miami-Dade Co.

Tree, to 30(40) m; bark dark reddish brown, deeply furrowed and forming irregular blocks. Leaves deciduous, the blade broadly ovate to elliptic, (5)6–15 cm long, 2.5–8 cm wide, the apex acute to acuminate, the base broadly cuneate to rounded, the margin entire, the lower surface sparsely pubescent or glabrous, the petiole 0.7–1 cm long. Flowers 2–3 in a cyme or solitary; sepals 4, ovate; petals 4, 1–2 cm long. Staminate flowers with the 16 stamens in 2 whorls of 4 pairs. Carpellate flowers usually with 8 staminodes; ovary 4- to 8-carpellate, glabrous except at the apex, the styles 4, basally connate. Fruit depressed-globose, globose, oblong, or conic, (2)3–5(7.5) cm long, yellow to orange or dark red, rarely purple, often glaucous; seeds ellipsoid, ca. 1.5 cm long, reddish brown.

Flatwoods, sandhills, and hammocks. Common; nearly throughout. New York south to Florida, west to Nebraska, Kansas, Oklahoma, and Texas. Spring–summer.

EXCLUDED TAXON

Diospyros caribaea (A. de Candolle) Standley—Reported by Menninger (1964), as "occasionally found growing wild in the Florida Keys." No Florida specimens known.

PRIMULACEAE Batsch ex Borkh. 1797. PRIMROSE FAMILY

Perennial herbs. Leaves in basal rosettes, alternate, simple, pinnate-veined, petiolate, estipulate. Flowers in terminal scapose umbels, bracteate, actinomorphic, bisexual; sepals 5, basally connate; petals 5, basally connate; stamens 5, epipetalous, basally connate or free, the anthers dehiscent by longitudinal slits; ovary superior, 5-carpellate, 1-loculate, the style 1. Fruit a valvate or operculate capsule.

A family of about 20 genera and about 500 species; North America, West Indies, Mexico, Central America, South America, Europe, and Asia.

Varying opinions exist regarding the Primulaceae complex. Kubitzki (2004), whom we follow here, recognizes Primulaceae, Myrsinaceae, Theophrastaceae, and Samolaceae as distinct families.

Selected reference: Channell and Wood (1959).

Primula L. 1753. PRIMROSE

Perennial herb. Leaves in basal rosettes, alternate, simple, pinnate-veined, petiolate, estipulate. Flowers in terminal, bracteate umbels; sepals 5, basally connate; petals 5, basally connate; stamens 5, free or somewhat connate, the anthers connivent. Ovary 5-carpellate, 1-loculate, the style 1, the stigma capitate. Fruit a valvate or operculate capsule; seeds numerous.

A genus of about 500 species; North America, Mexico, Central America, South America, Africa, Europe, and Asia. [*Primus*, first, and -*ulus*, diminutive, in reference to the early spring flowering.]

Dodecatheon L., 1753.

Selected reference: Reveal (2009).

Primula meadia (L.) A. R. Mast & Reveal [Between, apparently meant as a species intermediate between two other species.] SHOOTINGSTAR; PRIDE-OF-OHIO.

Dodecatheon meadia Linnaeus, Sp. Pl. 144. 1753. *Meadia dodecathea* Crantz, Inst. Rei. Herb. 2: 309. 1766. *Dodecatheon reflexum* Salisbury, Prodr. Stirp. Chap. Allerton 118. 1796, nom. illegit. *Dodecatheon meadia* Linnaeus subsp. *eumeadia* R. Knuth, in Engler, Pflanzenr. 4(Heft 22): 237. 1905, nom. inadmiss. *Dodecatheon meadia* Linnaeus var. *genuinum* Fassett, Amer. Midl. Naturalist 31: 463. 1944, nom. inadmiss. *Primula meadia* (Linnaeus) A. R. Mast & Reveal, Brittonia 59: 81. 2007.

Perennial herb, to 5 dm; stem erect, glabrous or glabrate. Leaves in a basal rosette, the blade oblanceolate to oblong or spatulate, the apex acute to rounded, the base narrowly cuneate, tapering to the petiole, the margin entire or coarsely toothed, the upper and lower surfaces glabrous, the petiole 3–5 cm long, winged nearly to the base. Flowers 1–25 in an umbel, the peduncle to 5 dm long, the pedicel (1)5–7 cm long, glabrous or glabrate; bracts lanceolate, 3–10 mm long, glabrous or glabrate; sepals green, 5–12 mm long, glabrous, the tube 2–4 mm long, the lobes 3–7 mm long; corolla tube maroon and yellow, with a wavy apical ring, the lobes spatulate, 1–2.5 cm long, strongly reflexed, white to lavender; stamens with the filaments usually connate into a tube 1–3 mm long, yellow, the anthers 4–10 mm long, yellow, the connective dark maroon or black. Fruit cylindric-ovoid, 0.7–2 cm long, 4–6(9) mm wide, dark reddish brown, glabrous; seeds 50–200, globose to ovoid or quadrate, irregularly alveolate, dark brown to black.

Calcareous prairies. Rare; Gadsden County. New York south to Florida, west to Manitoba, Minnesota, Iowa, Kansas, Oklahoma, and Texas. Spring.

Primula meadia is listed as endangered in Florida (Florida Administrative Code, Chapter 5B-40).

EXCLUDED TAXON

Hottonia inflata Elliott—Reported for Florida by Small (1903, 1913a, 1933), Correll and Johnston (1970), and Godfrey and Wooten (1981). No Florida specimens known.

MYRSINACEAE R. Br., nom. cons. 1810. MYRSINE FAMILY

Herbs, shrubs, or trees. Leaves alternate, opposite, or whorled, simple, pinnate-veined, petiolate or epetiolate, estipulate. Flowers in terminal or axillary racemes, panicles, cymes, or solitary, actinomorphic, bisexual or unisexual (plants dioecious), bracteate or ebracteate; sepals 4–6, basally connate; petals 4–6, basally connate; stamens 4–6, epipetalous, the filaments free or connate, the anthers dehiscent by longitudinal slits or apical pores; staminodes present in carpellate flowers; ovary 1, superior, 3- to 5-carpellate, 1-loculate, the style 1. Fruit a valvate or circumscissile capsule or a drupe.

A family of about 50 genera and about 1,400 species; nearly cosmopolitan.

Myrsinaceae is treated as a distinct family following Kubitzki (2004). See discussion under Primulaceae.

Ardisiaceae Juss., 1810.

Selected reference: Channell and Wood (1959).

1. Trees or shrub; fruit a drupe.
  2. Flowers borne in a terminal panicle or raceme or in an axillary cyme ....Ardisia
  2. Flowers borne on clustered, short, spurlike peduncles along the stem .... Myrsine
1. Herbs; fruit a capsule.
  3. Leaves to 2 cm long; plant annual........Anagallis
  3. Leaves more than 2 cm long; plant perennial........Lysimachia

Anagallis L. 1753. PIMPERNEL

Annual or perennial herbs. Leaves opposite, alternate, or whorled, simple, pinnate-veined, petiolate or epetiolate, estipulate. Flowers axillary, solitary, bisexual, ebracteate, pedicellate;

sepals (4)5, basally connate; petals (4)5, basally connate; stamens (4)5, the filaments basally connate, the anthers dehiscent by longitudinal slits. Fruit a circumscissile dehiscent capsule.

A genus of about 20 species; North America, West Indies, Mexico, Central America, South America, Europe, Africa, and Asia. [From the Greek *anagalao*, to laugh, referring to its fabled power to alleviate sadness.]

Centunculus L., 1753; *Micropyxis* Duby, 1844.

Selected reference: Cholewa (2009b).

1. Flowers subsessile .... **A. minima**
1. Flowers evidently pedicellate.
   2. Leaves all opposite or whorled .... **A. arvensis**
   2. Leaves alternate distally, opposite or subopposite proximally ....**A. pumila**

Anagallis arvensis L. [Pertaining to fields or cultivated land.] SCARLET PIMPERNEL.

Anagallis arvensis Linnaeus, Sp. Pl. 148. 1753. *Anagallis pulchella* Salisbury, Prodr. Stirp. Chap. Allerton 120. 1796, nom. illegit. *Anagallis punctifolia* Stokes, Bot. Mat. Med. 1: 305. 1812, nom. illegit. *Anagallis punctifolia* Stokes var. *rubra* Stokes, Bot. Mat. Med. 1: 305. 1812, nom. inadmiss. *Lysimachia arvensis* (Linnaeus) U. Manns & Anderberg, Willdenowia 39: 51. 2009.

Erect, ascending, or procumbent herb, to 3 dm; stem glabrous. Leaves opposite or sometimes whorled distally, the blade ovate to elliptic or lanceolate, 5–20 mm long, 4–10 mm wide, the apex acute to obtuse, the base broadly cuneate to rounded, the margin entire, the upper and lower surfaces glabrous, punctate, sessile. Flowers solitary, axillary, the pedicel to 2.5 cm long, longer than the subtending leaf; sepals 5, narrowly ovate-lanceolate, 4–5 mm long, somewhat keeled, the apex acuminate, basally connate, the margin entire or minutely crenulate; petals 5, obovate, 4–5 mm long, subequaling the sepals, the corolla salverform to rotate, 5–10 mm wide, salmon or red, rarely blue, the apex usually fringed with stipitate glands; stamens 5, the filaments pubescent. Fruit globose, 4–7 mm long; seeds 12–40, ca. 1 mm long, angled, dark reddish brown, the surface with fine excrescences.

Disturbed sites. Occasional; peninsula, central and western panhandle. Nearly throughout North America; Mexico; Europe, Africa, and Asia. Native to Europe and Asia. Spring–summer.

Anagallis minima (L.) E.H.L. Krause [Very small, in reference to the habit.] CHAFFWEED.

Centunculus minimus Linnaeus, Sp. Pl. 116. 1753. *Anagallis centunculus* Afzelius, Veg. Suec. 1: 1785, nom. illegit. *Anagallis pusilla* Salisbury, Prodr. Stirp. Chap. Allerton 121. 1796, nom. illegit. *Anagallidastrum exiguum* Bubani, Fl. Pyren. 1: 238. 1897, nom. illegit. *Anagallis minima* (Linnaeus) E.H.L. Krause, in Sturm, Deutschl. Fl., ed. 2. 9: 251. 1901. *Micropyxis exigua* Lunell, Amer. Midl. Naturalist 4: 506. 1916, nom. illegit. *Lysimachia minima* (Linnaeus) U. Manns & Anderberg, Willdenowia 39: 52. 2009.

Erect annual herb, to 4(10) cm; stem glabrous. Leaves alternate distally, the proximal ones sometimes opposite or subopposite, the blade elliptic, ovate, or obovate, 3–5 mm long, ca. 3 mm wide, the apex acute to obtuse, the base broadly cuneate, the margin entire, the upper and lower surfaces glabrous, sessile or subpetiolate. Flowers with the pedicel to 1 mm long or absent; sepals 4–5, narrowly ovate, 2–3 mm long, basally connate to ca. ⅓ their length, the lobes linear-oblong, the apex long-acuminate, the margin entire or minutely crenulate; petals 4–5, obovate, 1–2 mm long, the corolla salverform to rotate, 1–2 mm wide, white or pink; stamens 4–5, the filaments glabrous. Fruit globose, ca. 2 mm long; seeds 5–12, obpyramidal, brown, the surface with fine excrescences.

Marsh and pond margins. Occasional; peninsula, central and western panhandle. Nearly throughout North America; Mexico; Europe and Asia. Spring–fall.

Anagallis pumila Sw. [Small.] FLORIDA PIMPERNEL.

> *Anagallis pumila* Swartz, Prodr. 40. 1788. *Centunculus pentandrus* R. Brown, Prodr. 427. 1810, nom. illegit. *Micropyxis pumila* (Swartz) Duby, in de Candolle, Prodr. 8: 72. 1844. *Centunculus pumilus* (Swartz) Kuntze, Revis. Gen. Pl. 3(2): 193. 1898. *Centunculus pumilus* (Swartz) Kuntze var. *pentandrus* Kuntze, Revis. Gen. Pl. 3(2): 193. 1898, nom. inadmiss.

Procumbent to erect annual herb, to 3 dm; stem glabrous. Leaves alternate distally, the proximal ones opposite or subopposite, the blade elliptic to lanceolate, 4–8 mm long, 2–4 mm wide, the apex acute, the base cuneate, the margin entire, the upper and lower surfaces glabrate, sessile or short-petiolate. Flowers with the pedicel 3–6 mm long; sepals (4)5, lanceolate, 2–3 mm long, basally connate, the lobes with the margin entire, the apex acuminate; petals (4)5, ca. 2 mm long, the corolla salverform to rotate, greenish white; stamens 5, the filaments pubescent. Fruit globose, ca. 2 mm long; seeds 5–12, 3-angled, brown, the surface minutely rugose.

Wet flatwoods. Rare; Highlands, Lee, and Collier Counties. Florida; West Indies, Mexico, and Central America. Spring–fall.

Ardisia Sw., nom. cons. 1788. MARLBERRY

Subshrubs, shrubs, or trees. Leaves alternate, simple, pinnate-veined, petiolate, estipulate. Flowers in terminal or axillary panicles, racemes, or cymes, bisexual, ebracteate, pedicellate; sepals 4–6, basally connate; petals 4–6, basally connate; stamens 4–6, the filaments basally connate, the anthers dehiscent by longitudinal slits or terminal pores. Fruit a 1-seeded drupe.

A genus of about 450 species; North America, West Indies, Mexico, Central America, South America, Asia, Australia, and Pacific Islands. [From the Greek *ardis*, point of an arrow or spear, apparently in reference to the anther shape.]

Icacorea Aubl., nom. rej., 1775.

Selected reference: Pipoly and Ricketson (2009a).

1. Leaves serrulate; plant suffruticose, less than 1 m tall...**A. japonica**
1. Leaves entire or crenulate; plant a shrub, usually more than 1 m tall.
   2. Inflorescence a terminal panicle or raceme...**A. escallonioides**
   2. Inflorescence terminal on a short lateral branch or in an axillary subumbellate cyme or raceme.
      3. Leaf margins crenulate or undulate; fruit red or white ....**A. crenata**
      3. Leaf margins entire; fruit purplish or black.

4. Leaf blades subcoriaceous; peduncle longer than the pedicels; sepals ca. 1 mm long; fruit minutely black-punctate .... **A. elliptica**

4. Leaf blades chartaceous; peduncle as long as the pedicels; sepals ca. 3 mm long; fruit densely and coarsely black-punctate .... **A. solanacea**

Ardisia crenata Sims [With rounded teeth, in reference to the leaf margin.] SCRATCHTHROAT.

Ardisia crenata Sims, Bot. Mag. 45: pl. 1950. 1818. *Bladhia crenata* (Sims) H. Hara, Enum. Sperm. Jap. 1: 75. 1948.

Shrub, to 1.5(3) m; branchlets minutely reddish glandular-papillate, glabrous. Leaves with the blade elliptic, lanceolate, or oblanceolate, 7–15 cm long, 2–4 cm wide, subcoriaceous, the apex acute to acuminate, rarely obtuse to rounded, the base cuneate, the margin crenulate or undulate, slightly revolute, the upper and lower surfaces reddish glandular-papillate, glabrous, the petiole 6–10 mm long, glabrous. Flowers 5–18, terminal on a short 2- or 3-leaved lateral branch, in a subumbellate cyme, the pedicel 7–10 cm long, minutely glandular-papillate; sepals (4)5(6), oblong-ovate, 1–2 mm long, the margin entire; petals (4)5(6), ovate, 4–6 mm long, white or pinkish, the margin entire, glandular-papillose on the inner surface near the base; stamens (4)5(6), shorter than the petals, glandular-punctate dorsally, the anthers dehiscent by terminal pores. Fruit globose, 6–8 mm long, red, glandular-punctate; seed globose.

Moist hammocks. Occasional; northern and central peninsula, west to central panhandle. Escaped from cultivation. Georgia, Florida, Louisiana, and Texas; Africa, Asia, and Pacific Islands. Native to Asia. Spring.

Ardisia crenata is listed as a Category I invasive species in Florida by the Florida Exotic Pest Plant Council (FLEPPC, 2017).

Ardisia elliptica Thunb. [Shaped like an ellipse, in reference to the leaf shape.] SHOEBUTTON.

Ardisia elliptica Thunberg, Nov. Gen. Pl. 8: 119. 1798. *Bladhia elliptica* (Thunberg) Nakai, in Nakai & Honda, Nov. Fl. Jap. 9: 120. 1943.

Shrub, to 2 m; branchlets black punctate-lineate, glabrous. Leaves with the blade oblanceolate or obovate, 6–12(16) cm long, 3–5(7) cm wide, subcoriaceous, the apex obtuse to acute, the base cuneate, the margin entire, slightly revolute, the upper and lower surfaces glabrous, the petiole 5–10 mm long, glabrous. Flowers 5–10, lateral or subterminal on a short lateral branch, in a subumbellate cyme, the peduncle longer than the pedicel, the pedicel 1–2 cm long, glabrous; sepals 5, broadly ovate, ca. 1 mm long, the margin subentire, minutely ciliate, black-glandular-punctate, glabrous; petals 5, broadly ovate, 6–8 mm long, the margin entire, glandular-punctate, glabrous; stamens 5, subequaling the petals, the anthers glandular-punctate dorsally, dehiscent by longitudinal slits. Fruit subglobose, ca. 8 cm long, purplish black, minutely punctate.

Hammocks. Occasional; central and southern peninsula. Escaped from cultivation. Florida; West Indies; Africa, Asia, and Pacific Islands. Native to Asia. Spring.

Ardisia elliptica is listed as a Category I invasive species in Florida by the Florida Exotic Pest Plant Council (FLEPPC, 2017).

Ardisia escallonioides Schltdl. & Cham. [Resembling *Escallonia* (Escalloniaceae).] MARLBERRY.

Ardisia escallonioides Schlechtendal & Chamisso, Linnaea 6: 393. 1831. *Tinus escallonioides* (Schlechtendal & Chamisso) Kuntze, Revis. Gen. Pl. 2: 974. 1891.

Cyrilla paniculata Nuttall, Amer. J. Sci. Arts 5: 290. 1822. *Pickeringia paniculata* (Nuttall) Nuttall, J. Acad. Nat. Sci. Philadelphia 7: 95. 1834. *Ardisia pickeringia* Torrey & A. Gray ex A. de Candolle, in de Candolle, Prodr. 8: 124. 1844. *Bladhia paniculata* (Nuttall) Sudworth, Gard. & Forest 4: 239. 1891. *Tinus pickeringia* (Torrey & A. Gray ex A. de Candolle) Kuntze, Revis. Gen. Pl. 2: 974. 1891. *Icacorea paniculata* (Nuttall) Sudworth, Gard. & Forest 6: 324. 1893. *Ardisia paniculata* (Nuttall) Sargent, Man. Trees, ed. 2. 806. 1922; non Roxburgh, 1924. TYPE: FLORIDA: s.d., *Ware s.n.* (holotype: GH).

Shrub or tree, to 15 m; branchlets rufous-papillate, glabrous or glabrate. Leaves with the blade elliptic to oblanceolate, 4–17 cm long, 1.5–6 cm wide, subcoriaceous, the apex acute, the base cuneate, the margin entire, flat, the upper and lower surfaces glabrous, the petiole 5–12 mm long, glabrous. Flowers few to many in a terminal panicle or raceme, the peduncle ca. 2 cm long, the pedicel 4–6 mm long, rufous-papillate; sepals 5–6, ovate, ca. 2 mm long, the margin entire, glandular-ciliate, the apex minutely papillate; petals 5–6, lanceolate, 7–10 mm long, white to pink, the margin entire, glandular-punctate, with yellow papillae basally on the inner surface; stamens 5–6, shorter than the petals, the anthers glandular punctate dorsally, dehiscent by longitudinal slits. Fruit globose, 4–7 mm long, black, glandular-punctate.

Hammocks. Frequent; Flagler County, central and southern peninsula. Florida; West Indies, Mexico, and Central America. Spring.

Ardisia japonica (Thunb.) Blume [Of Japan.] JAPANESE ARDISIA.

Bladhia japonica Thunberg, Nov. Gen. Pl. 1: 7. 1781. *Ardisia japonica* (Thunberg) Blume, Bijdr. 690. 1826. *Tinus japonica* (Thunberg) Kuntze, Revis. Gen. Pl. 2: 405. 1891.

Stoloniferous subshrubs, to 3(4) dm; branchlets minutely rufous, puberulent, glabrescent. Leaves pseudoverticillate, the blade elliptic to obovate or lanceolate, 4–7 cm long, 1.5–4 cm wide, chartaceous, the apex acute, the base cuneate, the margin serrulate, the upper and lower surfaces minutely rufous-puberulent, glabrescent, the petiole 6–10 mm long, minutely puberulent. Flowers 3–5 in an axillary subumbellate cyme, the peduncle ca. 5 mm long, the pedicel 7–10 mm long, puberulent; sepals 5(6), ovate, 1–2 mm long, the margin entire, ciliate, sometimes glandular-punctate, glabrous; petals 5(6), broadly ovate, 4–5 mm long, the margin entire, glandular-punctate; stamens 5(6), shorter than the petals, the anthers glandular-punctate dorsally, dehiscent by longitudinal slits. Fruit globose, 5–6 mm long, red to blackish, glandular-punctate.

Disturbed sites. Rare; Alachua, Jackson, and Santa Rosa Counties. Escaped from cultivation. Florida, Louisiana, and Texas; Asia. Native to Asia. Summer.

Ardisia japonica is listed as a Category II invasive species in Florida by the Florida Exotic Pest Plant Council (FLEPPC, 2017).

Ardisia solanacea Roxb. [Resembling *Solanum* (Solanaceae).] CHINA-SHRUB.

> *Ardisia solanacea* Roxburgh, Pl. Coromandel 1: 27, pl. 27. 1795. *Anguillaria solanacea* (Roxburgh) Poiret, in Lamarck, Encycl. 7: 688. 1806. *Icacorea solanacea* (Roxburgh) Britton, Fl. Bermuda 284. 1918. *Bladhia solanacea* (Roxburgh) Nakai, in Nakai & Honda, Nov. Fl. Jap. 9: 120. 1943.

Shrub or tree, to 6 m; branchlets glabrous. Leaves with the blade elliptic to oblanceolate, 12–20 cm long, 4–7 cm wide, chartaceous, the apex acute, the base cuneate, the margin entire, sub-revolute, the upper and lower surfaces black-glandular-punctate, glabrous, the petiole 1–2 cm long, glabrous. Flowers 5–8 in an axillary subumbellate raceme, the peduncle 2–3 cm long, the pedicel 2–3 cm long, glabrous; sepals 5, ovate to reniform, ca. 3 mm long, the margin subentire to crenulate, ciliate, black-glandular-punctate, glabrous; petals 5, broadly ovate, ca. 9 mm long, the margin entire, entire, glandular-punctate; stamens 5, subequaling the petals, the anthers glandular-punctate dorsally, dehiscent by longitudinal slits. Fruit oblate, 7–9 mm long, black, glandular-punctate.

Disturbed hammocks. Rare; Hillsborough County. Escaped from cultivation. Florida; West Indies; Asia. Native to Asia. Spring.

EXCLUDED TAXON

> *Ardisia polycephala* Wallich ex A. de Candolle—Reported for Florida by Small (1933), the name misapplied to material of *A. elliptica*.

Lysimachia L. 1753. LOOSESTRIFE

Perennial herbs. Leaves opposite or whorled, simple, pinnate-veined, petiolate or epetiolate, estipulate. Flowers solitary, axillary, bisexual, bracteate, pedicellate; sepals 5, basally connate; petals 5, basally connate; stamens 5, the filaments basally connate or free, the anthers dehiscent by longitudinal slits; staminodes 5. Fruit a valvate dehiscent capsule; seeds few to many.

A genus of about 160 species; nearly cosmopolitan. [From the Greek *lysis*, dissolve, and *mache*, strife, in reference to its soothing properties.]

Steironema Raf., 1821.

Selected reference: Cholewa (2009c).

1. Principal leaves petiolate, the base rounded to cordate .... **L. ciliata**
1. Principal leaves sessile or the leaf base decurrent as a narrow wing .... **L. lanceolata**

Lysimachia ciliata L. [With cilia, in reference to the leaves.] FRINGED LOOSESTRIFE.

> *Lysimachia ciliata* Linnaeus, Sp. Pl. 147. 1753. *Steironema ciliatum* (Linnaeus) Baudo, Ann. Sci. Nat., Bot., ser. 2. 20: 346. 1843. *Nummularia ciliata* (Linnaeus) Kuntze, Revis. Gen. Pl. 2: 398. 1891.

Erect perennial, to 13 dm; stem glabrous. Leaves opposite, the blade lanceolate to ovate-lanceolate, 4–15(17) cm long, 1.5–6.5 cm wide, the apex acute to acuminate, the base rounded to cuneate, the margin entire, ciliolate, the upper and lower surfaces glabrous, the petiole 0.5–6 cm long, long-ciliate along its entire length. Flowers solitary, axillary in the distal leaves, the pedicel 1–7 cm long, usually stipitate-glandular; sepals 3–9 mm long, the lobes lanceolate, sometimes stipitate-glandular; petals 5–12 mm long, yellow, sometimes with a reddish base, the lobes obovate, stipitate-glandular; stamens connate basally, shorter than the corolla; staminodes 1–2 mm long. Fruit globose, 5–7 mm long, glabrous; seeds numerous, 1–2 mm long, 3-angled, reddish brown, reticulate.

Floodplain forests. Rare; Gadsden, Liberty, and Jackson Counties. Nearly throughout North America. Spring.

Lysimachia lanceolata Walter [Lance-shaped.] LANCELEAF LOOSESTRIFE.

Lysimachia lanceolata Walter, Fl. Carol. 92. 1788. *Steironema lanceolatum* (Walter) A. Gray, Proc. Amer. Acad. Arts 12: 63. 1876. *Nummularia lanceolata* (Walter) Kuntze, Revis. Gen. Pl. 2: 398. 1891.

Lysimachia heterophylla Michaux, Fl. Bor.-Amer. 1: 127. 1803. *Steironema heterophyllum* (Michaux) Baudo, Ann. Sci. Nat., Bot., ser. 2. 20: 347. 1843. *Lysimachia lanceolata* Walter var. *heterophylla* (Michaux) A. Gray, Manual 283. 1848. *Lysimachia ciliata* Linnaeus var. *heterophylla* (Michaux) Chapman, Fl. South. U.S. 280. 1860. *Lysimachia hybrida* Michaux var. *heterophylla* (Michaux) A. W. Wood, Amer. Bot. Fl. 213. 1870. *Steironema ciliatum* (Linnaeus) Baudo var. *heterophyllum* (Michaux) Chapman, Fl. South. U.S., ed. 3. 298. 1897.

Lysimachia hybrida Michaux, Fl. Bor.-Amer. 1: 126. 1803. *Lysimachia lanceolata* Walter var. *hybrida* (Michaux) A. Gray, Manual 283. 1848. *Lysimachia ciliata* Linnaeus var. *hybrida* (Michaux) Chapman, Fl. South. U.S. 280. 1860. *Steironema lanceolatum* (Walter) A. Gray var. *hybridum* (Michaux) A. Gray, Proc. Amer. Acad. Arts 12: 63. 1876. *Steironema ciliatum* (Linnaeus) Baudo var. *hybridum* (Michaux) Chapman, Fl. South. U.S., ed. 3. 298. 1897. *Nummularia hybrida* (Michaux) Farwell, Amer. Midl. Naturalist 11: 67. 1928. *Lysimachia lanceolata* Walter subsp. *hybrida* (Michaux) J. D. Ray, Illinois Biol. Monogr. 24(3–4): 39. 1956. *Steironema hybridum* (Michaux) Rafinesque ex Small, Fl. S.E. U.S. 904. 1903.

Erect perennial herb, to 1 m; stem glabrous, rarely stipitate-glandular or pubescent near the nodes. Leaves opposite or whorled, monomorphic or dimorphic, the basal rosette developed or not, the blade elliptic to lanceolate or linear-lanceolate, 2–18 cm long, 0.2–3 cm wide, the apex acute to acuminate or rounded, the base cuneate to rounded or obtuse, decurrent, the margin entire or rarely serrulate, ciliolate or eciliolate proximally, the upper and lower surfaces glabrous, punctate or epunctate, sessile or the leaves decurrent as a narrow wing. Flowers solitary, axillary in the distal leaves, the pedicel 1–5 cm long, glabrous to sparsely stipitate-glandular (rarely pubescent); sepals 4–8(10) mm long, glabrate, the lobes lanceolate to ovate; petals 4–12 mm long, yellow, the lobes broadly obovate, the apex apiculate, slightly erose distally, sessile or the petiole to 4 cm long, ciliate proximally, sparsely stipitate-glandular on the inner surface; stamens shorter than the corolla lobes; staminodes ca. 1 mm long. Fruit globose, 2–6 mm long, glabrous or sparsely stipitate-glandular distally; seeds few, 1–2 mm long, angular, dark brown, smooth.

Bluff forests, seepage areas, and floodplain forests. Occasional; Nassau County, eastern and central panhandle. New Brunswick and Quebec south to Florida, west to Alberta, North Dakota, South Dakota, Nebraska, Kansas, and Oklahoma, also Washington, Arizona, and New Mexico. Spring–summer.

Lysimachia hybrida has been recognized as a distinct species (e.g., Cholewa, 2009c) or variety (as *L. lanceolata* var. *hybrida*) (e.g., Wunderlin and Hansen, 2011), differing in some features such as stem thickness, degree of basal rosette development, leaf shape, and amount of marginal cilia. The two clearly intergrade and are best treated as a single taxon as we do here.

EXCLUDED TAXON

Lysimachia radicans Hooker—Reported for Florida by Gleason and Cronquist (1991) and Cholewa (2009c). No Florida specimens known.

Myrsine L. 1753. COLICWOOD

Trees or shrubs. Leaves alternate, simple, pinnate-veined, petiolate, estipulate. Flowers in fascicles, unisexual (plants dioecious), bracteate, pedicellate; sepals 5, basally connate; petals 5, basally connate; stamens 5, adnate to the corolla tube, the anthers dehiscent by longitudinal slits; staminodes 5 in the carpellate flowers; ovary superior, 3-carpellate, 1-loculate. Fruit a 1-seeded drupe.

A genus of about 300 species; nearly cosmopolitan. [Greek name for a type of Myrtle.]
Rapanea Aubl., 1775.
Selected reference: Pipoly and Ricketson (2009b).

Myrsine cubana A. DC. [Of Cuba.] MYRSINE; COLICWOOD.

Myrsine cubana A. de Candolle, Ann. Sci. Nat., Bot., ser. 2. 16: 86. 1841.
Sideroxylon punctatum Lamarck, Tabl. Encycl. 2: 42. 1794. *Bumelia punctata* (Lamarck) Roemer & Schultes, Syst. Veg. 4: 498. 1819. *Myrsine floridana* A. de Candolle, Trans. Linn. Soc. London 17: 107. 1834, nom. illegit. *Myrsine punctata* (Lamarck) Stearn, Bull. Brit. Mus. (Nat. Hist.), Bot. 4: 177. 1969; non. (H. Léveillé) Wilbur, 1965. *Rapanea punctata* (Lamarck) Lundell, Wrightia 4: 121. 1969. TYPE: FLORIDA: s.d., *Michaux s.n.* (lectotype: G-DC). Lectotypified by Ricketson & Pipoly (1997: 587).
Myrsine floridana Gandoger, Bull. Soc. Bot. France 65: 57. 1918; non. A. de Candolle, 1834. TYPE: FLORIDA.

Shrub or tree, to 15 m; branchlets glabrous. Leaves with the blade oblong, obovate, or oblanceolate, 4–13 cm long, 2.5–5.5 cm wide, the apex obtuse to rounded or obtuse, the base cuneate, the margin entire, revolute, the upper and lower surfaces glabrous, glandular-punctate, the petiole 4–7 mm long, glabrous. Flowers in a fascicle, subsessile; bracts ciliate. Staminate flowers with the sepals ovate, ca. 1 mm long, the margin entire, glandular-ciliate, the surface punctate; petals with the lobes lanceolate, 2–3 mm long, greenish white to pink, slightly connate at the base, the margin entire, glandular-granulose, the surface punctate, glabrous; stamens with the filaments rudimentary, the anthers attached to the apex of the corolla tube; pistillode glabrous. Carpellate flowers with the sepals and petals similar to the staminate; staminodes present;

ovary punctate, the style 1, the stigma conic, 3- to 5-lobed. Fruit globose, 4–5 mm long, black, punctate; seed globose.

Hammocks. Frequent; central and southern peninsula; Dixie and Wakulla Counties. Florida; West Indies, Mexico, and Central America. Fall–spring.

EXCLUDED TAXON

Myrsine guianensis (Aublet) Kuntze—Reported for Florida by Chapman (1897, as *M. rapanea* Roemer & Schultes), Small (1903, 1913a, 1913b, 1913c, 1913d, 1913e, 1933, all as *Rapanea guianensis* Aublet), Long and Lakela (1971), and Godfrey and Wooten (1981), all of whom misapplied the name to material of *M. cubana*.

EXCLUDED GENUS

Parathesis crenulata (Ventenat) Hooker f. ex Hemsley)—Reported for Florida by Clewell (1985, *Ardisia crenulata* Ventenat), who misapplied the name of this West Indian species to material of *A. crenata*.

THEOPHRASTACEAE D. Don, nom. cons. 1835.
THEOPHRASTA FAMILY

Shrubs or trees. Leaves alternate or pseudoverticillate, simple, pinnate-veined, petiolate, estipulate. Flowers in terminal racemes, bracteate, actinomorphic, bisexual; sepals 5, free; petals 5, basally connate; stamens 5, free, the anthers dehiscent by longitudinal slits; staminodes 5; ovary superior, 5-carpellate, 1-loculate, the style 1, the stigma 1. Fruit a berry.

A family of 5 genera; North America, West Indies, Mexico, Central America, and South America.

Theophrastaceae is treated as a distinct family following Kubitzki (2004). See discussion under Primulaceae above.

Selected reference: Channell and Wood (1959).

1. Young shoots puberulous, the trichomes uniseriate; corolla orange or yellow; seeds laterally compressed, partially covered by placental tissue ....**Bonellia**
1. Young shoots lepidote, the trichomes irregularly branched; corolla white or cream; seeds subglobose, completely covered by placental tissue ....**Jacquinia**

Bonellia Bertero ex Colla 1824.

Shrubs or trees. Leaves alternate or pseudoverticillate, simple, pinnate-veined, petiolate, estipulate. Flowers in terminal racemes, bracteate; sepals 5, free; petals 5, basally connate; stamens 5, borne at the base of the corolla tube, the filaments free; staminodes 5, borne at the apex of the corolla tube, petaloid; ovary superior, 5-carpellate, 1-loculate, the stigma capitate, lobed. Fruit a berry; seeds partly covered by placental tissue.

A genus of 22 species; North America, West Indies, Mexico, and Central America. [Commemorates Franco Andrea Bonelli (1784–1830), Italian zoologist.]

Selected reference: Wunderlin (2009).

Bonellia macrocarpa (Cav.) Ståhl & Källersjö [Large fruit.] CUDJOEWOOD.

Jacquinia macrocarpa Cavanilles, Icon 5: 55, t. 483. 1799. *Bonellia cavanillesii* Bertero ex Colla, Hort.
Ripul. 21. 1824, nom. illegit. *Bonellia macrocarpa* (Cavanilles) Ståhl & Källersjö, Novon 14: 117.
2004.

Shrub or tree, to 4 m; stem gray, smooth, the branchlets gray, puberulous when young, the trichomes uniseriate, glabrescent in age. Leaves alternate or sometimes inconspicuously pseudoverticillate, the blade elliptic, lanceolate, or oblanceolate, 3–6 cm long, 1–2 cm wide, the apex acute or obtuse, the base cuneate, the margin entire, the upper and lower surfaces glabrous, punctate, the petiole to 6 mm long, sparsely puberulent on the upper surface. Flowers in a terminal raceme to 3 cm long, the peduncle ca. 2 mm long, the pedicel ca. 1 mm long; bracts lanceolate, 3–7 mm long; sepals 3–4 mm long, the margin entire or slightly erose; petals yellow or orange, the lobes ovate to suborbiculate, 6–9 mm long; stamens 3–5 mm long; staminodes suborbiculate, longer than the stamens, petaloid, the apex slightly 3-lobed. Fruit ovoid or globose, 3–4 cm long, yellow or orange, the pericarp wrinkled; seeds 1–8, oblong or elliptic, ca. 1 cm long, dark brown, laterally compressed, the surface slightly alveolate, partially covered by the placental tissue.

Disturbed mangrove margins. Rare; Miami-Dade County. Escaped from cultivation. Florida; West Indies, Mexico, and Central America. Native to Mexico and Central America. All year.

Jacquinia L. 1759.

Shrubs or trees. Leaves alternate or pseudoverticillate, simple, pinnate-veined, petiolate, estipulate. Flowers in terminal racemes, bracteate; sepals 5, free; petals 5, basally connate; stamens 5, borne at the base of the corolla tube, the filaments free; staminodes 5, borne at the apex of the corolla tube, petaloid; ovary superior, 5-carpellate, 1-loculate, the stigma capitate, lobed. Fruit a berry; seeds covered by placental tissue.

A genus of 13 species; North America, West Indies, Central America, and South America. [Commemorates Nikolaus Joseph von Jacquin (1727–1817), Netherlands-born Austrian botanist.]

Selected reference: Whetstone and Wunderlin (2009).

1. Leaves alternate or indistinctly pseudoverticillate, the blade 1–4.5 cm long, 0.5–2.5 cm wide ....
   .... **J. keyensis**
1. Leaves usually distinctly pseudoverticillate, the blade 3–8(12) cm long, 1.5–5 cm wide .... **J. arborea**

Jacquinia arborea Vahl [Tree.] BRACELETWOOD

Jacquinia arborea Vahl, Eclog. Amer. 1: 26. 1796. *Jacquinia armillaris* Jacquin var. *arborea* (Vahl)
Grisebach, Fl. Brit. W.I. 397. 1861.

Shrub or tree, to 5 m; stem gray, nearly smooth, branchlets grayish brown, lepidote when young, glabrescent in age. Leaves alternate or distinctly pseudoverticillate, the blade oblong-obovate to spatulate, 3–8(12) cm long, 1.5–4(4.5) cm wide, coriaceous, the apex rounded or retuse, mucronate or not, the base cuneate, the margin entire, slightly revolute, the upper and lower surfaces glabrous, punctate, the petiole to 7 mm long, glabrous or sparsely puberulous. Flowers 7–25(40) in a terminal raceme to 6(12) cm long, usually exceeding the leaves, the peduncle ca. 2 mm long, the pedicel 7–13 mm long; bracts lanceolate, 1–1.5 cm long; sepals 2–3 mm long, the margin entire or slightly erose; petals 7–9 mm long, white or cream, the lobes slightly shorter to about as long as the tube; stamens 1–2 mm long, shorter than the staminodes; staminodes oblong to ovate, 2–3 mm long, the apex rounded or retuse. Fruit ovoid to globose, 7–11 mm long, orange-red to red, apiculate, the pericarp smooth; seeds oblong to elliptic, 3–5 mm long, light brown.

Disturbed mangrove margins. Rare; Broward and Miami-Dade Counties, Monroe County keys. Escaped from cultivation. Florida; West Indies, Mexico, and Central America. Native to West Indies and Central America. All year.

Jacquinia keyensis Mez [Of the Florida Keys.] JOEWOOD.

Jacquinia keyensis Mez, in Urban, Symb. Antill. 2: 44. 1901. TYPE: FLORIDA: Monroe Co.: Jewfish Key, 19 Jan 1895, *Curtiss 5447* (lectotype: G; isolectotypes: F, LE, US). Lectotypified by Ståhl (1992: 59).

Shrub or tree, to 6 m; stem light gray, nearly smooth, the branchlets gray, lepidote when young, glabrescent in age. Leaves alternate or indistinctly pseudoverticillate, the blade oblong-obovate to spatulate, 1–4.5 cm long, 0.5–2.5 cm wide, coriaceous, the apex obtuse to rounded or retuse, mucronate, the base cuneate, the margin entire, slightly to strongly revolute, the upper and lower surfaces glabrous, punctate, the petiole to 5 mm long, puberulous-lepidote. Flowers 4–30 in a terminal raceme to 6 cm long, usually exceeding the leaves, the peduncle 2–10 mm long, the pedicel 7–12 mm long; bracts lanceolate, less than 1 mm long; sepals 2–3(4) mm long, the margin entire or erose; petals 6–9 mm long, white or cream, the lobes subequaling or longer than the tube; stamens 3–4 mm long, shorter than the staminodes; staminodes oblong, 4–5 mm long, the apex obtuse, rounded, or truncate. Fruit ovoid or globose, 9–10 mm long, orange-red, apiculate, the pericarp smooth; seeds oblong to elliptic, 3–5 mm long, brown.

Coastal hammocks. Occasional; Charlotte and Lee Counties, southern peninsula. Florida; West Indies. All year.

Jacquinia keyensis is listed as threatened in Florida (Florida Administrative Code, Chapter 5B-40).

EXCLUDED TAXON

Jacquinia armillaris Jacquin—Reported for Florida by Chapman (1860, 1883, 1897), the name misapplied to material of *J. keyensis* (Ståhl, 1992).

SAMOLACEAE Raf. 1820. BROOKWEED FAMILY

Herbs. Leaves basal and cauline, alternate, simple, pinnate-veined, petiolate or epetiolate, estipulate. Flowers in terminal and axillary racemes or panicles, actinomorphic, bisexual; bracteate or ebracteate; sepals 5, basally connate; petals 5, basally connate; stamens 5, epipetalous, the anthers dehiscent by longitudinal slits; staminodes 5 or absent; ovary 1, semi-inferior, 5-carpellate, 1-loculate, the style 1. Fruit a capsule.

A family of 1 genus and about 12 species; North America, West Indies, Mexico, Central America, South America, Europe, Africa, and Asia.

Samolaceae is treated as a distinct family following Kubitzki (2004). See discussion under Primulaceae.

Selected references: Channell and Wood (1959); Ståhl (2004).

Samolus L. 1753. BROOKWEED

Perennial herbs. Leaves basal and cauline, simple, pinnate-veined, petiolate or epetiolate, estipulate. Flowers in terminal and axillary racemes or panicles, bracteate or ebracteate; sepals basally connate; petals basally connate; stamens borne at the apex of the corolla tube, opposite the corolla lobes, the filaments basally adnate to the corolla tube; staminodes borne at the apex of the corolla tube or absent, alternate with the corolla lobes; ovary semi-inferior, 5-carpellate, 1-loculate, the style 1, the stigma capitate. Fruit a capsule, loculicidally dehiscent; seeds numerous.

A genus of about 12 species; North America, West Indies, Mexico, Central America, South America, Europe, Africa, and Asia. [Ancient Celtic name of one of the European species.]

Selected reference: Cholewa (2009a).

1. Flowers 5–7 mm wide; pedicels ebracteate; cauline leaves not extending into the inflorescence............
....**S. ebracteatus**
1. Flowers 2–3 mm wide; pedicels with a minute bract near the middle; cauline leaves extending into the
inflorescence...**S. valerandi**

Samolus ebracteatus Kunth [Lacking bracts.] WATER PIMPERNEL; LIMEWATER BROOKWEED.

Samolus ebracteatus Kunth, in Humboldt et al., Nov. Gen. Sp. 2: 223, t. 129. 1818. *Samodia ebracteata* (Kunth) Baudo, Ann. Sci. Nat., Bot., ser. 2. 20: 350. 1843. *Samolus ebracteatus* Kunth subsp. *genuinus* R. Kunth, in Engler, Pflanzenr. 4(Heft 22): 340. 1905, nom. inadmiss.

Erect or ascending perennial herb, to 6 dm; stem glabrous. Leaves mostly basal, the blade spatulate, 2.5–16 cm long, 2–6 cm wide, the apex obtuse or rounded, sometimes apiculate, the base cuneate, decurrent, the margin entire, the upper and lower surfaces glabrous, punctate, sessile or short-petiolate. Flowers in a terminal raceme, the peduncle 5–10 mm long, the pedicel 2–20 mm long, glabrous or stipitate-glandular; bracts absent; sepals 2–3 mm long, the lobes triangular-ovate to triangular-lanceolate, subequaling or longer than the tube, the apex acute, glandular at least near the base; petals 3–7 mm long, pink or whitish, the lobes suborbicular, shorter than the tube, the apex rounded to truncate, sometimes erose, the base with a glandular tuft; staminodes absent. Fruit globose, 3–4 mm long, tan or brown; seeds 20–50, blackish brown to reddish brown.

Brackish and freshwater marshes. Frequent; central and southern peninsula west to central panhandle. Florida, Louisiana, Kansas, Oklahoma, Texas, New Mexico, and Nevada; West Indies and Central America. Spring–summer.

Samolus valerandi L. subsp. **parviflorus** (Raf.) Hultén. [Commemorates Dourez Valerand, sixteenth-century French pharmacist and botanist; small-flowered.] PINELAND PIMPERNEL; SEASIDE BROOKWEED.

Samolus parviflorus Rafinesque, Amer. Monthly Mag. & Crit. Rev. 2: 176. 1818. Samolus valerandi Linnaeus subsp. parviflorus (Rafinesque) Hultén, Kongl. Svenska Vetenskapsakad. Handl. 13: 148. 1971.

Samolus floribundus Kunth, in Humboldt et al., Nov. Gen. Sp. 2: 224. 1818. Samolus valerandi Linnaeus var. floribundus (Kunth) Britton et al., Prelim. Cat. 34. 1888, nom. illegit. Samolus valerandi Linnaeus var. americanus A. Gray, Manual, ed. 2. 274. 1856.

Samolus floridanus Rafinesque, Herb. Raf. 41. 1833. TYPE: "Florida to Carolina."

Erect, ascending, or prostrate perennial herb, to 8 dm; stem glabrous. Leaves basal and cauline or all cauline, the blade obovate or broadly spatulate to elliptic, 1.5–10 cm long, 1–4 cm wide, the apex rounded to obtuse, the base cuneate, decurrent, the margin entire, the upper and lower surface glabrous, punctate, sessile or short-petiolate. Flowers in a terminal or axillary raceme or panicle, sessile or the peduncle to 7 mm long, the pedicel 2–14 mm long, glabrous; bracts near the middle of the pedicel, minute; sepals 1–2 mm long, the lobes triangular to ovate, shorter than the tube, the apex acute, glabrous; petals 2–3 mm long, white, the lobes oblong, longer than the tube, the apex rounded or emarginate; staminodes 5. Fruit globose, 2–3 mm long, tan or brown; seeds 20–50, blackish brown to reddish brown.

Wet flatwoods and floodplain forests. Frequent, nearly throughout. New Brunswick and Quebec south to Florida, west to Ontario, Washington, Oregon, and California; West Indies, Mexico, Central America, and South America; Asia. Spring–summer.

THEACEAE Mirb. 1816. TEA FAMILY

Shrubs or trees. Leaves alternate, simple, pinnate-veined, petiolate, estipulate. Flowers solitary, axillary, bracteate, actinomorphic, bisexual; sepals 5, basally connate; petals 5, basally connate; stamens numerous, the anthers dehiscent by longitudinal slits; ovary superior, 5-carpellate and -loculate, the style 1, the stigmas lobed. Fruit a loculicidally dehiscent capsule.

A family of 9 genera and 450 species; North America, West Indies, Mexico, Central America, South America, and Asia.

Selected references: Prince (2009); Wood (1959).

1. Leaves coriaceous, the lower surface sparsely pubescent, appearing glabrous, the margin bluntly serrate; stamens yellow .... **Gordonia**
1. Leaves chartaceous, the lower surface evidently pubescent, the margin sharply denticulate-serrate; stamens purple .... **Stewartia**

Gordonia J. Ellis, nom. cons. 1771.

Trees. Leaves alternate, simple, pinnate-veined, petiolate, estipulate. Flowers solitary, axillary, bracteate; sepals 5, basally connate; petals 5, basally connate; stamens 75–125, basally connate; ovary 5-carpellate and -loculate, the style 1, the stigmas lobed. Fruit a loculicidally dehiscent capsule; seeds 10–20.

A genus of 2 species; North America, Mexico, Central America, and South America. [Commemorates James Gordon (1708–1781), London nurseryman and correspondent of Linnaeus.]

Gordonia lasianthus (L.) J. Ellis [From the Greek *lasios*, hairy, woolly, and *anthos*, flower, in reference to the stamens.] LOBLOLLY BAY.

> *Hypericum lasianthus* Linnaeus, Sp. Pl. 783. 1753. *Gordonia lasianthus* (Linnaeus) J. Ellis, Philos. Trans. 60: 519, 523. 1771. *Gordonia pyramidalis* Salisbury, Prodr. Stirp. Chap. Allerton 386. 1796, nom. illegit. *Lasianthus pyramidalis* Kuntze, Revis. Gen. Pl. 1: 63. 1891.

Tree, to 25 m; bark dark gray, roughened by coarse interlacing flat-topped ridges separated by rough narrow furrows, the branchlets dark brown to reddish, smooth, somewhat glaucous, glabrous or short-pubescent. Leaves with the blade elliptic to oblanceolate, 8–16(30) cm long, (2)3–5 cm wide, coriaceous, the apex obtuse to acute, sometimes retuse, the base cuneate-tapering, the margin bluntly serrate, the teeth deciduously gland-tipped, the upper surface dark green, glabrous, the lower surface olive-green, sparsely pubescent with 2- to 7-fascicled trichomes, the petiole short-pubescent proximally as on the leaves. Flowers solitary, axillary, the pedicel (3)5–8 cm long; bracts 2–4, scattered below the calyx, deciduous; sepals suborbicular to obovate, 9–11 mm long, short-clawed at the base, silky-pubescent on the outer surface; petals broadly obovate, 2–3 cm long, unequal, white, the apex rounded, the margin wavy-fringed, slightly curved upward, silky pubescent on the outer surface; stamens basally connate into a 5-lobed cup, each lobe with the free distal portion 4–6 mm long, yellow; style 1, stigma 5-lobed. Fruit ovate-oblong, 1.5–2 cm long, woody, brown, the surface silky-pubescent; seeds 10–20, 9–11 mm long, dark brown, apically winged.

Bayheads and swamps. Frequent; northern counties, central peninsula. North Carolina south to Florida, west to Mississippi. Summer.

Stewartia L. 1753.

Shrubs or trees. Leaves alternate, simple, pinnate-veined, petiolate, estipulate. Flowers solitary, axillary, bracteate; sepals basally connate; petals 5, basally connate; stamens 75–125, basally connate; ovary 5-carpellate and -loculate, the style 1, the stigma lobed. Fruit a loculicidally dehiscent capsule.

A genus of about 20 species; North America and Asia. [Commemorates John Stuart (1713–1792), 3rd earl of Bute, Scottish nobleman who served as prime minister of Great Britain.]

Stewartia malacodendron L. [From the Greek *malacos*, soft to the touch, and *dendron*, tree.] SILKY CAMELLIA.

> *Stewartia malacodendron* Linnaeus, Sp. Pl. 698. 1753. *Cavanilla florida* Salisbury, Stirp. Chap. Allerton 385. 1796, nom. illegit. *Malacodendron monogynum* Dumont de Courset, Bot. Cult., ed. 2. 5: 106. 1811.
>
> *Stewartia virginica* Cavanilles, Diss. 5: 303, t. 159(2). 1787. *Stewartia nobilis* Salisbury, Prodr. Stirp. Chap. Allerton 386. 1796, nom. illegit.

Shrub or tree, to 6 m; bark gray, closely fissured and exfoliating in longitudinal strips, the branchlets reddish brown, smooth, silky-pubescent. Leaves 2-ranked, the blade elliptic, oblong, ovate, or obovate, (3)5–10 cm long, 3–5 cm wide, chartaceous, the apex acute, the base rounded to cuneate, the margin denticulate-serrulate, somewhat ciliate, the upper surface glabrous, the lower surface silky-pubescent, the petiole 2–4(5) cm long, winged to the base, silky-pubescent on the lower surface. Flowers solitary, axillary, the pedicel ca. 5 mm long, silky-pubescent; bracts 2, 2–4 mm long, immediately below the calyx; sepals suborbicular to broadly ovate, 7–9 mm long, silky-pubescent on the outer surface; petals obovate, 3–3.5 cm long, unequal, creamy white, the margin crenulate to erose, slightly turned up at the apex, the outer surface silky-pubescent; stamens basally connate and gradually narrowed upward, the free portions 5–10 mm long, the filaments purple, the anthers bluish; style 1, the stigma 5-lobed. Fruit ovoid, 1–1.5 cm long, woody, brown, silky pubescent; seeds 2(4), lenticular or angular, 5–7 mm long, brown or reddish brown.

Bluff and riverine forests, steepheads, and bayheads. Occasional; central and western panhandle. Virginia south to Florida, west to Arkansas and Texas. Spring.

Stewartia malacodendron is listed as endangered in Florida (Florida Administrative Code, Chapter 5B-40).

EXCLUDED GENERA

> *Camellia japonica* Linnaeus—Reported for Florida by Diamond (2013), apparently based on unpublished and online resources. All Florida material seen is of cultivated material.
>
> *Camellia sasanqua* Thunberg—Reported for Florida by Serviss and Peck (2016), apparently based on unpublished and online resources. All Florida material seen is of cultivated material.
>
> *Franklinia alatamaha* W. Bartram—Reported for Florida by Chapman (1860, as *Gordonia pubescens* Cavanillas). No Florida specimens known.

SYMPLOCACEAE Desf., nom. cons. 1820. SWEETLEAF FAMILY

Shrubs or trees. Leaves alternate, simple, pinnate-veined, petiolate, estipulate. Flowers in axillary fascicles, actinomorphic, bisexual; sepals 5, basally connate; petals 5, basally connate;

nectariferous disk present; stamens numerous, the anthers dehiscent by longitudinal slits; ovary inferior, 3-carpellate and -loculate, the style 1. Fruit a drupe.

A family of 2 genera and about 320 species; North America, West Indies, South America, Asia, Australia, and Pacific Islands.

Selected references: Almeda and Fritsch (2009); Wood and Channell (1960).

Symplocos Jacq. 1760. SWEETLEAF

Shrubs or trees. Leaves alternate, simple, pinnate-veined, petiolate, estipulate. Flowers in axillary fascicles; sepals 5, basally connate; petals 5, basally connate; nectariferous disk present; stamens 40–100, epipetalous; ovary 1, 3-carpellate and -loculate, the style 1, the stigma lobed. Fruit a 1-seeded drupe.

A genus of about 318 species; North America, West Indies, South America, Asia, Australia, and Pacific Islands. [From the Greek *symplokos*, connected, entwined, apparently in reference to the stamens and petals in *S. martinicensis*.]

Symplocos tinctoria (Linnaeus) L'Her. [*Tinctorius*, used in dyeing.] COMMON SWEETLEAF; HORSE SUGAR.

> *Hopea tinctoria* Linnaeus, Mant. Pl. 105. 1767. *Symplocos tinctoria* (Linnaeus) L'Héritier de Brutelle, Trans. Linn. Soc. London 1: 176. 1791. *Protohopea tinctoria* (Linnaeus) Miers, J. Linn. Soc., Bot. 17: 290. 1879. *Eugeniodes tinctoria* (Linnaeus) Kuntze, Revis. Gen. Pl. 2: 976. 1891.

Shrub or tree, to 15 m; bark grayish pink, somewhat fissured and roughened, with warty excrescences, the branchlets grayish brown, sparsely pubescent. Leaves with the blade elliptic to oblong or oblanceolate, 5–12(15) cm long, 2–6(7) cm wide, subcoriaceous, the apex acuminate or acute, the base cuneate, the margin entire or crenulate-serrate, the upper surface glabrous or pubescent, the lower surface whitish green, pubescent, the petiole 8–12 mm long. Flowers 6–14 in an axillary fascicle on the branchlet of the previous year, enclosed in a bud with orange-red scales and appearing before the leaves; sepal lobes deltoid, small; petal lobes obovate, oblanceolate, or spatulate, 6–8 mm long, yellow to creamy white; stamens adnate to the corolla base in several series in fascicles; ovary surrounded by a nectariferous orange disk; style 5–6 mm long, the stigma 3-lobed. Fruit oblong, 10–14 mm long, dark orange to brown, the calyx lobes persistent, glabrous; seed ovoid, brown.

Hammocks, swamp margins, and wet flatwoods. Occasional; northern counties south to Hillsborough County. Maryland and Delaware south to Florida, west to Oklahoma and Texas. Spring–fall.

STYRACACEAE DC. & Spreng., nom. cons. 1929. STORAX FAMILY

Shrubs or trees. Leaves alternate, simple, pinnate-veined, petiolate, estipulate. Flowers in axillary racemes, false-terminal racemes, or solitary, actinomorphic, bisexual, bracteolate or ebracteolate; sepals (2)4–5(9), free or basally connate; petals 4–5(8), basally connate; stamens 2(4) times the number of petals, adnate to the corolla base, the anthers dehiscent by longitudinal slits; ovary partly or completely inferior, 1- to 4-carpellate, 1-loculate, the style 1. Fruit a loculicidally dehiscent capsule or nutlike and indehiscent.

A family of 11 genera and about 160 species; North America, West Indies, Mexico, Central America, South America, Europe, and Asia.

Selected references: Fritsch (2009); Wood and Channell (1960).

1. Corolla lobes 4; fruit oblanceolate, nutlike and indehiscent, longitudinally 2- or 4-winged.......**Halesia**
1. Corolla lobes 5–6(8); fruit subglobose, a 3-valved capsule or nutlike and indehiscent, not winged.......
   .... **Styrax**

Halesia J. Ellis ex L., nom. cons. 1759. SILVERBELL

Shrubs or trees. Leaves alternate, simple, pinnate-veined, petiolate, estipulate. Flowers in contracted axillary racemes or solitary, bracteolate or ebracteolate; sepals 4, free; petals 4, basally connate; stamens 7–16, the filaments basally connate; ovary inferior, 2- to 4-carpellate and -loculate, the style 1, the stigma capitate. Fruit nutlike, indehiscent, with 2 or 4 longitudinal wings.

A genus of 3 species; North America and Asia. [Commemorates Stephen Hales (1677–1761), English botanist.]

Mohrodendron Britton, 1893.

1. Petals united more than half their length; fruit 4-winged..**H. carolina**
1. Petals united only at the base; fruit 2-winged (if 4-winged, then with 2 broad wings alternate with 2 narrow ones) .... **H. diptera**

Halesia carolina L. [Of Carolina.] CAROLINA SILVERBELL.

> *Halesia carolina* Linnaeus, Syst. Nat., ed. 10. 1044. 1759. *Mohria carolina* (Linnaeus) Britton, Gard. & Forest 6: 434. 1893. *Mohrodendron carolinum* (Linnaeus) Britton, Gard. & Forest 6: 463. 1893. *Carlomohria carolina* (Linnaeus) Greene, Erythea 1: 246. 1893.
>
> *Halesia parviflora* Michaux, Fl. Bor.-Amer. 2: 40. 1803. *Mohria parviflora* (Michaux) Britton, Gard. & Forest 6: 434. 1893. *Mohrodendron parviflorum* (Michaux) Britton, Gard. & Forest 6: 463. 1893. *Carlomohria parviflora* (Michaux) Greene, Erythea 1: 246. 1893. *Halesia tetraptera* J. Ellis var. *parviflora* (Michaux) Schelle, in Beissner et al., Handb. Laubholzben. 405. 1903. *Halesia carolina* Linnaeus subsp. *parviflora* (Michaux) E. Murray, Kalmia 13: 7. 1983. *Halesia carolina* Linnaeus var. *parviflora* (Michaux) E. Murray, Kalmia 13: 7. 1983. TYPE: FLORIDA: St. Johns Co.: Matanzas, s.d., *Michaux s.n.* (holotype: P; isotype: P).

Shrub or tree, to 24 m; bark tightly striated, the branchlets sparsely pubescent. Leaves with the blade ovate-oblong, elliptic, or obovate to oblanceolate, 7–12(18) cm long, 3–7 cm wide, the apex abruptly short-acuminate to gradually tapered, the base rounded or broadly cuneate, the margin serrate or entire, the teeth gland-tipped, the upper surface sparsely pubescent or glabrous, the lower surface sparsely pubescent, sometimes only on the major veins, the petiole 1–2 cm long, sparsely pubescent or glabrous. Flowers 2–6 in a contracted axillary raceme and appearing fasciculate or solitary, the pedicel 2–5 mm long; ebracteolate; sepals 2–5 mm long, pubescent; petals 1–1.5 cm long, white, glabrous, the tube 0.7–2.5 cm long, the lobes broadly rounded, 3–7 mm long; stamens 12–16, 9–17 mm long, the filaments adnate to the corolla for 2–3 mm, the free portion 5–14 mm, basally connate for 1–3 mm, pubescent on the inner surface; style glabrous. Fruit oblanceolate to obovate, 2–4 cm long, 4-winged, beaked with the persistent style; seeds 1–4, fusiform.

Calcareous hammocks and floodplain forests. Occasional; northern counties, south to Citrus County. New York south to Florida, west to Michigan, Illinois, Arkansas, Oklahoma, and Mississippi. Spring.

Halesia diptera J. Ellis [Two-winged, in reference to the fruit.] TWO-WING SILVERBELL.

> *Halesia diptera* J. Ellis, Philos. Trans. 51: 931, t. 22(B). 1761. *Mohria diptera* (J. Ellis) Britton, Gard. & Forest 6: 434. 1893. *Mohrodendron dipterum* (J. Ellis) Britton, Gard. & Forest 6: 463. 1893. *Carlomohria diptera* (J. Ellis) Greene, Erythea 1: 246. 1893.
>
> *Halesia diptera* J. Ellis var. *magniflora* R. K. Godfrey, Rhodora 60: 88. 1958. TYPE: FLORIDA: Leon Co.: 1.5 mi. E of Tallahassee, s.d., *Godfrey 54434* (holotype: FSU).

Shrub or tree, to 10(15) m; bark striated, the branchlets sparsely pubescent. Leaves with the blade ovate, elliptic, or obovate, 9–16 cm long, 4–10 cm wide, the apex abruptly short-acuminate, the base rounded to cuneate, the margin irregularly dentate-serrate, the teeth glandtipped, the upper and lower surfaces sparsely pubescent, the petiole 1.5–3.5 cm long. Flowers 2–6 in a contracted axillary raceme or solitary, the pedicel 5–15 mm long; bracteolate; sepals 4–8 mm long; petals broadly elliptic, 1–3 cm long, the tube ca. 1 mm long, the lobes 1.5–3 cm long; stamens 7–10, 1.5–2.5 cm long, the filaments adnate to the corolla for ca. 1 mm, the free portion 4–12 mm long, basally connate 2.5–11 mm, pubescent on the inner surface; style pubescent. Fruit elliptic-oblong, oval, or oblanceolate, 2.5–5 cm long, 2-winged (if 4-winged, then with 2 broad wings alternate with 2 narrow ones), beaked with the persistent style; seeds 1–4, fusiform.

Hammocks and floodplain forests. Occasional; Suwannee County, panhandle. South Carolina south to Florida, west to Arkansas and Texas. Spring.

EXCLUDED TAXON

> *Halesia tetraptera* J. Ellis—Florida material reported under this name such as that by Chapman (1860) and Kurz and Godfrey (1962) is now referred to *H. carolina*, as shown by Reveal and Seldin (1976). However, Reveal and Seldin also mistakenly report this Appalachian species from Florida. No Florida specimens known.

Styrax L. 1753. SNOWBELL

Shrubs or trees. Leaves alternate, simple, pinnate-veined, petiolate, estipulate. Flowers in falseterminal racemes or solitary, ebracteolate; sepals (2)4–5(9), (2)4- to 5(9)-lobed; petals 5–6(8);

stamens 10(16), the filaments basally adnate to the corolla; ovary partly inferior, 3-carpellate, 1-loculate, the style 1, the stigma capitate. Fruit a 3-valved capsule or nutlike and indehiscent.

A genus of about 130 species; North America, Mexico, South America, Europe, and Asia. [From the Arabic *assthirak*, vernacular name for the type species.]

1. Leaves elliptic to obovate, to 8(10) cm long and 4(5.5) cm wide, the pubescence on the lower surface rust-colored (at least along the principal veins); flowers in the leaf axils on short shoots and the inflorescence appearing as a foliose raceme ....**S. americanus**
1. Leaves broadly elliptic to suborbicular, 8–12(15) cm long and 4–10(15) cm wide, the pubescence on the lower surface gray-green; flowers on short shoots with the upper leaves absent or vestigial and the inflorescence appearing as a drooping raceme .... **S. grandifolius**

Styrax americanus Lam. [Of America.] AMERICAN SNOWBELL.

Styrax americanus Lamarck, Encycl. 1: 82. 1783. *Styrax americanus* Lamarck forma *genuinus* J. R. Perkins, in Engler, Pflanzenr. 4(Heft 30): 76. 1907, nom. inadmiss.

Styrax pulverulentus Michaux, Fl. Bor.-Amer. 2: 41. 1803. *Styrax americanus* Lamarck forma *pulverulentus* (Michaux) J. R. Perkins, in Engler, Pflanzenr. 4(Heft 30): 76. 1907. *Styrax americanus* Lamarck var. *pulverulentus* (Michaux) Rehder, in L. H. Bailey, Stand. Cycl. Hort. 6: 3280. 1917.

Shrub or tree, to 5 m; branchlets glabrous or pubescent. Leaves with the blade elliptic to obovate, 1.5–8(10) cm long, 0.6–4(5.5) cm wide, the apex obtuse to rounded, sometimes cuspidate, the base cuneate, the margin entire, wavy, or obscurely and irregularly dentate or dentate-serrate, the upper surface glabrous or sparsely pubescent, the lower surface paler, glabrous or sparsely to densely pubescent, the trichomes rust-colored (at least along the principal veins), the petiole 2–6 mm long, pubescent or glabrous. Flowers solitary or 2–4(5) on short shoots and appearing as a foliose raceme 2–3.5 cm long, the pedicel 4–10(14) mm long, glabrous or sparsely pubescent; sepals 2.5–4 mm long, sparsely to densely stellate-pubescent; petals 11–16 mm long, white, the tube 2–3 mm long, the lobes 5(6), elliptic, 10–12 mm long, white, the outer surface sparsely stellate-pubescent. Fruit a capsule, 6–9 mm long, globose, loculicidally dehiscent nearly to the base, gray stellate-pubescent; seeds 1(3), subglobose, with 3(6) longitudinal grooves.

Calcareous hammocks and swamps. Occasional; northern counties, central peninsula. Virginia south to Florida, west to Illinois, Missouri, Oklahoma, and Texas. Spring.

Styrax grandifolius Aiton [Large-leaved.] BIGLEAF SNOWBELL.

Styrax grandifolius Aiton, Hort. Kew. 2: 75. 1789.

Shrub or tree, to 6 m; branchlets sparsely to densely pubescent. Leaves with the blade obovate to elliptic or rhombic, 8–12(20) cm long, 4–10(15) cm wide, the apex obtuse to rounded, abruptly short-acuminate or cuspidate, the base cuneate to rounded, the petiole 4–12 mm long, the margin entire or irregularly denticulate to serrate, rarely lobed, the upper surface glabrous or sparsely pubescent, especially along the major veins, the lower surface softly grayish pubescent. Flowers solitary or 2–12(15) in a drooping, false-terminal raceme-like inflorescence 3–15 cm long, the pedicel 4–9 mm long, stellate-pubescent; sepals 4–6 mm long, stellate-pubescent on the outer surface; petals 10–21 mm long, stellate-pubescent, the tube 3–5 mm long, the lobes 5(5), elliptic to oblong-elliptic, 15–20 mm long, spreading or reflexed. Fruit nutlike or at most with 1–3 narrow longitudinal fissures barely exposing the seeds, globose to ellipsoid, 7–12 mm long, gray to yellowish gray stellate-pubescent; seeds 1(3), subglobose, with 3(6) longitudinal grooves.

Calcareous hammocks and floodplain forests. Northern counties. Virginia south to Florida, west to Illinois, Arkansas, and Texas. Spring.

SARRACENIACEAE Dumort., nom. cons. 1829.
PITCHERPLANT FAMILY

Herbs. Leaves all basal, alternate, pinnate-veined, developing into hollow tubes (pitchers) with a hood, petiolate, estipulate. Flowers solitary on a scape, arising from the tip of the rhizome, actinomorphic, bisexual, bracteate; sepals 5, free; petals 5, free; stamens 50–100, basally connate into fascicles, the anthers dorsifixed, longitudinally dehiscent; ovary superior, 5-carpellate and -loculate. Fruit a loculicidally dehiscent capsule; seeds numerous.

A family of 3 genera and about 22 species; North America, South America, Europe, and Asia.

Selected reference: Wood (1960).

Sarracenia L. 1753. PITCHERPLANT

Perennial herbs. Leaves tubiform (pitchers), the orifice partly to completely covered with a hood arising dorsally from the orifice rim, the apex apiculate (with an apiculum) or not, with a ventral wing; phyllodia present or absent. Flowers solitary on a scape; bracts 3, subtending the sepals; sepals 5, free; petals 5, free, pendulous between the lobes of the style disk; stamens 50–100, slightly connate into 10–17 fascicles; ovary 5-lobed, the style distally expanded into a broad umbrella-like disk with 5 lobes, the stigmas simple, filiform, at the base of the style-disk notch. Fruit a capsule, basipetally or acropetally dehiscent; seeds numerous.

A genus of about 12 species; North America, Europe, and Asia. [Commemorates Michel Sarrazin de l'Etang (1659–1734), King's physician in New France, who sent specimens to Europe.]

Selected reference: Mellichamp and Case (2009).

1. Pitchers decumbent.
    2. Pitchers gradually tapering toward the base, marked with white, the hood curved and subglobose, concealing the aperture ....**S. psittacina**
    2. Pitchers markedly inflated in the upper portion, lacking white markings, the hood undulate, not concealing the aperture ....**S. rosea**
1. Pitchers erect or strongly ascending.
    3. Pitchers marked with white in the upper portion.
        4. Corolla pink; hood margin crenulate ....**S. leucophylla**

    4. Corolla yellow; hood margin entire .... **S. minor**

  3. Pitchers green, sometimes suffused with red or red-veined.

      5. Corolla red; aperture of the pitcher less than 3 cm wide; pitchers usually reddish brown with darker venation, the hood neck green within .... **S. rubra**

      5. Corolla yellow; aperture of the pitcher more than 4 cm wide, pitchers usually yellowish green, the hood neck maroon within .... **S. flava**

Sarracenia flava L. [Yellow.] YELLOW PITCHERPLANT.

Sarracenia flava Linnaeus, Sp. Pl. 510. 1753. *Sarracenia gronovii* A. W. Wood, Class-Book Bot., ed. 1861. 222. 1861, nom. illegit. *Sarracenia gronovii* A. W. Wood var. *flava* (Linnaeus) A. W. Wood, Class-Book Bot., ed. 1861. 222. 1861, nom. illegit.

Clump-forming perennial herb. Leaves (pitchers) erect, 2.4–9(10) dm long, yellow-green, often with dark red spots on the neck, dark red veins distally on the tube and hood, or the whole heavily suffused with bronze or purplish red, the outer surface glabrous, the orifice broadly ovate, 2–7(8) cm wide, the rim green, flaring and loosely revolute, often with a prominent, everted indentation immediately distal to the wing forming a spout over the wing, the ventral wing 0.5–1(2) cm wide, the hood orbiculate-reniform, 3–10 cm long, (3)5–15 cm wide, the proximal margin broadly cordate, recurved downward at the tip, opposite lobes reflexed upward, yellow-green, red-veined or suffused with bronze-red, the neck constricted, often red-spotted or -veined, 1–3 cm long, the margin revolute, the apiculum (2)3–12(18) mm long, the inner surface glabrous; phyllodia 2–4(5), erect, oblanceolate, (8)12–30 cm long, 1–3 cm wide. Flowering scape 1.5–6 dm long, shorter than the pitchers; bracts 1–2 cm long; sepals ovate-lanceolate, 3–5 cm long, the margin entire, yellow-green; petals obovate, 5–8 cm long, the margin entire, yellow-green; style-disk 6–8 cm wide, yellow-green. Fruit globose to ovoid, 1.5–2 cm long, tuberculate, basipetally dehiscent; seeds clavate to reniform-obovate, 2–3 mm long, laterally keeled.

Wet flatwoods, bogs, and acid swamps. Frequent; northern counties. Virginia south to Florida, west to Alabama, also Washington. Spring.

Sarracenia flava hybridizes in Florida with *S. leucophylla* (*S.* ×*mooreana*) and *S. rosea* (*S.* ×*naczii*).

Sarracenia leucophylla Raf. [With white leaves.] WHITETOP PITCHERPLANT.

Sarracenia leucophylla Rafinesque, Fl. Ludov. 14. 1817. TYPE: FLORIDA: Escambia Co.: near Pensacola. *Sarracenia drummondii* Croom, Ann. Lyceum Nat. Hist. New York 4: 100. 1848. *Sarracenia gronovii* A. W. Wood var. *drummondii* (Croom) A. W. Wood, Class-Book, ed. 1861. 222. 1861, nom. illegit. TYPE: FLORIDA: Franklin Co.: near Apalachicola, 1825, *Drummond 6* (holotype: ?; isotype: GH). *Sarracenia drummondii* Croom var. *rubra* Macfarlane, in L. H. Bailey, Stand. Cycl. Hort. 6: 3080. 1917. TYPE: FLORIDA. *Sarracenia drummondii* Croom var. *alba* Macfarlane, in L. H. Bailey, Stand. Cycl. Hort. 6: 3080. 1917. TYPE: "S. Ga. and W. Fla."

Clump-forming perennial herb. Leaves (pitchers) erect, 2.5–10 dm long, green proximally, with white areolae distally, with green to red or pink veins, green, often with dark red spots on the neck, dark red veins distally on the tube and hood, or the whole heavily suffused with bronze or purplish red, the surface glabrous or finely pubescent, the ventral wing 0.5–1.5(2) cm wide, the orifice ovate, 2–6(8) cm wide, the rim white to green or reddish, flaring and loosely revolute, often with a prominent, everted indentation immediately distal to the wing forming a spout over the wing, the hood recurved ventrally, held well beyond and covering the orifice, with white areolae bordered by white, green, or red, orbiculate-reniform to ovate reniform, 2.5–6.5 cm long, 2–6 cm wide, the proximal margin cordate, not narrowed to the neck, the apiculum (1)2–3 mm long, the inner surface densely retrorse strigose; phyllodia 5–6(8), erect, oblanceolate, 15–50 cm long, 2–3 cm wide. Flowering scape 3–8 dm long, shorter than the pitchers; bracts 0.5–1 cm long; sepals ovate-lanceolate, 3.5–5 cm long, the margin entire, maroon to red; petals orbiculate to rhombic, 4–5 cm long, the margin entire, maroon to red; style-disk 6–7 cm wide, red. Fruit globose to ovoid, 1.5–2 cm long, tuberculate, acropetally dehiscent; seeds clavate to reniform-obovate, ca. 2 mm long, laterally keeled.

Bogs and acid swamps. Frequent; central and western panhandle. Virginia south to Florida, west to Mississippi, also Washington. Spring.

Sarracenia leucophylla hybridizes in Florida with *S. flava* (*S.* ×*mooreana*), *S. psittacina* (*S.* ×*wrigleyana*), *S. rosea* (*S.* ×*mitchelliana*), and *S. rubra* subsp. *gulfensis* (*S.* ×*bellii*).

Sarracenia leucophylla is listed as endangered in Florida (Florida Administrative Code, Chapter 5B-40).

Sarracenia minor Walter [Small.] HOODED PITCHERPLANT.

Sarracenia minor Walter, Fl. Carol. 153. 1788.
Sarracenia variolaris Michaux, Fl. Bor.-Amer. 1: 310. 1803. TYPE: "Carolina ad Floridam," s.d., *Michaux s.n.* (holotype: P).
Sarracenia adunca Rafinesque, Autik. Bot. 34. 1840; non Smith, 1804. TYPE: FLORIDA.

Clump-forming perennial herb. Leaves (pitchers) erect, 1.2—4.5(10) dm long, green or suffused with red, with prominent, circular, white areolae on the dorsal and lateral sides, usually with a flush reddish pigment in the tissue surrounding the white, the surface glabrous or sparsely and unevenly short-pubescent, the ventral wing 1–3 cm wide, the orifice triangular-ovate, 1–3 cm wide, the rim red, strongly revolute, the hood strongly convex, arching and recurving ventrally, closely covering the orifice, green, flushed or streaked with bronze-red, 1.2–6 cm long and wide, the base continuous with the tube and not forming a neck, the apiculum ca. 1 mm long, the inner surface glabrate or with retrorse setae; phyllodia absent. Flowering scape 1.2–5.5 dm long, shorter than the pitchers; bracts 0.5–1.2 cm long; sepals ovate-lanceolate, 1.5–3.5 cm long, the margin entire, yellow-green; petals narrowly obovate, 3–5 cm long, the margin entire, yellow-green; style-disk 2.5–4.5 cm wide, pale yellow. Fruit globose to ovoid, 0.8–1.8 cm long, tuberculate, basipetally dehiscent; seeds clavate to reniform-obovate, ca. 1 mm long, laterally keeled.

Wet flatwoods, seepage areas, and bogs. Frequent; northern and central peninsula, west to central panhandle. North Carolina south to Florida. Spring.

Sarracenia minor hybridizes in Florida with *S. psittacina* (*S.* ×*formosa*).

Sarracenia minor is listed as threatened in Florida (Florida Administrative Code, Chapter 5B-40).

Sarracenia psittacina Michx. [Parrot-like, in reference to the leaf-shape.] PARROT PITCHERPLANT.

Sarracenia psittacina Michaux, Fl. Bor.-Amer. 1: 311. 1803. TYPE: "ad urbe Augusta ad Floridam," s.d, *Michaux s.n.* (holotype: P).
Sarracenia pulchella Croom, Amer. J. Sci. Arts 25: 75. 1834. TYPE: FLORIDA.
Sarracenia parviflora Rafinesque, Autik. Bot. 33. 1840. TYPE: FLORIDA.

Clump- or mat-forming perennial herb. Leaves (pitchers) prostrate, decumbent, or ascending, 8–30(40) cm long, red-purple reticulate, the distal ½ with white areolae, the surface glabrous, the ventral wing 1–4 cm wide, the orifice round, 0.5–1 cm wide, the rim green, turned inward, the hood recurved ventrally and expanded, forming a subglobose head surrounding the orifice, with white areolae surrounded by green to red-purple tissue, 2–4 cm long, the base and neck indistinct, the apiculum absent or to 1 mm long, the inner surface with retrorse setae; phyllodia absent. Flowering scape 1.5–3.5 dm long, usually longer than the pitchers; bracts 6–8 mm long; sepals ovate-lanceolate, 1.5–2.5 cm long, the margin entire, maroon; petals obovate, 2–4.5 cm long, the margin entire, maroon; style-disk 2–3 mm wide, maroon. Fruit globose to ovoid, ca. 1 cm long, tuberculate, acropetally dehiscent; seeds clavate to reniform-ovate, ca. 2 mm long, laterally keeled.

Bogs. Frequent; Nassau and Baker Counties, central and western panhandle. Georgia, Florida, west to Mississippi. Spring.

Sarracenia psittacina hybridizes in Florida with *S. leucophylla* (*S.* ×*wrigleyana*), *S. minor* (*S.* ×*formosa*), *S. rosea* (*S.* ×*courtii*), and *S. rubra* subsp. *gulfensis* (*S.* ×*galpinii*).

Sarracenia psittacina is listed as threatened in Florida (Florida Administrative Code, Chapter 5B-40).

Sarracenia rosea Naczi et al. [Reddish colored.] GULF PURPLE PITCHERPLANT; DECUMBENT PITCHERPLANT.

Sarracenia rosea Naczi et al., Sida 18: 1188. 1999. TYPE: FLORIDA: Liberty Co.: SW of Telogia, Apalachicola National Forest, 17 May 1993, *Naczi 3016* (holotype: MICH; isotypes NY, US).
Sarracenia purpurea Linnaeus var. *burkii* D. E. Schnell, Rhodora 95: 8. 1993.

Clump-forming perennial herb. Leaves (pitchers) decumbent to ascending, 6–28 cm long, green suffused with purple-red, frequently with darker reticulated red veins, inflated distally, the surface densely pubescent, the ventral wing 0.6–5(6) cm wide, the orifice round to oval, 2–5 cm wide, the rim red to maroon, revolute, sometimes slightly indented distally to the wing, the hood reniform, erect or with lobes variously arched together over the orifice, 2–6 cm long, the margin undulate or entire, the base lobes cordate, attached to the sides of the orifice without a neck, the distal portion outwardly recurved and apically notched, the apiculum absent, the inner surface with retrorse setae; phyllodia absent. Flowering scape 16–35 cm long, longer than the pitchers; bracts 5–8 mm long; sepals ovate-lanceolate, 3–4.5 cm long, purple red, the margin entire; petals ovate to obovate,(4)5–6.5 cm long, the margin entire, pale pink to nearly white; style-disk 4.5–7 cm wide, pale green to nearly white. Fruit globose to ovoid, 1.5–2 cm long, tuberculate, acropetally dehiscent; seeds clavate to reniform-obovate, ca. 2 mm long, laterally keeled.

Bogs. Occasional; central and western panhandle. Georgia, Florida, Alabama, and Mississippi. Spring.

Sarracenia rosea hybridizes in Florida with *S. flava* (*S. ×naczii*), *S. leucophylla* (*S. ×mitchelliana*), *S. psittacina* (*S. ×courtii*), and *S. rubra* subsp. *gulfensis* (*S. ×chelsonii*).

Sarracenia rosea is listed as threatened in Florida (Florida Administrative Code, Chapter 5B-40).

Sarracenia rubra Walter [Red, in reference to the flower color.] REDFLOWER PITCHERPLANT.

Clump-forming perennial herb. Leaves (pitchers) 2.8–6.3 dm long, green, often suffused with bronze or red, gradually tapering from the base to the orifice with a slight distal bulge, the ventral wing 0.3–1.5 cm wide, the surface finely pubescent, the orifice oval, 2.4–5.3 cm wide, the hood broadly ovate, recurved and held well above the orifice, slightly undulate, 3.6–6.3 cm long, the neck indistinct, the apiculum 1–3 mm long; phyllodia absent. Flowering scape 1.4–7.5 dm long, subequaling or exceeding the pitchers; bracts 0.4–1 cm long; sepals obovate, 2–3 cm long, maroon; petals obovate, 2.5–4 cm long, maroon; style-disk 2.5–3.5 cm wide, green. Fruit globose to ovoid, 0.5–1.5 cm long, tuberculate, acropetally dehiscent; seeds clavate to reniform-obovate, 1–2 mm long, laterally keeled.

1. Pitcher 4.3–6.1 dm tall, the orifice 2.4–3.5 cm wide, the hood 3.6–4.7 cm long..........................
....**S. rubra** subsp. **gulfensis**
1. Pitcher 2.8–4.3 dm tall, the orifice 3.4–5.3 cm wide, the hood 4.6–6.3 cm long..........................
....**S. rubra** subsp. **wherryi**

Sarracenia rubra subsp. **gulfensis** D. E. Schnell [Of the Gulf of Mexico region.] GULF COAST REDFLOWER PITCHERPLANT.

Sarracenia rubra Walter subsp. *gulfensis* D. E. Schnell, Castanea 44: 218. 1979. TYPE: FLORIDA: Okaloosa Co.: ca. 20 km N of Niceville, Eglin AFB, Jun 1974, *Schnell s.n.* (holotype: NCU).

Pitcher 4.3–6.1 dm tall, the orifice 2.4–3.5 cm wide, the hood 3.6–4.7 cm long.

Bogs. Rare; western panhandle. Georgia, Florida, and Alabama. Spring.

Sarracenia rubra subsp. *gulfensis* hybridizes in Florida with *S. leucophylla* (*S. ×bellii*), *S. psittacina* (*S. ×galpinii*), and *S. rosea* (*S. ×chelsonii*).

Sarracenia rubra subsp. *gulfensis* is listed as threatened in Florida (Florida Administrative Code, Chapter 5B-40).

Sarracenia rubra subsp. **wherryi** (Case & R. B. Case) D. E. Schnell [Commemorates Edgar Theodore Wherry (1885–1982), American mineralogist, soil scientist, and botanist with a special interest in ferns and *Sarracenia*.] WHERRY'S REDFLOWER PITCHERPLANT.

Sarracenia alabamensis Case & R. B. Case subsp. *wherryi* Case & R. B. Case, Rhodora 78: 315. 1976. *Sarracenia rubra* Walter subsp. *wherryi* (Case & R. B. Case) D. E. Schnell, Castanea 43: 261. 1978.

Pitcher 2.8–4.3 dm tall, the aperture 3.4–5.3 cm wide, the hood 4.6–6.3 cm long.

Bogs. Rare; Escambia County. Florida, Alabama, and Mississippi. Spring.

Sarracenia rubra subsp. *wherryi* is listed as threatened in Florida (Florida Administrative Code, Chapter 5B-40).

HYBRIDS

Sarracenia ×bellii Mellich. (*S. leucophylla* × *S. rubra* subsp. *gulfensis*). [Commemorates Clyde Ritchie Bell (1921–2013), specialist on the southeastern U.S. flora.]

Sarracenia ×bellii Mellichamp, Carniv. Pl. Newslett. 37: 114. 2008. TYPE: Santa Rosa Co.: W side of Fl. 87, just N of Yellow River, ca. 1952, *Case P-42*; cult. as *Mellichamp s.n.* (holotype: UNCC).

Rare; Santa Rosa County. Spring.

Sarracenia ×chelsonii Mast. (*S. rosea* × *S. rubra*). [Of the Chelsea District of central London, an affluent area where the Veitch Nursery was located and where many *Sarracenia* hybrids were made.]

Sarracenia ×chelsonii Masters, Gard. Chron., ser. 2. 8: 600. 1877.

Rare; Walton and Santa Rosa Counties. Spring.

Sarracenia ×courtii Anon. (*S. psittacina* × *S. rosea*). [Commemorates William Court (1843–1888), *Sarracenia* hybridizer at the Veitch Nurseries.]

Sarracenia ×courtii Anonymous, Gard. Chron., ser. 2. 16: 381. 1881.

Rare; Liberty County. Spring.

Sarracenia ×formosa Veitch ex Mast. (*S. minor* × *S. psittacina*). [Finely formed, beautiful.]

Sarracenia ×formosa Veitch ex Masters, Gard. Chron., ser. 2. 16: 41. 1881.

Rare; northern peninsula. Spring.

Sarracenia ×galpinii C. R. Bell & Case (*S. psittacina* × *S. rubra*). [Commemorates Robert Galpin of North Carolina, who assisted Bell and Case in their studies.]

Sarracenia ×galpinii C. R. Bell & Case, J. Elisha Mitchell Sci. Soc. 72: 149. 1956. TYPE: FLORIDA: Santa Rosa Co.: Savannah N of Yellow River and E of FL 87, Apr. 1955, *Bell 1523* (holotype: NCU).

Rare; Okaloosa and Santa Rosa Counties. Spring.

Sarracenia ×mitchelliana G. Nicholson (*S. leucophylla* × *S. rosea*). [Commemorates Mr. Mitchell.]

Sarracenia ×mitchelliana G. Nicholson, Ill. Dict. Gard. 3: 365. 1886.

Rare; central and western panhandle. Spring.

Sarracenia ×mooreana Veitch (*S. flava* × *S. leucophylla*). [Commemorates David Moore (1808–1879), director of the Glasnevin Botanic Gardens.]

Sarracenia ×mooreana Veitch, Gard. Chron., ser. 2. 7: 425. 1877.

Occasional; central and western panhandle. Spring.

Sarracenia ×naczii Mellich. (*S. flava* × *S. rosea*). [Commemorates Robert F. C. Naczi (1963–), Arthur J. Conquist Curator of North American Botany, New York Botanical Garden, specialist in Cyperaceae and Sarraceniaceae.]

Sarracenia ×naczii Mellichamp, Carniv. Pl. Newslett. 37: 113. 2008. TYPE: FLORIDA: Escambia Co.: S side of Ocie Phillips Road, ca. 3 mi. E of Perdido River, 3 Sep 2007, *Mellichamp s.n.* (holotype: UNCC; isotype: NCU).

Occasional; central and western panhandle. Spring.

Sarracenia ×wrigleyana S. G. (*S. leucophylla* × *S. psittacina*). [Commemorates O. O. Wrigley, Esq. of Bury, England.]

Sarracenia ×wrigleyana S. G., Garden (London) 28: 219. 1885.

Rare; western panhandle. Spring.

EXCLUDED TAXA

Sarracenia purpurea Linnaeus—All reports of this species for Florida, such as Chapman (1860), Small (1903, 1913a, 1933), Radford et al. (1964, 1968), Godfrey and Wooten (1981), Clewell (1985), and Wunderlin (1998), are misapplications of the name to material of *S. rosea*, fide Naczi et al. (1999).

Sarracenia purpurea Linnaeus subsp. *venosa* (Rafinesque) Wherry—Although the original description and Wilhelm (1984) both indicate Florida as part of this taxon's range, our plants represent *S. rosea*, fide Naczi et al. (1999).

Sarracenia rubra Walter—Because infraspecific categories were not recognized, the typical subspecies was reported for Florida by implication by Small (1903, 1913a, 1933), Radford et al. (1964, 1968), Godfrey and Wooten (1981), Wilhelm (1984), and Clewell (1985). All Florida material except one Escambia County collection is subsp. *gulfensis*.

Sarracenia rubra subsp. *jonesii* (Wherry) Wherry—Reported for Florida by Small (1933) as *S. jonesii* Wherry. This taxon is confined to the western Carolinas.

Sarracenia ×catesbaei Elliott (*S. flava* × *S. purpurea*)—Reported from Florida by Wilhelm (1984) and Wunderlin (1998). With the recognition of Florida material formerly called *S. purpurea* L. as the separate species *S. rosea*, this name, whose type is from South Carolina, becomes misapplied to our material of *S. ×naczii*.

Sarracenia ×readii C. R. Bell (*S. leucophylla* × *S. alabamensis* Case & R. B. Case subsp. *wherryi* (D. E. Schnell) Case & R. B. Case)—The type of this name is from Alabama and refers to a hybrid not found in Florida. The correct name for our plants formerly treated as this by Wilhelm (1984), Wunderlin (1998), and Wunderlin and Hansen (2003) is *S. ×bellii*.

CLETHRACEAE Klotzsch, nom. cons. 1851. WHITE ALDER FAMILY

Shrubs. Leaves alternate, simple, pinnate-veined, petiolate, estipulate. Flowers in terminal racemes, actinomorphic, bisexual; sepals 5, basally connate; petals 5, basally connate; stamens 10, in 2 whorls of 5 each, the outer whorl antipetalous, the anthers dehiscent by apical, pore-like slits; ovary superior, 3-carpellate and -loculate, the style 1, the stigmas 3. Fruit a loculicidally dehiscent capsule.

A family of 2 genera and about 75 species; North America, West Indies, Mexico, Central America, South America, and Asia.

Selected references: Thomas (1961); Tucker and Jones (2009).

Clethra L. 1753. SWEETPEPPERBUSH

Shrubs. Leaves alternate, simple, pinnate-veined, petiolate, estipulate. Flowers borne on the new growth in racemes, bracteate; sepals 5, basally connate; petals 5, basally connate; stamens with the filaments basally epipetalous, the anthers reflexed in bud, erect at anthesis, the theca divergent distally. Fruit a loculicidally dehiscent capsule; seeds numerous.

A genus of about 65 species; North America, West Indies, Mexico, Central America, and Asia. [From the Greek *klethra*, alder, in reference to the resemblance of the leaves to those of some *Alder* (Betulaceae).]

Clethra alnifolia L. [With leaves like *Alnus* (Betulaceae).] COASTAL SWEETPEPPERBUSH.

Clethra alnifolia Linnaeus, Sp. Pl. 396. 1753. *Clethra alnifolia* Linnaeus var. *denudata* Aiton, Hort. Kew. 2: 73. 1789, nom. inadmss. *Clethra alnifolia* Linnaeus var. *glabella* Michaux, Fl. Bor.-Amer. 1: 260. 1803, nom. inadmiss.
Clethra tomentosa Lamarck, Encycl. 2: 46. 1786. *Clethra incana* Persoon, Syn. Pl. 1: 483. 1805, nom. illegit.
Clethra alnifolia Linnaeus var. *tomentosa* Michaux, Fl. Bor.-Amer. 1: 260. 1803.
Clethra angustifolia Rafinesque, Autik. Bot. 6. 1840; non de Candolle, 1838. TYPE: FLORIDA.
Clethra bracteata Rafinesque, Autik. Bot. 7. 1840. TYPE: FLORIDA.

Shrub, to 3 m; branchlets puberulent to woolly-tomentose. Leaves with the blade obovate to oblong, 5–9(12) cm long, 2–4(6) cm wide, the apex obtuse to acute, the base cuneate to rounded, the margin serrate, especially distally, the upper surface glabrous or with sparse, stellate trichomes (especially when young), the lower surface woolly-tomentose, puberulent, or glabrescent, the petiole 0.5–3.5(6) cm long. Flowers in a solitary or cluster of 2–4 racemes (6)9–15(19) cm long, the axis stellate-pubescent, the pedicel 2–4 mm long, short-stellate-puberulent; bract 1, basal, to 7 mm long, shorter than the flower, short-stellate-puberulent; sepals 4–5 mm long, the outer surface stellate-pubescent; petals obovate, 5–8 mm long, white or rarely pink; stamens 4–7 mm long; style 6–7 mm long, pubescent in the proximal ½, straight.

Fruit subglobose, 3–5 mm long; seeds 40–100, oblong-ovoid, 1–2 mm long, brown, strongly flattened, slightly winged.

Acid swamps, bogs, and creek margins. Occasional; northern counties, central peninsula. Maine south to Florida, west to Texas. Summer.

CYRILLACEAE Lindl., nom. cons. 1846. TITI FAMILY

Shrubs or trees. Leaves alternate, simple, pinnate-veined, petiolate or epetiolate, estipulate. Flowers in terminal or axillary racemes, bracteate, bracteolate, actinomorphic, bisexual; sepals (4)5(8), basally connate; petals (4)5(8), free or basally connate; nectariferous disk present; stamens 5 or 10; ovary superior, 2- to 5-carpellate and -loculate, the style 1, the stigma lobed. Fruit a drupe.

A family of 3 genera and 14 species; North America, West Indies, Mexico, Central America, and South America.

Selected references: Lemke (2009); Thomas (1960, 1961).

1. Flowers in terminal racemes of the current season; fruit broadly 2- to 4-winged; stamens 10 ....
   ....**Cliftonia**
1. Flowers in clustered racemes at the summit of the previous season; fruit not winged; stamens 5 ....
   .... **Cyrilla**

Cliftonia Banks ex C. F. Gaertner 1807.

Shrubs or trees. Leaves alternate, simple, pinnate-veined, petiolate or epetiolate, estipulate. Flowers in racemes on the distal ends of branches of the previous season, bracteate, bracteolate; sepals (4)5(8), basally connate; petals (4)5(8), free; stamens 10, free; ovary 3- to 5-carpellate and -loculate, the stigma 3- to 5-lobed. Fruit a 2- to 4-winged dry drupe.

A monotypic genus; North America. [Commemorates British botanist William Clifton (fl. 1760s), attorney general of Georgia, later chief justice of West Florida.]

Cliftonia monophylla (Lam.) Sarg. [With simple leaves.] BLACK TITI; BUCKWHEAT TREE.

> *Ptelea monophylla* Lamarck, Tabl. Encycl. 1: 336. 1792. *Cliftonia monophylla* (Lamarck) Sargent, Silva 2: 7. 1892.
> *Cliftonia nitida* C. F. Gaertner, Suppl. Carp. 247. 1807. TYPE: FLORIDA: "habitat in Florida occidentali."
> *Mylocaryum ligustrinum* Willdenow, Enum. Pl. 454. 1809. *Cliftonia ligustrina* (Willdenow) Sims ex Sprengel, Syst. Veg. 2: 316. 1825.

Shrub or tree, to 15 m; branchlets glabrous, glaucous. Leaves with the blade elliptic to oblanceolate, 2.5–10 cm long, 1.2–1.8 cm wide, coriaceous, evergreen, the apex acute to obtuse or emarginate, the base cuneate, the margin entire, the upper surface glabrous, the lower surface glabrous, glaucous, the petiole relatively short or absent. Flowers in a terminal raceme 2–6 cm long; sepal lobes broadly deltate to oblong, 2–3 mm long, the apex obtuse to rounded, glabrous; the petals spatulate to obovate, 6–8 mm long, white or pink, weakly clawed, the apex rounded; stamens in 2 whorls of 5 each, the outer whorl antisepalous, the filament basally flattened and petaloid. Fruit 2- to 5-winged, broadly ellipsoid, 5–7 mm long, brown; seeds 1–5 or sometimes absent.

Acid swamps and bogs. Frequent; northern counties, Marion County. South Carolina south to Florida, west to Louisiana. Spring.

Cyrilla Garden ex L. 1767. TITI

Shrubs or trees. Leaves alternate, simple, pinnate-veined, petiolate, estipulate. Flowers in racemes on the distal ends of branches of the previous season, bracteate, bracteolate; sepals (4)5(8), connate proximally; petals (4)5(8), free; stamens 5, free; ovary 2–3(4)-carpellate and -loculate, the stigma 2–3(4)-lobed. Fruit a dry drupe.

A monotypic genus; North America, West Indies, Mexico, Central America, and South America. [Commemorates Dominico Cirillo (1739–1799), Italian physician and professor of natural history at the University of Naples.]

Cyrilla racemiflora L. [Flowers in racemes]. TITI.

> *Cyrilla racemiflora* Linnaeus, Mant. Pl. 50. 1767. *Itea cyrilla* Swartz, Prodr. 50. 1788, nom. illegit. *Cyrilla caroliniana* Michaux, Fl. Bor.-Amer. 1: 158. 1803, nom. illegit. *Itea caroliniana* Lamarck, Encycl. 3: 315. 1789, nom. illegit.
> *Andromeda plumata* Marshall, Arbust. Amer. 9. 1785. TYPE: "Carolina and Florida."
> *Cyrilla parvifolia* Rafinesque, Autik. Bot. 8. 1840. *Cyrilla racemiflora* Linnaeus subsp. *parvifolia* (Rafinesque) E. Murray, Kalmia 13: 5. 1983. *Cyrilla racemiflora* Linnaeus var. *parvifolia* (Rafinesque) E. Murray, Kalmia 13: 5. 1983; non Sargent, 1921. TYPE: "Florida and Alabama."
> *Cyrilla polystachia* Rafinesque, Autik. Bot. 8. 1840. TYPE: "Louisiana and Florida."
> *Cyrilla parvifolia* Shuttleworth ex Nash, Bull. Torrey Bot. Club 23: 101. 1896; non Rafinesque, 1840. *Cyrilla racemiflora* Linnaeus var. *parvifolia* Sargent, J. Arnold Arbor. 2: 166. 1921. TYPE: FLORIDA: Wakulla Co.
> *Cyrilla arida* Small, Bull. Torrey Bot. Club 51: 383. 1924. TYPE: FLORIDA: Highlands Co.: between Avon Park and Sebring, 13 Dec. 1920, *Small et al. 11486* (holotype: NY; isotypes: GH, MICH, MO, TEX).

Shrub or tree, to 10 m; branchlets glabrous. Leaves with the blade oblanceolate to elliptic, 1–10 cm long, 0.5–2.5 cm wide, chartaceous to coriaceous, deciduous or semievergreen, the apex acute, rounded, or emarginate, the base cuneate, the margin entire, the upper and lower surfaces glabrous, the petiole short. Flowers in terminal or subterminal racemes 6–18 cm long; sepal lobes lanceolate-ovate, 2–3 mm long, the apex acute to acuminate; petals ovate to ovate-lanceolate, 2–4 mm long, white or creamy white, the apex acute. Fruit ovoid to subglobose, 2–3 mm long; seeds 1–4 or sometimes absent.

Swamps and bogs, rarely scrub. Occasional; northern counties, central peninsula. Virginia south to Florida west to Texas; West Indies, Mexico, Central America, and South America. Spring–Summer.

ERICACEAE Juss., nom. cons. 1789. HEATH FAMILY

Herbs, vines, shrubs, or trees. Leaves alternate, opposite, or whorled, simple, pinnate-veined, petiolate or epetiolate, estipulate. Flowers in terminal axillary racemes, umbels, corymbs, panicles, fascicles, spikes, or solitary, actinomorphic or zygomorphic, bisexual or unisexual (plants dioecious), bracteate or ebracteate, bracteolate or ebracteolate; sepals 2–7, free or basally connate; petals 2–7 or absent, free or basally connate; stamens 2–10(14), free, the anthers dehiscent by pores or longitudinal slits; ovary superior or inferior, 2- to 5-carpellate, 2- to 10-loculate, the style 1. Fruit a loculidally or septicidally dehiscent capsule or an indehiscent drupe or a berry.

A family of about 120 genera and about 4,100 species; nearly cosmopolitan.

The taxonomy of the Ericaceae may certainly change with further study. For example *Vaccinium* is not monophyletic and *Gaylussacia* is embedded within it (W. S. Judd, pers. comm.).

Empetraceae Hook. & Lindl., nom. cons., 1821; *Monotropaceae* Nutt., nom. cons., 1818; *Pyrolaceae* Lindl., nom. cons., 1829; *Vacciniaceae* DC. ex Perleb, nom. cons., 1818.

Selected references: Wood (1961); Wood and Channell (1959).

1. Plant an achlorophyllous herb.
　　2. Petals free to their base, deciduous; fruit a dehiscent capsule .... **Monotropa**
　　2. Petals united more than ½ their length; fruit indehiscent and fleshy............................**Monotropsis**
1. Plant a chlorophyllous tree, shrub, woody vine, or subwoody herb.
　　3. Leaves needlelike; flowers much reduced, unisexual; calyx and corolla 2-merous.............**Ceratiola**
　　3. Leaves other than needlelike; flowers not reduced, bisexual; calyx and corolla 4- to 7-merous.
　　　　4. Ovary inferior; fruit fleshy.
　　　　　　5. Ovary 4- to 5-loculate; fruit a many seeded berry...**Vaccinium**
　　　　　　5. Ovary 10-loculate; fruit a berrylike drupe with 10 hard nutlets............................**Gaylussacia**
　　　　4. Ovary superior; fruit dry.
　　　　　　6. Plant trailing on the ground, or if ascending, then only to 20 cm tall.
　　　　　　　　7. Leaves lanceolate, dark green with whitish veins; stem glabrous....................**Chimaphila**
　　　　　　　　7. Leaves elliptic, elliptic-ovate, or suborbicular, solid green; stem hirsute................**Epigaea**
　　　　　　6. Plant erect or ascending, sometimes weak vines but not trailing on the ground.
　　　　　　　　8. Petals free..**Bejaria**
　　　　　　　　8. Petals united.
　　　　　　　　　　9. Corolla campanulate, funnelform, or rotate; capsules septicidal.
　　　　　　　　　　　　10. Corolla campanulate or funnelform; anthers long-exserted; capsules longer than wide... **Rhododendron**
　　　　　　　　　　　　10. Corolla rotate; anthers included, inserted into a circle of 10 pouches in the corolla; capsules shorter than wide or the length/width subequal............................ **Kalmia**
　　　　　　　　　　9. Corolla urceolate; capsules loculicidal.
　　　　　　　　　　　　11. Corolla densely strigose..**Oxydendrum**
　　　　　　　　　　　　11. Corolla glabrous, lepidote, or sparsely strigulose.
　　　　　　　　　　　　　　12. Inflorescences of corymbose clusters; fruit with the carpel midrib differentiated (colored or thickened) from the valve margins... **Lyonia**
　　　　　　　　　　　　　　12. Inflorescence clearly racemose or paniculate; fruit with the carpel midrib not differentiated from the valve margins.

13. Weak shrub or vine; anthers appendaged on the side with 2 stout refl
spurs just above the junction with the filament..F

13. Erect shrub; anthers not appendaged as above.

14. Filaments sigmoid-curved above; anthers obscurely 2-mucronate at the
apex; pith chambered; inflorescences emerging in the spring....**Agarista**

14. Filaments straight or nearly so; anthers awned or at least prominently
2-mucronate at the apex; pith solid; inflorescences emerging in the fall.

15. Leaves coriaceous, persistent; racemes axillary; pedicel with the
bracts subequal..**Leucothoe**

15. Leaves chartaceous, deciduous; racemes terminal; peduncle bearing
2 apical bracts that closely subtend the calyx ....**Eubotrys**

Agarista D. Don ex G. Don 1834. HOBBLEBUSH

Shrubs or trees. Leaves alternate, simple, pinnate-veined, petiolate, estipulate. Flowers in axillary racemes, actinomorphic, bisexual, bracteolate; sepals 5, basally connate; petals 5, basally connate; stamens 10, free, the anthers dehiscent by elliptic pores; ovary superior, 5-carpellate and -loculate, the style 1, the stigma capitate. Fruit a septicidally dehiscent capsule; seeds numerous.

A genus of 31 species; North America, Mexico, Central America, South America, and Africa. [From the Greek mythological daughter of Clisthenes, in reference to the beauty of the flowers.]

Selected reference: Judd (2009c).

Agarista populifolia (Lam.) Judd [With leaves like *Populus* (Salicaceae).] FLORIDA HOBBLEBUSH; PIPESTEM.

Andromeda populifolia Lamarck, Encycl. 1: 159. 1783. *Andromeda laurina* Michaux, Fl. Bor.-Amer. 1:
253. 1803, nom. illegit. *Lyonia populifolia* (Lamarck) K. Koch, Dendrologie 2: 123. 1872. *Leucothoe
populifolia* (Lamarck) Dippel, Handb. Laubholzk. 1: 356. 1889. *Agarista populifolia* (Lamarck) Judd,
J. Arnold Arbor. 60: 495. 1979.

Andromeda lucida Jacquin, Collectanea 1: 95. 1786; non Lamarck, 1783. *Andromeda acuminata* Aiton,
Hort. Kew. 2: 70. 1789. *Leucothoe acuminata* (Aiton) G. Don, Gen. Hist. 3: 832. 1834.

Shrubs or trees, to 7 m; branchlets glabrous, with chambered pith. Leaves with the blade ovate, 2.5–9(11) cm long, 0.9–4(5) cm wide, coriaceous, the apex acuminate, the base cuneate to rounded, the margin entire or serrate, flat, the upper surface glabrous, the lower surface glabrous or with stipitate-glandular trichomes on the midvein. Flowers 10–20 in an axillary raceme; bracteoles 2, near the base to the middle of the pedicel; sepal lobes 1–2 mm long, glabrous; corolla cylindric to urceolate, the petals 6–10 mm long, connate for most of their length (the lobes ca. 1 mm long), white, glabrous; stamens 4–5 mm long, the filaments geniculate, pubescent. Fruit subglobose to pyramidal, 3–4 mm long; seeds oblong to angular-obovoid, ca. 2 mm long, brown.

Swamps and wet hammocks. Occasional; northern peninsula south to Lake and Orange Counties. South Carolina, Georgia, and Florida. Spring.

Bejaria Mutis ex L., nom. et orth. cons. 1771.

Shrubs. Leaves alternate, simple, pinnate-veined, petiolate, estipulate. Flowers in terminal racemes or panicles, actinomorphic, bisexual, ebracteate, ebracteolate; sepals 7, basally connate; petals 7, free; stamens 14, free, the anthers dehiscent by terminal pores; ovary superior, 7-carpellate and -loculate, the style 1, the stigma 7-lobed. Fruit a septicidally dehiscent capsule; seeds numerous.

A genus of 15 species; North America, West Indies, Mexico, Central America, and South America. [Commemorates José Béjar, eighteenth-century Spanish physician.]

Selected reference: Clemants (2009).

Bejaria racemosa Vent. [Flowers in a raceme.] TARFLOWER.

Bejaria racemosa Ventenat, Descr. Pl. Nouv. pl. 51. 1801. TYPE: FLORIDA.
Bejaria paniculata Michaux, Fl. Bor.-Amer. 1: 280, t. 26. 1803; non Cels ex Dumont de Courset, 1802.

Shrub, to 3 m; branchlets glabrous. Leaves with the blade ovate, elliptic, obovate, or lanceolate, 1.8–5.2 dm long, 0.6–2.5 cm wide, chartaceous, the apex acute to rounded, the base cuneate to rounded, the margin entire, the upper and lower surfaces glabrous or reddish-tomentose when young, the petiole ca. 1 mm long, glabrous or hispid to tomentose. Flowers 2–5 in a terminal raceme or panicle, sometimes solitary; sepals 3–5 mm long, slightly unequal in size (outer ones larger), the lobes ovate, 2–3 mm long, glabrous or rusty-tomentose; petals spatulate, 18–33 cm long, spreading to reflexed, white, pink, or white with pink lines, with a sticky exudate, glabrous; stamens 12–25 cm long, tomentose; ovary glabrous. Fruit depressed oblong, 4–6 mm long; seeds ellipsoid, 1–2 mm long, the surface reticulate, brown.

Flatwoods. Common; peninsula, eastern panhandle. Georgia, Florida, and Alabama. Summer.

Ceratiola Michx. 1803. FLORIDA ROSEMARY; SAND HEATH.

Shrubs. Leaves whorled, simple, pinnate-veined, petiolate, estipulate. Flowers in axillary fascicles, actinomorphic, unisexual (plants dioecious), bracteolate; sepals 2(3), free; petals 2(3), free; stamens 2, free, the anthers longitudinally dehiscent; ovary superior, 2-carpellate and -loculate, the style 1, the stigma linear. Fruit a 2-seeded drupe.

A monotypic genus; North America. [From the Greek *keration*, little horn, in reference to the style branches.]

Ceratiola is sometimes placed in the Empetraceae by various authors.

Selected reference: Elisens (2009b).

Ceratiola ericoides Michx. [Resembling *Erica*.] FLORIDA ROSEMARY; SAND HEATH.

Ceratiola ericoides Michaux, Fl. Bor.-Amer. 2: 222. 1803. TYPE: "Georgiae et Floridae."
Ceratiola falcatula Gandoger, Bull. Soc. Bot. France 66: 232. 1919. TYPE: FLORIDA: Manatee Co.: Palma Sola.

Shrub, to 2.5 m; stem bark gray, exfoliating, rough with the persistent leaf bases, the branchlets tomentose. Leaves persistent, 4(6) per node, the blade linear to aciculate, 5–15 mm long, coriaceous, the apex acute, the base cuneate to rounded, the margin entire, the surfaces minutely glandular to glandular-pubescent, the lower surface with a prominent longitudinal groove, the petiole ca. 1 mm long, tan. Flowers 2–3 per fascicle, sessile; bracteoles 2(4), ovate, ca. 1 mm long, sessile, the margin erose; sepals ovate to orbicular, ca. 1 mm long, reddish brown, the margin fimbrillate, persistent; petals ovate, ca. 1 mm long, reddish brown, the margin fimbrillate; stamens 3–4 mm long, reddish brown; ovary ovoid, ca. 1 mm long, reddish purple. Fruit globose, ca. 2 mm long, red or tan to yellow, dry; pyrenes 2, 1–2 mm long, reddish brown.

Scrub. Frequent; nearly throughout. South Carolina south to Florida, west to Mississippi. Spring–summer.

Chimaphila Pursh 1814. PRINCE'S PINE

Subwoody herbs. Leaves alternate or pseudovertillate, simple, pinnate-veined, petiolate, estipulate. Flowers in corymbs, subumbels, or solitary, bracteate, ebracteolate; sepals 5, basally connate; petals 5, free; stamens 10, free, the anthers dehiscent by 2 crescent-shaped to round pores; ovary superior, 5-carpellate, imperfectly 5-locular, the style 1, the stigma capitate. Fruit a loculicidally dehiscent capsule; seeds numerous.

A genus of 5 species; North America, West Indies, Mexico, Central America, Europe, and Asia. [From the Greek *cheima*, winter, and *phila*, love, in reference to the evergreen habit.]

Chimaphila is sometimes placed in the Pyrolaceae by various authors.

Selected reference: Freeman (2009).

Chimaphila maculata (L.) Pursh [Spotted, in reference to the leaves.] SPOTTED WINTERGREEN; STRIPED PRINCE'S PINE.

Pyrola maculata Linnaeus, Sp. Pl. 396. 1753. *Chimaphila maculata* (Linnaeus) Pursh, Fl. Amer. Sept. 300. 1814.

Erect to decumbent, subwoody, rhizomatous herb, to 5 dm; stem papillose to hispidulous. Leaves with the blade lanceolate to ovate-lanceolate or ovate, 2–10 cm long, 0.8–3 cm wide, the apex acute to acuminate, the base rounded to cuneate, the margin coarsely serrate, the upper surface glossy green to dark green with white achlorophyllous areas bordering the larger veins, the lower surface dull and light green, maculate, the petiole 3–13 mm long, glabrous. Flowers solitary or 2–5 in a corymb or subumbel, the peduncle 4–19 cm long, papillose to hispidulous, the pedicel 0.5–2.5 cm long; bracts linear-lanceolate, basally adnate to the peduncle, the free portion 4–6 mm long, membranous, the margin entire; sepal lobes ovate, 1–4 mm long, spreading or reflexed, greenish, the margin sometimes whitish green, the apex rounded, the margin erose-dentate; petals orbicular, 6–12 mm long, white or pink, often tinged with violet, the margin fimbriate to erose-denticulate; stamens 6–8 mm long, the base dilated, villous, the anthers white to tan or pinkish brown, abruptly narrowed into tubules above the theca. Fruit depressed-globose, 5–10 mm long; seeds fusiform.

Dry hammocks. Rare; Leon County. Quebec south to Florida, west to Michigan, Illinois, Tennessee, and Mississippi, also Arizona; Mexico and Central America; Europe. Native to North America, Mexico, and Central America. Spring–summer.

Epigaea L. 1753. TRAILING ARBUTUS

Subwoody vines. Leaves alternate, simple, pinnate-veined, petiolate, estipulate. Flowers in axillary or terminal racemes, actinomorphic, bisexual or unisexual (functionally dioecious), bracteolate; sepals 5, basally connate; petals 5, basally connate; stamens 10, the filaments epipetalous, the anthers dehiscent by longitudinal slits; ovary superior, 5-carpellate and -loculate, the style 1, the stigma 5-lobed. Fruit a septicidally dehiscent capsule; seeds numerous.

A genus of 3 species; North America and Asia. [From the Greek *epi-*, upon, and *gaia*, earth, in reference to its creeping habit].

Selected reference: Judd and Kron (2009b).

Epigaea repens L. [Creeping.] TRAILING ARBUTUS.

Epigaea repens Linnaeus, Sp. Pl. 395. 1753.

Creeping or prostrate subwoody vine; stem to 3(5) dm long, usually hirsute-hispid, glabrescent in age. Leaves with the blade ovate-elliptic, (2)3–8(10) cm long, 1.5–5.5 cm wide, the apex acute to mucronate, the base rounded or cordate, the margin ciliate, the upper and lower surfaces hirsute, becoming glabrate, the petiole 1–5 cm long, hirsute. Flowers 2–6(10) in an axillary or terminal cluster 2–5 cm long, bisexual or functionally unisexual due to undeveloped or sterile stamens or ovary, the staminate larger than the carpellate; bracteoles 2, 4–7 mm long, hirsute; sepal lobes lanceolate, 5–6 mm long, hirsute; petals rose or pink to white, the tube 6–10(15) mm long, the lobes 6–10 mm long, spreading; stamens with the filaments hirsute; ovary glandular-pubescent. Fruit depressed-globose, 5–8 mm long, glandular-hirsute; seeds ovoid to globose, ca. 0.5 mm long, the surface foveolate, brown.

Dry hammocks. Rare; Liberty County, western panhandle. Newfoundland and Labrador south to Florida, west to Manitoba, Minnesota, Iowa, Illinois, Tennessee, and Mississippi. Spring.

Epigaea repens is listed as endangered in Florida (Florida Administrative Code, Chapter 5B-40).

Eubotrys Nutt., nom. cons. 1842.

Shrubs. Leaves alternate, simple, pinnate-veined, petiolate, estipulate. Flowers in axillary racemes, actinomorphic, bisexual, bracteate, bracteolate; sepals 5, basally connate; petals 5, basally connate; stamens 8(10), free, the anthers dehiscent by terminal pores; ovary superior, 5-carpellate, falsely 10-locular, the style 1, the stigma 5-lobed. Fruit a loculicidally dehiscent capsule; seeds 5–10.

A genus of 2 species; North America. [From the Greek *eu-*, good, well-developed, or true, and *botrys*, bunch of grapes, in reference to the capsules in a tight raceme.]

Selected references: Judd et al. (2012); Tucker (2009b).

Eubotrys racemosa (L.) Nutt. [In a raceme, in reference to the inflorescence.] SWAMP DOGHOBBLE.

> *Andromeda racemosa* Linnaeus, Sp. Pl. 394. 1753. *Lyonia racemosa* (Linnaeus) D. Don, Edinburgh New Philos. J. 17: 159. 1834. *Zenobia racemosa* (Linnaeus) de Candolle, Prodr. 7: 598. 1839. *Cassandra racemosa* (Linnaeus) Spach, Hist. Nat. Vég. 9: 478. 1840. *Eubotrys racemosa* (Linnaeus) Nuttall, Trans. Amer. Philos. Soc., ser. 2. 8: 269. 1843. *Leucothoe racemosa* (Linnaeus) A. Gray, Manual, ed. 2. 252. 1856.
>
> *Leucothoe elongata* Small, Bull. New York Bot. Gard. 1: 284. 1899. *Eubotrys elongata* (Small) Small, Shrubs Florida 95, 133. 1913. *Leucothoe racemosa* (Linnaeus) A. Gray var. *elongata* (Small) Fernald, Rhodora 41: 554. 1939. TYPE: FLORIDA.

Shrub, to 4 m; branchlets glabrous. Leaves with the blade oblong to lanceolate or obovate, membranous, the apex acute to acuminate, the base cuneate, the margin spinulose-serrulate, the upper surface with sparse stipitate-glandular trichomes and sometimes with unicellular trichomes on the veins, the lower surface glabrous or sometimes with sparse to moderate unicellular trichomes on the major veins, the petiole 1–3 mm long, glabrous. Flowers in a raceme 0.5–21 cm long; bracts lanceolate, 4–5 mm long, deciduous, bracteoles 2, small, distal; calyx campanulate, the sepals lanceolate, 2–3(4) mm long, the apex acute, with sparse to dense unicellular trichomes; corolla cylindric, 5.5–9 mm long, white, the lobes much shorter than the tube, recurved, glabrous; stamens 2–4 mm long, the anthers 1–2 mm long, 4-awned, the theca divergent distally; ovary glabrous. Fruit depressed-globose, 2–3.5 mm long, glabrous; seeds wedge- to crescent-shaped, ca. 1 mm long, the surface reticulate, glossy.

Flatwoods, acid swamps, and riverbanks. Frequent; northern counties, central peninsula. New York and Massachusetts south to Florida, west to Tennessee and Texas. Spring.

Gaylussacia Kunth, nom. cons. 1819. HUCKLEBERRY

Shrubs. Leaves alternate, simple, pinnate-veined, petiolate, estipulate. Flowers in axillary or terminal racemes, actinomorphic, bisexual, bracteate, bracteolate; sepals 5, basally connate; petals 4–5, basally connate; stamens 10, free, the anthers dehiscent by terminal pores; ovary inferior, 5- or 10-carpellate and -loculate, the style 1, the stigma capitate. Fruit a drupe; pyrenes 10.

A genus of about 50 species; North America and South America. [Commemorates Louis Joseph Gay-Lussac (1778–1850), French chemist and physicist.]

Decachaena (Hook.) Torr. & A. Gray ex Lindl., 1846; *Lasiococcus* Small, 1933.

Selected reference: Sorrie et al. (2009).

1. Leaves, hypanthium, and fruit without stipitate glands; leaves with sessile glands on the lower surface only .... **G. frondosa**
1. Leaves, hypanthium, and fruit stipitate-glandular; leaves with or without sessile glands.

2. Hypanthium and fruit with short stipitate-glandular trichomes, some of the glands subsessile; lower surface of the leaves uniformly dotted with sessile glands .... **G. dumosa**

2. Hypanthium and fruit with long stipitate-glandular trichomes; lower surface of the leaves not gland-dotted, but sometimes with short stipitate glands...**G. mosieri**

Gaylussacia dumosa (Andrews) Torr. & A. Gray [With a bushy habit.] DWARF HUCKLEBERRY.

Vaccinium dumosum Andrews, Bot. Repos. 2: pl. 112. 1800. *Decamerium dumosum* (Andrews) Nuttall, Trans. Amer. Philos. Soc., ser. 2. 2: 260. 1843. *Gaylussacia dumosa* (Andrews) A. Gray, Proc. Amer. Acad. Arts, ser. 2. 3: 50. 1846. *Adnaria dumosa* (Andrews) Kuntze, Revis. Gen. Pl. 2: 382. 1891. *Lasiococcus dumosus* (Andrews) Small, Man. S.E. Fl. 1008, 1506. 1933.

Vaccinium hirtellum W. T. Aiton, Hortus Kew. 2: 357. 1811. *Decamerium hirtellum* (W. T. Aiton) Nuttall, Trans. Amer. Philos. Soc., ser. 2. 2: 260. 1843. *Gaylussacia hirtella* (W. T. Aiton) Torrey & A. Gray ex Torrey, Fl. New York 1: 448. 1843.

Vaccinium dumosum Andrews var. *humile* P. Watson, Dendrol. Brit. 1: t. 32. 1823–1825. *Gaylussacia dumosa* (Andrews) Torrey & A. Gray var. *humilis* (P. Watson) Zabel, in Beissner et al., Handb. Laubholzben. 395. 1903. TYPE: "New Jersey to Florida."

Gaylussacia dumosa (Andrews) A. Gray var. *hirtella* A. Gray, Manual 259. 1848.

Gaylussacia dumosa (Andrews) A. Gray var. *hirtella* Chapman, Fl. South. U.S. 259. 1860; non. A. Gray, 1848.

Decamerium hirtellum (W. T. Aiton) Nuttall var. *griseum* Ashe, Rhodora 33: 198. 1931. TYPE: FLORIDA: Okaloosa Co.: SW part of county, 19 Apr 1923, *Ashe s.n.* (holotype: NCU?).

Decamerium hirtellum (W. T. Aiton) Nuttall forma *minimum* Ashe, Rhodora 33: 198. 1931. TYPE: FLORIDA: Okaloosa Co.: 19 Apr 1923, *Ashe s.n.* (holotype: NCU?).

Shrub, to 3(4) dm; branchlets of current season pale green, stipitate-glandular-pubescent, glabrescent in age. Leaves with the blade oblanceolate to obovate, 2.5–4 cm long, 3–10 mm wide, subcoriaceous, the apex obtuse to subacute, mucronate, the base cuneate, the margin entire, with scattered stipitate-glandular trichomes and cilia, the upper and lower surfaces with scattered stipitate-glandular and sessile-glandular trichomes, the upper surface sometimes glabrescent, the petiole 1–2 mm long. Flowers 5–8 in a raceme 3–6 cm long, the pedicel 2–3 mm long, stipitate-glandular pubescent; bracts foliate, 5–12 mm long, equaling or shorter than the pedicels, pubescent with stipitate-glandular or nonglandular trichomes; bracteoles 1–2; sepal lobes ca. 2 mm long, pubescent with stipitate-glandular and nonglandular trichomes; corolla campanulate, 3–5 mm long, the petals 5, white to pink or reddish; stamens with the filaments ca. 0.5 mm long, sparsely pubescent, the anthers 3–4 mm long, the theca divergent distally; ovary stipitate-glandular-pubescent. Fruit subglobose, 6–8 mm long, black, stipitate-glandular; pyrenes ellipsoid, 1–2 mm long, the surface papillose.

Acid swamps, sandhills, flatwoods, and scrub. Frequent; nearly throughout. Virginia south to Florida, west to West Virginia, Tennessee, Mississippi, and Louisiana. Spring.

Gaylussacia frondosa (L.) Torr. & A. Gray ex Torr. [Leafy.] BLUE HUCKLEBERRY.

Vaccinium frondosum Linnaeus, Sp. Pl. 351. 1753. *Gaylussacia frondosa* (Linnaeus) Torrey & A. Gray, Fl. New York 1: 449. 1843. *Decamerium frondosum* (Linnaeus) Nuttall, Trans. Amer. Philos. Soc., n.s. 8: 260. 1843. *Adnaria frondosa* (Linnaeus) Kuntze, Revis. Gen. Pl. 2: 382. 1891. *Decachaena frondosa* (Linnaeus) Torrey & A. Gray ex Small, Man. S.E. Fl. 1007. 1933.

Gaylussacia frondosa (Linnaeus) Torrey & A. Gray ex Torrey var. *tomentosa* A. Gray, Syn. Fl. N. Amer. 2(1): 19. 1878. *Gaylussacia tomentosa* (A. Gray) Pursh ex Small, Bull. Torrey Bot. Club 24: 442. 1897. *Decamerium tomentosum* (A. Gray) Ashe, Rhodora 33: 198. 1931. *Decachaena tomentosa* (A. Gray) Small, Man. S.E. Fl. 1007, 1506. 1933. TYPE: FLORIDA/GEORGIA.

Gaylussacia frondosa (Linnaeus) Torrey & A. Gray ex Torrey var. *nana* A. Gray, Syn. Fl. N. Amer., ed. 2. 2(1): 396. 1886. *Gaylussacia nana* (A. Gray) Small, Bull. Torrey Bot. Club 24: 442. 1897. *Decamarium nanum* (A. Gray) Ashe, Rhodora 33: 198. 1931. *Decachaena nana* (A. Gray) Small, Man. S.E. FL. 1008, 1506. 1933. TYPE: FLORIDA.

Colonial shrub, to 20 dm; branchlets pale green, densely pubescent, often with sessile glands, glabrescent in age. Leaves with the blade ovate to oblong, 2.5–6 cm long, 1–3 cm wide, subcoriaceous, the apex rounded to obtuse, the base cuneate, the margin entire, the upper surface glabrous or puberulent, the lower surface glabrous or puberulent, sometimes sessile-glandular, glaucous, the petiole 2–3 mm long. Flowers 1–4 in a raceme 1–2.5 cm long, sometimes solitary, glabrous or pilose, sometimes sessile-glandular, the pedicel 8–15(20) mm long, glabrous, sessile-glandular; bracts foliate or not, 5–6 mm long, shorter than the pedicel, glabrous or pubescent, sessile-glandular; bracteoles 1–2 mm long; sepal lobes ca. 1 mm long, glabrous or sparsely puberulent, sessile-glandular or not; corolla campanulate-conic, 3–5 mm long, greenish white, the petals 5, the lobes deltate, ca. 1 mm long; filaments ca. 1 mm long, the anthers 2–3 mm long; ovary glabrous. Fruit subglobose, 5–8 mm long, dark blue to black, rarely white, glabrous, glaucous; pyrenes ellipsoid, 1–2 mm long, the surface papillose.

Flatwoods. Frequent; northern counties, central peninsula. North Carolina south to Florida, west to Louisiana. Spring.

Other authors (e.g., Gajdeczka et al., 2010; Sorrie et al., 2009) have segregated taxa at the specific level (e.g., *G. nana, G. tomentosa*) from this complex based on leaf and plant size, degree of pubescence, and glaucescence. We treat *G. frondosa* here in the broad sense as a polymorphic species as these characters show much overlap. More study needs to be done.

If you choose to recognize them at the species level, the three taxa can be distinquished best by the characters given in the following key.

1. Branchlets of the current season glabrous or glabrate; lower leaf surface glabrous or moderately pubescent, sometimes glaucescent .... **G. frondosa**
1. Branchlets of the current season densely short-pubescent; lower leaf surface sparsely to densely short-pubescent, glaucous or glaucescent.
    2. Larger leaves mostly 2–4 cm long, 1–2 cm wide, the lower leaf surface glabrous to sparsely short-pubescent (longer trichomes ca. 1 mm long), glaucous; calyx glaucous; plants 2–6(10) dm tall........ ....**G. nana**
    2. Larger leaves mostly 3–6 cm long, 2–3.5 cm wide, the lower leaf surface densely short-pubescent (longer trichomes ca. 2 mm long), glaucescent; calyx not glaucous; plants 7.5–20 dm tall............... ....**G. tomentosa**

In Florida, *G. frondosa* is a rare panhandle species, *G. nana* is common in the northern counties and central peninsula, while *G. tomentosa* is common in the northern and central peninsula, west to the central panhandle.

Gaylussacia mosieri Small [Commemorates Charles A. Mosier (1871–1936), first superintendent of Royal Palm Hammock State Park, who often collaborated with John Kunkel Small in his explorations in Florida.] WOOLLY HUCKLEBERRY.

Gaylussacia mosieri Small, Torreya 27: 36. 1927. TYPE: FLORIDA.

Shrub, to 1(1.5) m; branchlets of current season pale green, often finely glandular-pubescent with reddish trichomes. Leaves with the blade ovate to oblong, 1.5–5 cm long, 0.7–2.5 cm wide, subcoriaceous, the apex rounded to obtuse, the base cuneate, the margin entire, the upper surface stipitate-glandular and eglandular-pubescent, the lower surface similar and also sessile-glandular, the petiole 1–2 mm long. Flowers 4–8 in a raceme 3–6 cm long, often glandular pubescent, the pedicel 5–7 mm long, stipitate-glandular-pubescent; bracts foliaceous, 5–12 mm long, longer than the pedicels, stipitate-glandular- and eglandular-pubescent; bracteoles 1–2, 3–6 mm long; sepal lobes 2–3 mm long, stipitate-glandular-pubescent; corolla campanulate, 7–8 mm long, white or pinkish, the petals 4–5, the lobes triangular, 1–2 mm long; filaments 3–4 mm long, glabrous, the anthers 2–4 mm long, the thecae divergent distally; ovary stipitate-glandular-pubescent. Fruit subglobose, 6–9 mm long, black, stipitate-glandular-pubescent; pyrenes ellipsoid, ca. 2 mm long, the surface papillose.

Flatwoods, bogs, titi swamps, and cypress swamp margins. Frequent; Duval and Volusia Counties, panhandle. Georgia south to Florida, west to Louisiana. Spring.

EXCLUDED TAXON

Gaylussacia baccata (Wangenheim) K. Koch—Reported for Florida by Nuttall (1843, as *Decamerium resinosum* (Aiton) Nuttall). No Florida specimens known.

Kalmia L. 1753. LAUREL

Shrubs. Leaves alternate, simple, pinnate-veined, petiolate, estipulate. Flowers in terminal panicles, axillary fascicles, racemes, or solitary, actinomorphic, bisexual; sepals 5, basally connate; petals 5, basally connate, with saccate anther pockets; stamens 10, free, the anthers dehiscent by apical slits; ovary superior, 5-carpellate and -loculate, the style 1, the stigma capitate. Fruit a septicidally dehiscent capsule; seeds numerous.

A genus of 10 species; North America, West Indies, Europe, and Asia. [Commemorates Peter Kalm (1715–1779), Swedish botanist, collector in North America.]

Kalmiella Small, 1903.

Selected reference: Liu et al. (2009).

1. Plant hirsute; corolla deep pink; leaves chartaceous ....**K. hirsuta**
1. Plant glabrous; corolla white or pink-tinged; leaves coriaceous...**K. latifolia**

Kalmia hirsuta Walter [With coarse trichomes.] WICKY; HAIRY LAUREL.

Kalmia hirsuta Walter, Fl. Carol. 138. 1788. *Chamaedaphne hirsuta* (Walter) Kuntze, Revis. Gen. Pl. 2: 399. 1891. *Kalmiella hirsuta* (Walter) Small, Fl. S.E. U.S. 886, 1336. 1903.

Shrub, to 1 m; branchlets puberulent-hispid, viscid. Leaves with the blade elliptic to ovate, 5–14 mm long, 1–8 mm wide, the apex acute to rounded-apiculate, the base cuneate, the margin entire, revolute, the upper and lower surfaces puberulent-hispid, stipitate-glandular, the petiole ca. 1 mm long, puberulent-hispid. The flowers axillary, solitary or 2–5 in a fascicle or compact raceme, the pedicel 1–1.5 cm long; sepal lobes lanceolate, 3–8 mm long, green, puberulent-hispid, stipitate-glandular; corolla saucer-shaped, the petals 8–10 mm long, connate their entire length, pink, with a ring of red spots near the anther pockets and a ring of red spots lower down, the outer surface often sparsely hirsute and stipitate-glandular on the keels, the inner surface puberulent basally; filaments 3–4 mm long, puberulent basally; ovary stipitate-glandular, the style 5–7 mm long. Fruit subglobose, 2–4 mm long, sparsely stipitate-glandular; seeds ovoid, ca. 0.5 mm long, the surface reticulate.

Flatwoods, scrub, and coastal swales. Frequent; northern counties south to Lake and Hernando Counties. South Carolina Georgia, Florida, and Alabama. Spring–summer.

Kalmia latifolia L. [With wide leaves.] MOUNTAIN LAUREL.

> *Kalmia latifolia* Linnaeus, Sp. Pl. 391. 1753. *Chamaedaphne latifolia* (Linnaeus) Kuntze, Revis. Gen. Pl. 2: 388. 1891.

Shrub, to 8 m; branchlets stipitate-glandular or glabrescent, viscid. Leaves with the blade elliptic to elliptic-lanceolate, 4–12 cm long, 1.5–5 cm wide, the apex acute, the base cuneate, the margin entire, the upper surface glabrous, the lower surface glabrous or sometimes stipitate-glandular, the petiole 1–3 cm long, glabrous or puberulent or stipitate-glandular. Flowers (12)20–40 in a terminal panicle, the pedicel 2–4 cm long; sepals oblong, 3–4 mm long, green to reddish, glabrous or stipitate-glandular; corolla saucer-shaped, the petals connate nearly their length, 2–2.5 cm long, pink to red or white, with purple spots around the anther pocket, the inner surface lightly stipitate-glandular, the outer surface puberulent; filaments 4–5 mm long, glabrous; ovary stipitate-glandular, the style 10–18 mm long. Fruit subglobose, 3–5 mm long, stipitate-glandular; seeds ovoid, ca. 1 mm long, winged, the surface reticulate.

Bluff forests and creek swamps. Occasional; Suwannee County, central and western panhandle. Maine south to Florida, west to Michigan, Indiana, Kentucky, Tennessee, Mississippi, and Louisiana; Europe. Native to North America. Spring.

Kalmia latifolia is listed as threatened in Florida (Florida Administrative Code, Chapter 5B-40).

Leucothoe D. Don 1834. DOGHOBBLE

Shrubs. Leaves alternate, simple, pinnate-veined, petiolate, estipulate. Flowers axillary, in fascicles or solitary, bracteate; sepals 5, basally connate; petals 5, basally connate; stamens 10, free, the anthers awned, dehiscent by pores; ovary superior, 5-carpellate, falsely 10-loculate, the style 1, the stigma 5-lobed. Fruit a loculicidally dehiscent capsule; seeds numerous.

A genus of 5 species; North America and Asia. [From the Greek, name of the daughter of Babylonian king Orchamus.]

Selected references: Judd et al. (2013); Tucker (2009a).

Leucothoe axillaris (Lam.) D. Don [Axillary, in reference to the inflorescence.]
COASTAL DOGHOBBLE.

Andromeda axillaris Lamarck, Encycl. 1: 157. 1783. *Andromeda walteri* Willdenow, Enum. Pl. 453.
   1809, nom. illegit. *Leucothoe axillaris* (Lamarck) D. Don, Edinburgh New Philos. J. 17: 159. 1834.
   Lyonia axillaris (Lamarck) K. Koch, Dendrologie 2: 124. 1872.
Eubotrys bracteata Nuttall, Trans. Amer. Philos. Soc., ser. 2. 8: 269. 1843. TYPE: FLORIDA: "East
   Florida," s.d., *Ware s.n.* (holotype: PH?).

Shrub, to 2 m; branchlets with scattered multicellular-glandular and sparse to moderate short
to elongate trichomes or glabrous. Leaves with the blade oblong, lanceolate, oblanceolate, or
elliptic, 3–14 mm long, 1–5(8) cm wide, the apex obtuse or acute to short-acuminate, the base
rounded or obtuse to narrowly cuneate, the margin obscurely to clearly serrate, the serra-
tions all along the margin to restricted to the distal ⅓, the upper surface with sparse unicel-
lar trichomes on the midveins and sometimes the secondary veins and/or the extreme basal
portion of the blade near the margin, the lower surface with scattered multicellular-glandular
trichomes and sparse unicellular trichomes on the proximal portion of the midvein, the petiole
5–10 mm long. Flowers 8–30(44) in an axillary raceme 1–5.5 cm long, the pedicel 2–3 mm long;
bracts ovate, 2–3(7) mm long; sepal lobes ovate, ca. 2 mm long, whitish; corolla oblong-urce-
olate, 5–8 mm long, the lobes short-deltoid; stamens 4–6 mm long, the filaments pubescent,
papillate, the anther theca divergent, each with 2 awns; ovary glabrous. Fruit subglobose, 3–4
mm long, glabrous; seeds angular, papillose.

Moist hammocks and creek swamps. Occasional; northern counties, central peninsula. Virginia south to Florida, west to Louisiana. Spring.

Lyonia Nutt., nom. cons. 1818. STAGGERBUSH

Shrubs or trees. Leaves alternate, simple, pinnate-veined, petiolate, estipulate. Flowers in axillary fascicles, panicles, racemes, or solitary, bracteate, actinomorphic, bisexual; sepals 5, basally
connate; petals 5, basally connate; stamens 10, free, dehiscent by elliptic pores; ovary 5-carpel-
late and -loculate, the style 1, the stigma capitate. Fruit a loculicidally dehiscent capsule; seeds
numerous.

A genus of about 35 species; North America, West Indies, Mexico, and Asia. [Commemo-
rates John Lyon (1765–1814), English nurseryman and American plant explorer.]

Arsenococcus Small, 1913; *Desmothamnus* Small, 1913; *Neopieris* Britton, 1913; *Xolisma* Raf.,
1919.

Selected references: Judd (1981, 2009d).

1. Corolla about twice as long as the calyx.
    2. Corolla urceolate; sepals persistent; leaves persistent, the blade with intermarginal veins ....
    .... **L. lucida**
    2. Corolla tubular; sepals deciduous during the winter; leaves deciduous, the blade lacking intermar-
    ginal veins ....**L. mariana**
1. Corolla more than 2 times as long as the calyx.

3. Inflorescence a panicle; capsule depressed, not angled; leaves deciduous, pubescent...**L. ligustrina**

3. Inflorescence an axillary cluster; capsule longer than wide, prominently angled; leaves persistent, lepidote.

    4. Irregular shrub or small tree; leaves distinctly revolute, not reduced toward the end of the flowering shoot ....**L. ferruginea**

    4. Strict shrub; leaves not revolute, reduced toward the end of the flowering shoot.....**L. fruticosa**

Lyonia ferruginea (Walter) Nutt. [Rusty-colored, in reference to the leaf surface.] RUSTY STAGGERBUSH.

Andromeda ferruginea Walter, Fl. Carol. 138. 1788. *Lyonia ferruginea* (Walter) Nuttall, Gen. N. Amer. Pl. 1: 266. 1818. *Cassandra ferruginea* (Walter) Niedenzu, Bot. Jahrb. Syst. 11: 146, 148. 1889. *Xolisma ferruginea* (Walter) A. Heller, Cat. N. Amer. Pl. 6. 1898.

Andromeda ferruginea Walter var. *arborescens* Michaux, Fl. Bor.-Amer. 1: 252. 1803. *Lyonia ferruginea* (Walter) Nuttall var. *arborescens* (Michaux) Rehder, in L. H. Bailey, Cycl. Amer. Hort. 3: 960. 1900. TYPE: FLORIDA: St. Johns Co.: St. Augustine, s.d., *Michaux s.n.* (Lectotype: P). Lectotypified by Judd (1981: 411).

Andromeda rigida Pursh, Fl. Amer. Sept. 292. 1814. *Lyonia rigida* (Pursh) Nuttall, Gen. N. Amer. Pl. 1: 266. 1818. TYPE: "Carolina to Florida."

Shrub or tree, to 6(12) m; branchlets ferruginous-pubescent when young, glabrescent in age. Leaves with the blade elliptic to obovate or ovate, 1–9 cm long, 0.5–4.5 cm wide, not reduced toward the end of the flowering shoot, coriaceous, persistent, the apex acute or obtuse to mucronate, sometimes rounded, the base cuneate, the margin entire or undulate, revolute, the upper surface at first ferruginous-lepidote, soon glabrous, the lower surface ferruginous-lepidote and -pubescent, the petiole 1–5 mm long. Flowers in a fascicle, the pedicel lepidote; bract 1, linear-lanceolate, 1–2 mm long; sepal lobes lanceolate, 1–2 mm long, lepidote; corolla urceolate, 2–4 mm long; stamen filaments 1–2 mm long, geniculate, roughened, lacking appendages or with 2 short spurs below the anther. Fruit ovoid to ellipsoid, 3–6 mm long, lepidote, pubescent; seeds ellipsoid.

Scrub and scrubby flatwoods. Frequent; northern counties, central peninsula. South Carolina, Georgia, and Florida. Spring.

Lyonia fruticosa (Michx.) G. S. Torr. [*Frutex*, shrub or bush.] COASTALPLAINS STAGGERBUSH.

Andromeda ferruginea Walter var. *fruticosa* Michaux, Fl. Bor.-Amer. 1: 252. 1803. *Xolisma fruticosa* (Michaux) Nash, Bull. Torrey Bot. Club 22: 153. 1895. *Lyonia ferruginea* (Walter) Nuttall var. *fruticosa* (Michaux) Rehder, in L. H. Bailey, Cycl. Amer. Hort. 3: 960. 1900. *Lyonia fruticosa* (Michaux) G. S. Torrey, in B. L. Robinson, Proc. Amer. Acad. Arts 51: 527. 1916.

Shrub, to 1.3(3) m; branchlets ferruginous-pubescent when young, glabrescent in age. Leaves with the blade obovate to elliptic, 0.5–6 cm long, 0.3–4 cm wide, reduced toward the end of the flowering shoot, coriaceous, persistent, the apex acute to rounded, the base cuneate, the margin entire or undulate, flat or sometimes slightly revolute distally, the upper side lepidote (soon deciduous), pubescent along the midvein, the lower surface ferruginous-lepidote, pubescent, soon glabrous, the petiole 1–5 mm long. Flowers in an axillary fascicle, the pedicel lepidote, glabrous; bract 1, linear-lanceolate; sepal lobes triangular-lanceolate, 1-2 mm long, glabrous or pubescent, lepidote; corolla urceolate, 3–5 mm long, white; stamen filaments 1–3 mm long, geniculate, roughened, lacking appendages or with 2 short spurs below the anther. Fruit ovoid to ellipsoid, 3–5 mm long, lepidote, pubescent; seeds ellipsoid.

Scrub and flatwoods. Frequent; peninsula, west to central panhandle. South Carolina, Georgia, and Florida. Spring.

Lyonia ligustrina (L.) DC. [Resembling *ligustrum* (Oleaceae).] MALEBERRY.

Vaccinium ligustrinum Linnaeus, Sp. Pl. 351. 1753. *Andromeda ligustrina* (Linnaeus) Muhlenberg, Cat. Pl. Amer. Sept. 44. 1813. *Lyonia ligustrina* (Linnaeus) de Candolle, Prodr. 7(2): 599. 1839. *Xolisma ligustrina* (Linnaeus) Britton, Mem. Torrey Bot. Club 4(2C): 135. 1894. *Arsenococcus ligustrinus* (Linnaeus) Small, Fl. Lancaster Co. 218. 1913. *Andromeda ligustrina* (Linnaeus) Muhlenberg var. *typica* Fernald, Rhodora 43: 625. 1941.
Andromeda paniculata Linnaeus var. *foliosiflora* Michaux, Fl. Bor.-Amer. 1: 255. 1803. *Andromeda frondosa* Pursh, Fl. Amer. Sept. 295. 1814. *Lyonia frondosa* (Pursh) Nuttall, Gen. N. Amer. Pl. 1: 267. 1818. *Andromeda ligustrina* (Linnaeus) Muhlenberg var. *frondosa* (Pursh) A. W. Wood, Class-Book, ed. 1861. 488. 1861, nom. illegit. *Andromeda ligustrina* (Linnaeus) Muhlenberg var. *pubescens* A. Gray, Syn. Fl. N. Amer. 2(1): 33. 1878, nom. illegit. *Andromeda paniculata* Linnaeus var. *pubescens* Dippel, Handb. Laubholzk. 1: 370. 1889, nom. illegit. *Xolisma ligustrina* (Linnaeus) Britton var. *foliosiflora* (Michaux) C. Mohr, Bull. Torrey Bot. Club 24: 24. 1897. *Xolisma foliosiflora* (Michaux) Small, Fl. S.E. U.S. 889, 1336. 1903, nom. illegit. *Lyonia ligustrina* (Linnaeus) de Candolle var. *foliosiflora* (Michaux) Fernald, Rhodora 10: 53. 1903. *Andromeda ligustrina* (Linnaeus) Muhlenberg var. *foliosiflora* (Michaux) C. K. Schneider, Ill. Handbr. Laubholzk. 2: 533. 1911. *Xolisma ligustrina* (Linnaeus) Britton var. *pubescens* Millspaugh, W. Virginia Geol. Surv. Rep. 5A: 324. 1913, nom. illegit. *Arsenococcus frondosus* (Pursh) Small, Shrubs Florida 97, 133. 1913. *Lyonia ligustrina* (Linnaeus) de Candolle var. *pubescens* Bean, Trees Shrubs Brit. Isl. 2: 64. 1914, nom. illegit.
Andromeda parabolica Veillard, in Duhamel du Monceau, Traite Arbr. Arbust. 2: 191. 1804. *Lyonia parabolica* (Veillard) K. Koch, Dendrologie 2: 119. 1872. TYPE: "dans la Georgie et la Floride."

Shrub, to 4 m; branchlets short-pubescent when young, glabrous in age. Leaves with the blade elliptic, obovate, or ovate, (1.5)2–10 cm long, 1–4(5) cm wide, membranous, deciduous to semi-persistent, the apex acute or mucronate, the base cuneate or rounded, the margin serrulate, flat to slightly revolute, the upper and lower surfaces sparsely pubescent, especially on the main veins, the petiole 1–5 mm long. Flowers in an axillary raceme of fascicles or a short raceme to somewhat paniculate, the pedicel 3–10 mm long; bracts 1–6, variable, some foliaceous, 5–30 mm long, others 1–2 mm long; sepal lobes ca. 1 mm long, sparsely pubescent; corolla urceolate, 2–4(5) mm long, white; stamen filaments 1–2 mm long, geniculate, pubescent basally, with 2 short spurs below the anther. Fruit globose to subglobose, 2–3 mm long, glabrous or pubescent; seeds ellipsoid.

Flatwoods and bogs. Frequent; northern counties, central peninsula. Maine south to Florida, west to Oklahoma and Texas. Spring–summer.

Specimens from the southeastern U.S. Coastal Plain (including Florida) have more bracteate inflorescences and have been referred to as var. *foliosiflora*. As this character intergrades with northern populations where the ranges of the two overlap in six states, we prefer to subsume the infraspecific taxon.

Lyonia lucida (Lam.) K.Koch [Shining, in reference to the upper leaf surface.] FETTERBUSH.

Andromeda lucida Lamarck, Encycl. 157. 1783. *Lyonia lucida* (Lamarck) K. Koch, Dendrologie 2: 118. 1872. *Desmothamnus lucidus* (Lamarck) Small, in Britton, N. Amer. Fl. 29(1): 64. 1914. *Pieris lucida* (Lamarck) Rehder, Mitt. Deutsch. Dendrol. Ges. 24: 266. 1915; non Léveillé, 1906. *Xolisma lucida* (Lamarck) Rehder, J. Arnold Arbor. 5: 50. 1924.

Andromeda nitida W. Bartram ex Marshall, Arbust. Amer. 8. 1785. *Pieris nitida* (W. Bartram ex Marshall) Hooker f., in Bentham & Hooker f., Gen. Pl. 2: 588. 1876. *Lyonia nitida* (W. Bartram ex Marshall) Fernald, Rhodora 10: 53. 1908. *Neopieris nitida* (W. Bartram ex Marshall) Britton, in Britton & A. Brown, Ill. Fl. N. U.S., ed 2. 2: 690. 1913. *Desmothamnus nitidus* (W. Bartram) Small, Shrubs Florida 96, 133. 1913. TYPE: "Carolina and Florida."

Andromeda marginata Veillard, in Duhamel du Monceau, Traite Arbr. Arbust. 2: 188. 1804. *Lyonia marginata* (Veillard) D. Don, Edinburgh New Philos. J. 17: 159. 1834. *Leucothoe marginata* (Veillard) Spach, Hist. Nat. Vég. 9: 482. 1840. TYPE: "La Caroline et la Floride."

Shrub, to 3(5) m; branchlets sparsely lepidote when young, soon glabrous. Leaves with the blade elliptic, obovate, or ovate, 1–8.5(10) cm long, 0.5-4(5) cm wide, coriaceous, persistent, the apex acuminate to acute or rounded, the base cuneate to rounded, the margin entire, usually revolute, the upper and lower surfaces glabrous, glandular-punctate, the petiole 1–5 mm long. Flowers in an axillary fascicle, the pedicel glandular-puberulent; bract 1, linear-lanceolate, 1–4 mm long; sepal lobes 2–8 mm long, glandular-puberulent; corolla cylindric, 5–9 mm long, pink, red, or white; stamen filaments 3–5 mm long, geniculate, roughened, with 2 well-developed spurs below the anther. Fruit ovoid to globose, 3–5 mm long, glabrous; seeds ellipsoid.

Flatwoods, bogs, and cypress ponds. Common; nearly throughout. Virginia south to Florida, west to Louisiana. Spring.

Lyonia mariana (L.) D. Don [Marine, by the sea]. PIEDMONT STAGGERBUSH.

Andromeda mariana Linnaeus, Sp. Pl. 393. 1753. *Andromeda pulchella* Salisbury, Prodr. Stirp. Chap. Allerton 289. 1796, nom. illegit. *Lyonia mariana* (Linnaeus) D. Don, Edinburgh New Philos. J. 17: 159. 1834. *Leucothoe mariana* (Linnaeus) de Candolle, Prodr. 7: 602. 1839. *Pieris mariana* (Linnaeus) Hooker f., in Bentham & Hooker f., Gen. Pl. 2: 588. 1876. *Neopieris mariana* (Linnaeus) Britton, in Britton & A. Brown, Ill. Fl. N. U.S., ed. 2. 2: 691. 1913. *Xolisma mariana* (Linnaeus) Rehder, J. Arnold Arbor. 5: 51. 1924.

Shrub, to 1.5 m; branchlets glabrous. Leaves with the blade elliptic, ovate, or obovate, (2.5)3–8 (10) cm long, 1–4(5) cm wide, membranous, deciduous, the apex acute to rounded-mucronate, the base cuneate to rounded, the margin entire, flat to slightly revolute, the upper surface glabrous, the lower surface glabrous or short-pubescent on the main veins, the petiole 1–5 mm long. Flowers in an axillary fascicle; pedicel glandular-pubescent; bract 1, linear-lanceolate, 2–4 mm long; sepal lobes 3–9 mm long, glandular-pubescent; corolla cylindric, white or pink, 7–14 mm long; stamen filaments 4–7 mm long, geniculate, pubescent, especially basally, with 2 spurs below the anther. Fruit ovoid, 4–6 mm long, pubescent; seeds ellipsoid.

Flatwoods and creek swamps. Occasional; northern and central peninsula, west to central panhandle. New York and Connecticut south to Florida, also Missouri south to Louisiana, west to Oklahoma and Texas. Spring.

Monotropa L. 1753. INDIANPIPE

Achlorophyllous, heterotrophic herbs; stems and leaves absent. Flowers in racemes or solitary, bracteate, actinomorphic, bisexual; sepals 3–6, free; petals 3–6, free; nectariferous disk present; stamens 8–14, free, the anthers longitudinally dehiscent by 1 slit; ovary superior, 4- to 6-carpellate and -loculate, the style 1, the stigma umbilicate or funnelform. Fruit a basipetally, loculicidally dehiscent capsule; seeds numerous.

A genus of 2 species; North America, Mexico, Central America, and Europe. [From the Greek *monos*, one, and *tropos*, turn or direction, in reference to the flowers turned in one direction].

Monotropa is sometimes placed in the Monotropaceae by various authors.

Hypopitys J. Hill, 1756.

Selected reference: Wallace (2009a).

1. Flowers 2–11, rarely solitary; sepals and bracts similar ....**M. hypopithys**
1. Flowers solitary; sepals and bracts dissimilar ....**M. uniflora**

Monotropa hypopithys L. [*Hypo*, under, beneath, and *pitys*, pine, in reference to it growing under pines.] PINESAP; FALSE BEECHDROPS.

> *Monotropa hypopithys* Linnaeus, Sp. Pl. 387. 1753. *Hypopitys monotropa* Crantz, Inst. Rei. Herb. 2: 467. 1766. *Hypopitys multiflora* Scopoli, Fl. Carniol., ed. 2. 1: 285. 1771, nom. illegit. *Monotropa hypopithys* Linnaeus var. *hirsuta* Roth, Tent. Fl. Germ. 1: 180. 1788, nom. inadmiss. *Hypopitys europea* Nuttall, Gen. N. Amer. Pl. 1: 271. 1818, nom. illegit. *Hypopitys lutea* Gray, Nat. Arr. Brit. Pl. 2: 404. 1821, nom. illegit. *Monotropa hirsuta* Hornemann, Fors. Oecon. Plantel., ed. 3. 2: 179. 1835, nom. illegit. *Monotropa europea* Eaton & J. Wright, Man. Bot., ed. 8. 323. 1840, nom. illegit. *Hypopitys multiflora* Scopoli var. *hirsuta* Ledebour, Fl. Ross. 2: 934. 1846, nom. inadmiss. *Monotropa squamiformis* Dulac, Fl. Hautes-pyrénées 421. 1867. *Hypopitys hypopithys* (Linnaeus) Small, Mem. Torrey Bot. Club 4: 137. 1893, nom. inadmiss. *Monotropa multiflora* Fritsch, Excursionsfl. Oesterreich 426. 1897, nom. illegit.
>
> *Monotropa lanuginosa* Michaux, Fl. Bor.-Amer. 1: 266. 1803. *Hypopitys lanuginosa* (Michaux) Rafinesque, Med. Repos., ser. 3. 1: 297. 1810. *Monotropa hypopithys* Linnaeus subsp. *lanuginosa* (Michaux) H. Hara, J. Fac. Sci. Univ. Tokyo, Sect. 3, Bot. 6: 348. 1956.

Inflorescence an erect, secund raceme, 5–32 cm long, the axis yellowish to orange or reddish, flowers 2–11, rarely solitary, the pedicel nodding at anthesis, erect in fruit, finely pubescent or glabrous, sometimes glandular-pubescent; bracts differing from the sepals; flowers with the sepals absent or 4–5, spatulate to elliptic, 7–12 mm long; petals 4–5, oblong, yellowish to orange or reddish, the apex acute to rounded, the base narrowly saccate, the margin ciliate or erose, the inner surface glabrous or pubescent; lobes of the nectariferous disk 8–10, paired; stamens 8–10, the filaments sparsely pubescent, the anthers horizontal at anthesis, horseshoe-shaped, the theca of equal size; ovary 4–8 mm long, pubescent or glabrous, the style 2–10 mm long, sparsely pubescent, the stigma umbilicate, 2–3 mm in diameter, sometimes subtended by a ring of trichomes. Fruit 4- to 5-segmented, 6–10 mm long; seeds oblong-fusiform, ca. 1 mm long, usually membranous-winged.

Mesic hammocks; parasitic on roots of various trees. Rare; Marion, Lake, Walton, and Oka-loosa Counties. Nearly throughout North America; Mexico and Central America; Europe and Asia. Summer–fall.

Monotropa hypopithys is listed as endangered in Florida (Florida Administrative Code, Chapter 5B-40).

Monotropa uniflora L. [One-flowered.] INDIANPIPE.

Monotropa uniflora Linnaeus, Sp. Pl. 387. 1753. *Hypopitys uniflora* (Linnaeus) Crantz, Inst. Rei Herb. 2: 476. 1766. *Monotropa uniflora* Linnaeus var. *typica* Domin, Sitzungsber. Königl. Böhm. Ges. Wiss., Math.-Naturwiss. Cl. 1915: 4. 1915, nom. inadmiss.

Monotropa brittonii Small, J. New York Bot. Gard. 28: 7. 1927. TYPE: FLORIDA: Palm Beach Co.: near Pompano, 2 Dec 1919, *Small et al. 9219* (lectotype: NY; isolectotype: NY. Lectotypified by Wilbur and Luteyn (1978: 102).

Inflorescence an erect, solitary flower, 5–30 cm long, the axis white, the pedicel nodding at anthesis, erect in fruit, glabrous; bracts similar to the sepals; flowers with the sepals 3–6, lanceolate to oblong, 7–10 mm long; petals 3–6, obovate, 1–2 cm long, white, pinkish, or reddish, the apex rounded or slightly lacerate, the base slightly saccate, the margin entire, the inner surface with scattered trichomes; lobes of the nectariferous disk 10, elongate, curved-cylindric; stamens 8–14, the filaments glabrous or sparsely pubescent, the anthers horizontal on anthesis, transversely ellipsoid to depressed ovoid, the abaxial pair of theca smaller; ovary 6–12 mm long, glabrous or sparsely pubescent, the style 2–7 mm long, glabrous or sparsely pubescent, the stigma broadly funnelform, 2–6 mm in diameter, not subtended by a ring of trichomes. Fruit 5-segmented, 7–11 mm long; seeds ca. 1 mm long, usually membranous-winged.

Mesic hammocks and scrub; parasitic on roots of various trees. Occasional; peninsula, central and western panhandle. Nearly throughout North America; Mexico and Central America; Europe and Asia. Summer–fall.

Monotropsis Schwein. 1817. PIGMYPIPES

Achlorophyllous, heterotrophic herbs; stems and leaves absent. Flowers in racemes, bracteate, bracteolate, actinomorphic, bisexual; sepals 5, free; petals 5, basally connate; nectariferous disk present; stamens 10, free, the anthers longitudinally dehiscent by 2 slits; ovary superior, 5-carpellate, 1-loculate, the style 1, the stigma capitate, angular. Fruit indehiscent and fleshy; seeds numerous.

A genus of 2 species; North America. [*Monotropa* and the Greek *-opsis*, to resemble, in reference to its similarity to the genus *Monotropa*.]

Although recently treated as monotypic (e.g., Wunderlin, 1998), Wunderlin and Hansen (2003, 2011), and Wallace (2009b), we follow Rose (2012) in recognizing two species in the genus.

Monotropsis is sometimes placed in the Monotropaceae by various authors.

Schweinitzia Elliott, 1817.

Selected references: Rose (2012); Wallace (2009b).

Monotropsis reynoldsiae (A. Gray) A. Heller [Commemorates Mary Collins Reynolds (1852–1936), who collected around St. Augustine.] PIGMYPIPES.

> *Schweinitzia reynoldsiae* A. Gray, Proc. Amer. Acad. Arts 20: 301. 1885. *Monotropsis reynoldsiae* (A. Gray) A. Heller, Cat. N. Amer. Pl. 5. 1898. TYPE: FLORIDA: St. Johns Co.: St. Augustine, 1 Dec 1884, *Reynolds s.n.* (holotype: GH; isotype: GH, NY, P).

Inflorescence an erect, secund raceme, 4–13 cm long, the axis violet to purple, flowers 4–16(19), the pedicel nodding at anthesis, erect in fruit, glabrous; bracts ovate to deltoid, 3–7 mm long, glabrous, the margin entire or dentate; bracteoles 2, subtending the flower; sepals lanceolate-ovate, 2–6 mm long, the margin entire or dentate; corolla campanulate, 5–9 mm long, white or tinged with pink basally; stamens with the filaments pinkish, 2–5 mm long, the anthers ca. 1 mm long, yellow, usually with a network of reddish reticulations; ovary flask-shaped, 2–5 mm long, pink, the stigma white. Fruit 4–8 mm long; seeds ovoid, ca. 1 mm long.

Mesic hammocks; parasitic on roots. Rare; St. Johns, Volusia, Marion, Citrus, Hernando, and Pasco Counties. Endemic. Winter.

Monotropsis reynoldsiae is listed as endangered in Florida (Florida Administrative Code, Chapter 5B-40).

EXCLUDED TAXON

> *Monotropsis odorata* Schweinitz—Reported for Florida by Wunderlin (1998), Wunderlin and Hansen (2003, 2011), and Wallace (2009b), based on misapplication of the name to material of *M. reynoldsiae*.

Oxydendrum DC. 1839. SWAMP CRANBERRY

Trees. Leaves alternate, simple, pinnate-veined, petiolate, estipulate. Flowers in terminal panicles, bracteolate, actinomorphic, bisexual; sepals 5, basally connate; petals 5, basally connate; stamens 10, free, the anthers dehiscent by longitudinal slit-like pores; ovary superior, 5-carpellate and -loculate, the style 1, the stigma capitate. Fruit a loculicidally dehiscent capsule; seeds numerous.

A monotypic genus; North America. [From the Greek *oxys*, sour, and *dendron*, tree, in reference to the taste of the twigs and leaves.]

Selected reference: Judd (2009a).

Oxydendrum arboreum (L.) DC. [*Arboreus*, treelike.] SOURWOOD.

> *Andromeda arborea* Linnaeus, Sp. Pl. 394. 1753. *Lyonia arborea* (Linnaeus) D. Don, Edinburgh New Philos. J. 17: 159. 1834. *Oxydendrum arboreum* (Linnaeus) de Candolle, Prodr. 7: 601. 1839.

Tree, to 25(35) m; branchlets puberulent. Leaves with the blade elliptic-oblong to elliptic, ovate, or obovate, 6–23 cm long, 2–8 cm wide, chartaceous, deciduous, the apex acute to acuminate, the base cuneate or rounded, the margin irregularly serrate or serrulate, at least distally, or entire, sometimes fringed with trichomes when young, the upper surface with coarse trichomes on the midvein and lamina, the lower surface with coarse trichomes on the main veins, the petiole 1–3 cm long. Flowers 15–50 in a terminal panicle of racemes or secondary panicles, the pedicel 2–4 mm long; bracteoles 2, medial or distal, subulate; sepals connate to ca. ½ their length, the lobes lanceolate, 1–2 mm long; corolla urceolate, 4–7 mm long, the tube longer than the lobes, the surfaces puberulent; stamens with the filaments 2–4 mm long, straight, flat, pubescent; style strongly impressed into the ovary apex, the stigma capitate-truncate. Fruit narrowly ovoid, 4–9 mm long, with slightly thickened sutures; seeds narrowly oblong, tailed.

Hammocks, bluff forests, and bayheads. Occasional; panhandle. New York south to Florida, west to Indiana, Kentucky, Tennessee, Mississippi, and Louisiana. Spring.

Pieris D. Don 1834. FETTERBUSH

Shrubs or vines. Leaves alternate, simple, pinnate-veined, petiolate, estipulate. Flowers in axillary racemes, bracteate, bracteolate, actinomorphic, bisexual; sepals 5, basally connate; petals 5, basally connate; stamens 10, free, the anthers dehiscent by elliptic pores; ovary superior, 5-carpellate and -loculate, the style 1, the stigma capitate. Fruit a loculicidally dehiscent capsule; seeds numerous.

A genus of 7 species; North America, West Indies, and Asia. [Latin name for one of the Muses.]

Ampelothamnus Small, 1913.

Selected reference: Judd (2009b).

Pieris phyllyreifolia (Hook.) DC. [With leaves like *Phillyrea* (Oleaceae), a genus native to the Mediterranean region.] FETTERBUSH.

Andromeda phyllyreifolia Hooker, Icon. Pl. 2: t. 122. 1837. *Pieris phyllyreifolia* (Hooker) de Candolle, Prodr. 7: 599. 1839. *Ampelothamnus phyllyreifolius* (Hooker) Small, Shrubs Florida 96, 133. 1913. TYPE: FLORIDA: Franklin Co.: Apalachicola, s.d., *Drummond 27* (holotype: E; isotypes: E, GH?).
Andromeda croomia Torrey ex A. W. Wood, Class-Book Bot., ed. 1861. 487. 1861. TYPE: FLORIDA: Gadsden Co.: Quincy, s.d., *Wood s.n.* (holotype: NY).

Shrub or vine, to 1 m or climbing to 10 m by a flattened rhizome; branchlets with stipitate-glandular and nonglandular trichomes. Leaves ovate, elliptic, or obovate, 1.5–7 cm long, 0.5–2 cm wide, coriaceous, deciduous or persistent, the apex acute to rounded, the base cuneate to rounded, the margin entire or serrate, at least near the apex, revolute, the upper and lower surfaces short stipitate-glandular, the petiole 1–5 mm long. Flowers 3–9 in an axillary raceme, the pedicel 3–5 mm long, stipitate-glandular; bract 1, subulate, basal; bracteoles 2, subulate, a little below the calyx; sepal lobes lanceolate-triangular, 4–5 mm long, stipitate-glandular above the middle; corolla urceolate, 6–9, glabrous, the petal lobes short, recurved; stamen filaments 4–6 mm long, geniculate, glabrous; style strongly sunken into the ovary apex. Fruit subglobose, 3–4 mm long, glabrous; seeds angular-ovoid to obovoid, ca. 1 mm long, brown.

Cypress and white cedar swamps and cypress ponds. Occasional; northern counties, south to Lake County. South Carolina south to Florida, west to Mississippi. Winter–spring.

Rhododendron L. 1753.

Shrubs or trees. Leaves alternate, simple, pinnate-veined, petiolate, estipulate. Flowers in terminal racemes, zygomorphic, bisexual, bracteate, bracteolate; sepals 5, basally connate; petals 5, basally connate; stamens 5 or 10, free, the filaments unequal, the anthers dehiscent by terminal pores; ovary superior, 5-carpellate and -loculate, the style 1, the stigma capitate. Fruit a septicidally dehiscent capsule; seeds numerous.

A genus of about 1,000 species; North America, Europe, Asia, and Australia. [From the Greek *rhodon*, rose, and *dendron*, tree.]

Azalea L., nom. rej., 1753.

Selected reference: Judd and Kron (2009a).

1. Leaves with the lower surface lepidote; stamens 10 ....**R. minus**
1. Leaves with the lower surface glabrous or pubescent; stamens 5.
    2. Corolla yellow or orange-yellow ....**R. austrinum**
    2. Corolla white or pink.
        3. Calyx and pedicel eglandular (rarely with stipitate-glandular trichomes ....**R. canescens**
        3. Calyx and pedicel stipitate-glandular.
            4. Calyx, corolla tube, and pedicel with sparse short-stipitate glands ....**R. alabamense**
            4. Calyx, corolla tube, and pedicel with dense long-stipitate glands ....**R. viscosum**

Rhododendron alabamense Rehder [Of Alabama]. ALABAMA AZALEA.

Rhododendron alabamense Rehder, in E. H. Wilson & Rehder, Monogr. Azaleas 141. 1921. *Azalea alabamensis* (Rehder) Ashe, J. Elisha Mitchell Sci. Soc. 38: 90. 1922.

Shrub, to 3(5) m; bark smooth or vertically furrowed, shredding, the branchlets pubescent. Leaves with the blade ovate to obovate, 4–7(9) cm long, 1.2–2.5(3) cm wide, membranous, deciduous, the apex acute to obtuse, often mucronate, the base cuneate, the margin entire, flat, ciliate, the upper surface sparsely pubescent, the lower surface glabrous or pubescent, the petiole 2–5 mm long, with stipitate-glandular trichomes. Floral bud scales glabrous or glabrate, the margin ciliate. Flowers 6–7 in a raceme, the pedicel 4–12 mm long, with eglandular and stipitate-glandular trichomes; bracts and bracteoles similar to the bud scales; sepal lobes 0.5–3 mm long, with eglandular and stipitate-glandular trichomes, the margin pubescent; corolla funnelform, the petals 2.5–4 cm long, the tube 1.5–3 cm long, the lobes 1–2 cm long, white, sometimes pink-tinged, with a yellow blotch on the upper lip, the outer surface with stipitate-glandular and eglandular trichomes; stamens 5, 3–7 cm long. Fruit ovoid to cylindric, 1.5–2 cm long, sparsely to moderately pubescent with eglandular trichomes; seeds dorsiventrally flattened.

Hammocks. Rare; eastern and central panhandle. Tennessee, Georgia, Florida, Alabama, and Mississippi. Spring.

Rhododendron alabamense is listed as endangered in Florida (Florida Administrative Code, Chapter 5B-40).

Rhododendron austrinum (Small) Rehder [Southern.] FLORIDA FLAME AZALEA; ORANGE AZALEA.

Azalea austrina Small, FL. S.E. U.S., ed. 2. 1356, 1375. 1913. *Rhododendron austrinum* (Small) Rehder, in L. H. Bailey, Stand. Cycl. Hort. 6: 3571. 1917. TYPE: FLORIDA: Gadsden Co.: Chattahoochee, Apr & Oct, *Curtis 1718* (holotype: NY).

Shrub or tree, to 3(5) m; bark smooth or vertically furrowed, shredding, the branchlets with stipitate-glandular or eglandular trichomes. Leaves with the blade ovate to obovate, 3–11 cm long, 1.5–4.5 cm wide, membranous, deciduous, the apex acute to obtuse, often mucronate, the base cuneate, the margin entire, flat, ciliate, with stipitate-glandular and eglandular trichomes, the upper surface with sparse stipitate-glandular and eglandular trichomes, the lower surface with eglandular trichomes, the petiole 2–5 mm long, with stipitate-glandular or eglandular trichomes. Floral bud scales pubescent, the margin ciliate, partly glandular-serrate. Flowers 9–24 in a raceme; bracts and bracteoles similar to the bud scales; sepal lobes 0.5–2.5 mm long, with sparse stipitate-glandular trichomes, the margin with long stipitate-glandular trichomes; corolla funnelform, 2.5–4.5 cm long, the tube 1.5–2.5 cm long, the lobes 1–2 cm long, the petals yellow to orange, the tube often red to orange-red, with an inconspicuous darker yellow, orange, or red spot on the upper lip, the outer surface with sparse stipitate-glandular and eglandular trichomes; stamens 5, 5–7.5 cm long. Fruit ovoid to cylindric, 1.5–2.5 cm long, sparsely to densely pubescent with stipitate-glandular and eglandular trichomes; seeds dorsiventrally flattened.

Hammocks and floodplain forests. Occasional; Baker County, central and western panhandle. Georgia, Florida, Alabama, and Mississippi. Spring.

Rhododendron austrinum is listed as endangered in Florida (Florida Administrative Code, Chapter 5B-40).

Rhododendron canescens (Michx.) Sweet [Becoming gray, in reference to the leaves.] SWEET PINXTER AZALEA; MOUNTAIN AZALEA.

Azalea canescens Michaux, Fl. Bor.-Amer. 1: 150. 1803. *Azalea rosea* Loiseleur-Deslongchamps, in Duhamel du Monceau, Traite Arbr. Arbust. 5: 224. 1812, nom Illegit. *Azalea nudiflora* Linnaeus var. *rosea* Sweet, Hort. Brit. 265. 1826. *Rhododendron nudiflorum* (Linnaeus) Torrey var. *roseum* (Sweet) Sweet, Hort. Brit., ed. 2. 344. 1830. *Rhododendron canescens* (Michaux) Sweet, Hort. Brit., ed. 2. 343. 1830. *Azalea nudiflora* Linnaeus var. *canescens* (Michaux) Rehder, in L. H. Bailey, Cycl. Amer. Hort. 1: 121. 1900, nom. illegit. *Rhododendron roseum* (Sweet) Rehder, in E. H. Wilson & Rehder, Monogr. Azaleas 138. 1921, nom. illegit.

Shrub or tree, to 6 m; bark smooth or vertically furrowed, shredding, the branchlets with scattered eglandular trichomes. Leaves with the blade ovate to obovate, 2.5–10(13) cm long, 1–3(4) cm wide, membranous to chartaceous, deciduous, the apex acute to obtuse, usually mucronate, the base cuneate, the margin entire or minutely serrulate, inconspicuously ciliate, the upper and lower surfaces with scattered eglandular trichomes or glabrous, the petiole 2–5 mm long, with eglandular trichomes. Floral bud scales pubescent, the margin often ciliate. Flowers 6–19 in a raceme, the pedicel 4–17 mm long, with eglandular (rarely with stipitate-glandular) trichomes; bracts and bracteoles similar to the bud scales; sepal lobes 0.5–4 mm long, the surface and margin with eglandular (rarely with glandular) trichomes; corolla funnelform, 2.5–4.5 cm long, the tube 1.5–2.5 cm long, the lobes 1–2 cm long, pink to white with a pink tube, the outer surface with scattered stipitate-glandular and eglandular trichomes; stamens 5, 3–6 cm long. Fruit ovoid to cylindric, 1.5–3 cm long, with eglandular trichomes; seeds dorsiventrally flattened.

Flatwoods, bay swamps, hammocks, and floodplain forests. Frequent; northern counties, Marion County. North Carolina south to Florida, west to Illinois, Kentucky, Arkansas, Oklahoma, and Texas. Spring.

Rhododendron minus Michx. var. **chapmanii** (A. Gray) W. H. Duncan & Pullen [Less, in reference to its small stature; commemorates Alvan Wentworth Chapman (1809–1899), physician and botanist.] CHAPMAN'S RHODODENDRON.

> *Rhododendron chapmanii* A. Gray, Proc. Amer. Acad. Arts 12: 61. 1876. *Azalea chapmanii* (A. Gray) Kuntze, Revis. Gen. Pl. 2: 387. 1891. *Rhododendron minus* Michaux var. *chapmanii* (A. Gray) W. H. Duncan & Pullen, Brittonia 14: 297. 1962. TYPE: FLORIDA: s.d., *Chapman s.n.* (holotype: GH; isotype: NY).
>
> *Rhododendron punctatum* Andrews var. *chapmanii* A. W. Wood, Amer. Bot. Fl. 204. 1870. TYPE: FLORIDA.

Shrub, to 3 m; bark smooth or vertically furrowed, shredding, the branchlets ferruginous-lepidote. Leaves with the blade elliptic or obovate, 1.5–6.5 cm long, 1–3 cm wide, coriaceous, persistent, the apex obtuse to rounded, rarely notched or acute, the base cuneate, the margin entire, flat or revolute, the upper surface scattered ferruginous-lepidote, soon glabrous, dark-dotted, the lower surface ferruginous-lepidote, the petiole 2–6(7) mm long, ferruginous-lepidote and with simple trichomes. Floral bud scales ferruginous-lepidote, sometimes also with simple trichomes, the margin ciliate. Flowers 5–10 in a raceme, the pedicel 5–15 mm long, ferruginous-lepidote; bracts and bracteoles similar to the bud scales; sepal lobes 1–2 mm long, ferruginous-lepidote; corolla campanulate to funnelform, 2.3–3.7 cm long, the tube 1–2 cm long, the lobes 0.5–1.5 cm long, dark to pale pink or white, often with greenish spots, the outer surface scattered ferruginous-lepidote; stamens 10, 1.5–2.5 cm long. Fruit ovoid to cylindric, 0.5–1.5 cm long, ferruginous-lepidote; seeds dorsiventrally flattened, each end with a short, truncate tail.

Flatwoods. Rare; Clay County, central panhandle. Endemic. Spring.

Rhododendron minus var. *chapmanii* is listed as endangered in Florida (Florida Administrative Code, Chapter 5B-40) and in the United States (U.S. Fish and Wildlife Service, 50 CFR 23).

Rhododendron viscosum (L.) Torr. [Glutinous, in reference to the stipitate-glandular trichomes.] SWAMP AZALEA.

> *Azalea viscosa* Linnaeus, Sp. Pl. 151. 1753. *Rhododendron viscosum* (Linnaeus) Torrey, Fl. N. Middle United States 424. 1824. *Anthodendron viscosum* (Linnaeus) Reichenbach, in Moessler, Handb. Gewaechsk., ed. 2. 1: 309. 1827.
>
> *Azalea serrulata* Small, Fl. S.E. U.S. 883, 1336. 1903. *Rhododendron serrulatum* (Small) Millais,

Rhododendrons 241. 1917. *Rhododendron viscosum* (Linnaeus) Torrey var. *serrulatum* (Small) Ahles, J. Elisha Mitchell Sci. Soc. 80: 173. 1964. TYPE: FLORIDA: Lake Co.: vicinity of Eustis, 1–15 Jun 1894, *Nash 967* (holotype: NY).

Rhododendron serrulatum (Small) Millais forma *molliculum* Rehder, in E. H. Wilson & Rehder, Monogr. Azaleas 155. 1921. TYPE: FLORIDA: Lake Co.: Eustis, 23 June 1919, *Harbison 17* (holotype: A?).

Shrub or tree, to 7 m; bark smooth or vertically furrowed, shredding, the branchlets pubescent. Leaves with the blade ovate to obovate, 2–7(9) cm long, 1–3(3.5) cm wide, membranous to chartaceous, deciduous, the apex acute to obtuse, usually mucronate, the base cuneate, the margin entire or minutely serrulate, ciliate, the upper surface glabrous or with scattered trichomes, the lower surface glabrous or pubescent, the petiole 3–10 mm long, pubescent. Floral bud scales glabrous or pubescent, the margin eglandular- or glandular-ciliate. Flowers 3–14 in a raceme, the pedicel 0.5–2.5 cm long, with stipitate-glandular or eglandular trichomes; bracts and bracteoles similar to the bud scales; sepal lobes 0.5–5 mm long, with sparse stipitate-glandular or eglandular trichomes; corolla funnelform, 2–5.5 cm long, the tube 1.3–3.5 cm long, the lobes 0.7–2 cm long, white or pink, the outer surface with long stipitate-glandular trichomes; stamens 5, 2–6.5 cm long. Fruit ovoid to cylindric, 0.7–2 cm long, with stipitate-glandular trichomes; seeds dorsiventrally flattened.

Wet flatwoods and swamps. Frequent; northern counties, central peninsula. Maine south to Florida, west to Oklahoma and Texas. Spring–summer.

EXCLUDED TAXA

Rhododendron canescens (Michaux) Sweet var. *candidum* (Small) Rehder—Reported for Florida by Small (1933, as *Azalea candida* Small), probably based on a misidentification. All known Florida material is of var. *canescens*.

Rhododendron nudiflorum (Linnaeus) Torrey—Reported for Florida by Chapman (1860) and Small (1903, 1913a, 1913e, as *Azalea nudiflora* Linnaeus). No Florida specimens known. Small's 1933 work excludes Florida from the plant's range.

Rhododendron minus Michaux—Reported for Florida by Chapman (1860, as *R. punctatum* Andrews), the name misapplied to material of *R. minus* var. *chapmanii*.

Vaccinium L. 1753. BLUEBERRY

Shrubs or trees. Leaves alternate, simple, pinnate-veined, petiolate or epetiolate, estipulate. Flowers in terminal or axillary racemes or fascicles, sometimes solitary, bracteate, actinomorphic, bisexual; sepals 5, basally connate; petals (4)5, basally connate; stamens 8 or 10, free, the anthers dehiscent by pores; ovary inferior, (4)5-carpellate and -loculate or falsely 10-loculate, the style 1, the stigma capitate. Fruit a berry; seeds 5–many.

A genus of about 500 species; North America, West Indies, Mexico, Central America, Europe, and Asia. [Ancient Latin name for the blueberry, perhaps derived from *bacca*, berry.]

Vaccinium is not monophyletic and *Gaylussacia* is apparently embedded within it (W. S. Judd, pers. comm.). More study is needed.

Batodendron Nutt., 1842; *Cyanococcus* (A. Gray) Rydb., 1918; non Hansg., 1905; *Picrococcus* Nutt., 1843; *Polycodium* Raf., 1818.

Selected reference: Vander Kloet (2009).

1. Young branchlets verrucose.
   2. Leaves deciduous, the largest ones more than 1.5 cm long...**V. corymbosum**
   2. Leaves persistent, the largest ones less than 1.5 cm long.
      3. Leaves with stipitate glands on the lower surface...**V. myrsinites**
      3. Leaves without stipitate glands on the lower surface...**V. darrowii**
1. Young branchlets not verrucose.
   4. Leaves rarely glandular, truncate at the base, the upper surface dull, the lower surface strongly glaucous; corolla open in bud; anthers exserted...**V. stamineum**
   4. Leaves often stipitate-glandular, cuneate at the base, the upper surface lustrous, the lower surface light green; corolla closed in bud; anthers included...**V. arboreum**

Vaccinium arboreum Marshall [*Arboreus*, treelike.] SPARKLEBERRY; FARKLEBERRY.

Vaccinium arboreum Marshall, Arbust. Amer. 157. 1785. *Batodendron arboreum* (Marshall) Nuttall, Trans. Amer. Philos. Soc., ser. 2. 8: 261. 1843.
Batodendron speciosum Greene, Pittonia 3: 326. 1898. TYPE: FLORIDA.

Shrub or tree, to 10(14) m; bark thin, grayish brown, shredding in irregular, thin plates or flakes, exposing the smooth, reddish brown inner bark, the branchlets reddish green, glabrous, glaucous, or glandular-pubescent. Leaves with the blade obovate, elliptic, ovate, or suborbicular, 2–4(7) cm long, 1–2.5 cm wide, coriaceous, tardily deciduous, the apex rounded to obtuse, mucronate, the base cuneate to rounded, the margin entire, with small glands, the upper surface glabrous or pubescent along the midrib, the lower surface pubescent along the midrib and the major veins or over throughout, short-petiolate or subsessile. Flowers in racemes 2.5–7.5 cm long or solitary, the pedicel ca. 1 mm long; bracts 2, near the middle of the pedicel; calyx campanulate, closed in bud, 1–2 mm long, glabrous, the sepal lobes deltoid, pubescent distally; corolla campanulate, the tube 3–4 mm long, the lobes oblong, 2–3 mm long, usually recurved; stamens shorter than the petals, the filaments pubescent, the anthers with dorsal awns. Fruit subglobose, 6–9 mm long, black, lustrous; seeds ellipsoid, the surface reticulate.

Hammocks, flatwoods, and scrub. Frequent; northern counties, central peninsula. Virginia south to Florida, west to Kansas, Oklahoma, and Texas. Spring.

Vaccinium corymbosum L. [Flowers and/or fruits in a corymb.] HIGHBUSH BLUEBERRY.

Vaccinium corymbosum Linnaeus, Sp. Pl. 350. 1753. *Cyanococcus corymbosus* (Linnaeus) Rydberg, Brittonia 1: 94. 1931. *Vaccinium corymbosum* Linnaeus forma *typicum* Camp, Amer. Midl. Naturalist 23: 177. 1940, nom. inadmiss.
Vaccinium amoenum Aiton, Hort. Kew. 2: 12. 1789. *Vaccinium corymbosum* Linnaeus var. *amoenum* (Aiton) A. Gray, Manual, ed. 2. 250. 1856. *Cyanococcus amoenus* (Aiton) Small, Man. S.E. Fl. 1014, 1506. 1933.

Vaccinium fuscatum Aiton, Hort. Kew. 2: 11. 1789. *Vaccinium corymbosum* Linnaeus var. *fuscatum* (Aiton) Hooker, Bot. Mag. 62: t. 3433. 1835. *Cyanococcus fuscatus* (Aiton) Small, Man. S.E. U.S. 1013, 1506. 1933.

Vaccinium virgatum Aiton, Hort. Kew. 2: 12. 1789. *Cyanococcus virgatus* (Aiton) Small, Man. S.E. Fl. 1014, 1506. 1933.

Vaccinium formosum Andrews, Bot. Repos. 2: pl. 97. 1800.

Vaccinium corymbosum Linnaeus var. *atrocarpum* A. Gray, Manual, ed. 2. 250. 1856. *Vaccinium corymbosum* Linnaeus var. *atrococcum* A. Gray, Manual, ed. 2. rev. 250. 1857, nom. illegit. *Cyanococcus atrococcus* Small, Man. S.E. Fl. 1014, 1507. 1933.

Vaccinium elliottii Chapman, Fl. South. U.S. 260. 1860. *Vaccinium virgatum* Aiton var. *parvifolium* A. Gray, Syn. Fl. N. Amer. 2(1): 22. 1878. *Cyanococcus elliottii* (Chapman) Small, Man. S.E. Fl. 1014, 1506. 1933. TYPE: "Florida to South Carolina."

Vaccinium atrococcum A. Heller, Bull. Torrey Bot. Club 21: 24. 1894.

Vaccinium australe Small, Fl. S.E. U.S. 895, 1335. 1903.

Vaccinium caesariense Mackenzie, Torreya 10: 230. 1910. *Vaccinium corymbosum* Linnaeus forma *caesariense* (Mackenzie) Camp, Amer. Midl. Naturalist 23: 177. 1940.

Vaccinium arkansanum Ashe, Rhodora 33: 195. 1931.

Vaccinium fuscatum Aiton var. *pullum* Ashe, Rhodora 33: 196. 1931. TYPE: FLORIDA: Marion Co.: vicinity of Sellers Lake, 14 May 1923, *Ashe s.n.* (holotype: NCU?).

Vaccinium ashei Reade, Torreya 31: 71. 1931. TYPE: FLORIDA: Okaloosa Co.: near Niceville.

Cyanococcus holophyllus Small, Man. S.E. Fl. 1015, 1506. 1933. *Vaccinium holophyllum* (Small) Uphof, Mitt. Deutsch. Dendrol. Ges. 48: 19. 1936. TYPE: FLORIDA: Highlands Co.

Shrub, to 3(5) m; branchlets green, yellowish, or reddish, glabrous or pubescent, sometimes glaucous, rarely glandular. Leaves with the blade ovate, oblong, elliptic, lanceolate, or spatulate, 3–7(8) cm long, 1–2.5(4.5) cm wide, subcoriaceous, deciduous or tardily deciduous, the apex acute to obtuse, the base cuneate, the margin entire or serrate, the upper surface glabrous, the lower surface glabrous or pubescent, sometimes glandular, short-petiolate to subsessile. Flowers in a short raceme or a fascicle, the pedicel 1–2 mm long; bracts 2, basal; calyx campanulate, 3–4 mm long, glabrous, the sepal lobes deltoid, 1–2 mm long; corolla subcylindric to urceolate, closed in bud, 5–12 mm long, pink to white, the lobes short, ca. 1 mm long; stamens shorter than the petals, the filaments pubescent. Fruit subglobose, 4–12 mm long, blue to black, glabrous, glaucous; seeds ellipsoid, the surface reticulate.

Hammocks, flatwoods, swamps, and bayheads. Frequent; nearly throughout. Nova Scotia, New Brunswick, and Quebec south to Florida, west to Ontario, Wisconsin, Missouri, Oklahoma, and Texas, also British Columbia and Washington; Europe. Native to North America. Spring.

Vaccinium elliottii is diploid and apparently reproductively isolated from the rest of *V. corymbosum*, which is tetrapoid. It may represent a distinct species, but more study is needed. It is tentatively placed in synonymy here until the situation can be resolved.

Vaccinium darrowii Camp [Commemorates George McMillan Darrow (1889–1983), United States Department of Agriculture.] DARROW'S BLUEBERRY.

Vaccinium myrsinites Lamarck var. *glaucum* A. Gray, Syn. Fl. N. Amer. 2(1): 21. 1878. *Vaccinium darrowii* Camp, Bull. Torrey Bot. Club 69: 240. 1942.

Shrub, to 6(15) dm; branchlets green, puberulent, glaucous. Leaves with the blade oblanceolate or elliptic, 0.5–1.5 cm long, 2–10 mm wide, coriaceous, persistent, the apex acute to obtuse, the base cuneate, the margin entire or with low, appressed-ascending, gland-tipped teeth, often revolute, the upper surface short-pubescent along the midrib or the entire surface with short, appressed, gray trichomes, the lower surface sparsely pubescent along the midrib, sessile or subsessile. Flowers 2–8 in a fascicle or short raceme; the pedicel 3–5 mm long; bracts 2, basal; calyx greenish, glabrous, sometimes glaucous, the tube ca. 2 mm long, the sepal lobes triangular, deltoid, or broadly rounded, ca. 1 mm long, the margin sometimes short-ciliate; corolla urceolate, 4–6(8) mm long, closed in bud, white or pink, the lobes short-triangular or oblong-triangular; stamens shorter than the petals, the filaments pubescent. Fruit subglobose, 6–8 mm long, blue, glabrous, glaucous; seeds ellipsoid, the surface reticulate.

Flatwoods and sandhills. Common; nearly throughout. Georgia south to Florida, west to Louisiana. Spring.

Vaccinium myrsinites Lam. [Resembling some species of *Myrsine* (Myrsinaceae).] SHINY BLUEBERRY.

Vaccinium myrsinites Lamarck, Encycl. 1: 73. 1783. *Vaccinium corymbosum* Linnaeus var. *myrsinites* (Lamarck) Castiglioni, Viagg. Stati Uniti 2: 391. 1790. *Cyanococcus myrsinites* (Lamarck) Small, Man. S.E. Fl. 1013, 1506. 1933. TYPE: FLORIDA: St. Johns Co.: near St. Augustine, 1779, *Lamarck 116* (holotype: P).
Vaccinium nitidum Andrews, Bot. Repos. 7: pl. 480. 1807.

Shrub, to 6(10) dm; branchlets green, glabrous, or puberulent. Leaves with the blade oblanceolate or elliptic, 0.5–1.5 cm long, 2–10 mm wide, coriaceous, persistent, the apex acute to obtuse, the base cuneate, the margin entire or with low, appressed-ascending, gland-tipped teeth, often revolute, the upper surface with sparse stipitate-glandular trichomes or glabrous, the lower surface sparsely pubescent along the midrib, with stipitate-glandular trichomes, sessile or subsessile. Flowers 2–8 in a fascicle or short raceme; the pedicel 3–5 mm long; bracts 2, basal; calyx greenish, glabrous, sometimes glaucous, the tube ca. 2 mm long, the sepal lobes triangular, deltoid, or broadly rounded, ca. 1 mm long, the margin sometimes short-ciliate; corolla urceolate, 4–6(8) mm long, closed in bud, white or pink, the lobes short-triangular or oblong-triangular; stamens shorter than the petals, the filaments pubescent. Fruit subglobose, 6–8 mm long, black or blue-black, glabrous, glaucous; seeds ellipsoid, the surface reticulate.

Flatwoods and sandhills. Common; nearly throughout. South Carolina south to Florida, west to Alabama. Spring.

Vaccinium stamineum L. [Stamens, in reference to the stamens being exserted in the flower.] DEERBERRY.

Vaccinium stamineum Linnaeus, Sp. Pl. 350. 1753. *Picrococcus stamineus* (Linnaeus) Nuttall, Trans. Amer. Philos. Soc., ser. 2. 8: 261. 1843. *Polycodium stamineum* (Linnaeus) Greene, Pittonia 3: 324. 1898.
Picrococcus floridanus Nuttall, Trans. Amer. Philos. Soc., ser. 2. 8: 262. 1843. *Polycodium floridanum*

(Nuttall) Greene, Pittonia 3: 325. 1898. *Vaccinium floridanum* (Nuttall) Sleumer, Bot. Jahrb. Syst. 71: 425. 1941. TYPE: FLORIDA: s.d., *Cooper s.n.*

Vaccinium caesium Greene, Pittonia 3: 249. 1897. *Polycodium caesium* (Greene) Greene, Pittonia 3: 325. 1898. *Polycodium floridanum* (Nuttall) Greene var. *caesium* (Greene) Ashe, J. Elisha Mitchell Sci. Soc. 46: 203. 1931. *Vaccinium floridanum* (Nuttall) Sleumer var. *caesium* (Greene) Sleumer, Bot. Jahrb. Syst. 71: 425. 1941. *Vaccinium stamineum* Linnaeus var. *caesium* (Greene) D. B. Ward, Castanea 39: 200. 1974. TYPE: FLORIDA: Lake Co.: vicinity of Eustis, 1–15 May 1894, *Nash 473* (holotype: ND-G).

Vaccinium revolutum Greene, Pittonia 3: 250. 1897. *Polycodium revolutum* (Greene) Greene, Pittonia 3: 325. 1898. *Polycodium floridanum* (Nuttall) Greene var. *revolutum* (Greene) Ashe, J. Elisha Mitchell Sci. Soc. 46: 203. 1931. *Vaccinium floridanum* (Nuttall) Sleumer var. *revolutum* (Greene) Sleumer, Bot. Jahrb. Syst. 71: 425. 1941. TYPE: FLORIDA: Lake Co.: Eustis, 12–31 Mar 1894, *Nash 53* (Holotype: ND-G).

Vaccinium stamineum Linnaeus var. *melanocarpum* C. Mohr, Bull. Torrey Bot. Club 24: 25. 1897. *Vaccinium melanocarpum* (C. Mohr) C. Mohr ex Kearney, Bull. Torrey Bot. Club 24: 570. 1897. *Polycodium melanocarpum* (C. Mohr) Small, Fl. S.E. U.S. 894, 1336. 1903.

Vaccinium melanocarpum (C. Mohr) C. Mohr ex Kearney var. *sericeum* C. Mohr, Contr. U.S. Natl. Herb. 6: 658. 1901. *Polycodium sericeum* (C. Mohr) C. B. Robinson, Bull. Torrey Bot. Club 39: 559. 1912. *Vaccinium sericeum* (C. Mohr) E. J. Palmer, J. Arnold Arbor. 13: 429. 1932. *Vaccinium stamineum* Linnaeus var. *sericeum* (C. Mohr) D. B. Ward, Castanea 39: 201. 1974.

Polycodium neglectum Small, Fl. S.E. U.S. 893, 1336. 1903. *Vaccinium neglectum* (Small) Fernald, Rhodora 10: 53. 1908. *Vaccinium stamineum* Linnaeus var. *neglectum* (Small) Deam, Shrubs Indiana, ed. 2. 288. 1932.

Polycodium oliganthum Greene, Leafl. Bot. Observ. Crit. 2: 226. 1912. TYPE: FLORIDA: Miami-Dade Co.: Lemon City, 23 May 1901, *Tracy 7264* (holotype: ND-G).

Polycodium quercinum Ashe, Bull. Torrey Bot. Club 54: 580. 1927. *Vaccinium quercinum* (Ashe) Sleumer, Bot. Jahrb. Syst. 71: 425. 1941. TYPE: FLORIDA: Marion Co.: Lynne, 10 Jun 1925, *Ashe s.n.* (holotype: NCU?).

Polycodium depressum Small, Torreya 28: 5. 1928. *Vaccinium depressum* (Small) Sleumer, Bot. Jahrb. Syst. 71: 426. 1941.

Polycodium bellum Ashe, J. Elisha Mitchell Sci. Soc. 46: 209. 1931. *Vaccinium bellum* (Ashe) Sleumer, Bot. Jahrb. Syst. 71: 426. 1941.

Polycodium concoloratum Ashe, J. Elisha Mitchell Sci. Soc. 46: 204. 1931. *Vaccinium concoloratum* (Ashe) Sleumer, Bot. Jahrb. Syst. 71: 425. 1941. TYPE: FLORIDA: Seminole Co.: near Sanford, s.d., *Ashe s.n.* (holotype: NCU?).

Polycodium glandulosum Ashe, J. Elisha Mitchell Sci. Soc. 46: 201. 1931. *Vaccinium glandulosum* (Ashe) Sleumer. Bot. Jahrb. Syst. 71: 425. 1941. *Vaccinium stamineum* Linnaeus var. *glandulosum* (Ashe) D. B. Ward, Castanea 39: 201. 1974. TYPE: FLORIDA: Wakulla Co.

Polycodium lautum Ashe, J. Elisha Mitchell Sci. Soc. 46: 205. 1931. *Vaccinium lautum* (Ashe) Sleumer, Bot. Jahrb. Syst. 71: 425. 1941. TYPE: FLORIDA: Washington Co.: E of Holmes Valley, 6 Jun 1929, *Ashe s.n.* (holotype: NCU?).

Polycodium multiflorum Ashe, J. Elisha Mitchell Sci. Soc. 46: 204. 1931. *Vaccinium semipersistens* Sleumer, Bot. Jahrb. Syst. 71: 425. 1941. *Vaccinium stamineum* Linnaeus var. *multiflorum* (Ashe) D. B. Ward, Castanea 39: 201. 1974. TYPE: FLORIDA: Okaloosa Co.

Polycodium multiflorum Ashe var. *uniquum* Ashe, J. Elisha Mitchell Sci. Soc. 46: 205. 1931. *Vaccinium semipersistens* Sleumer var. *uniquum* (Ashe) Sleumer, Bot. Jahrb. Syst. 71: 425. 1941. TYPE: FLORIDA.

Shrubs, to 5 m; bark brownish gray, shredding in irregular flakes, the branchlets glabrous or pubescent, sometimes stipitate-glandular, usually glaucous. Leaves with the blade membranous, deciduous or tardily deciduous, elliptic, ovate, oblong, obovate, or suborbicular, 2–8 cm long, 1–3 cm wide, the apex acute to acuminate, mucronate, the base cuneate to rounded or sub-cordate, the margin entire or irregularly serrulate, often revolute, usually stipitate-glandular proximally, the upper and lower surfaces glabrous or variously pubescent with simple and sometimes stipitate-glandular trichomes, often glaucous, sessile or subsessile. Flowers 2–7 in a fascicle or short raceme, the pedicel 1–2 cm long; bracts 2, basal; calyx greenish, glabrous, the tube 2–4 mm long, the sepal lobes triangular or deltoid, ca. 1 mm long, the margin ciliate; corolla campanulate, 4–6(8) mm long, open in bud, white to greenish white, the lobes oblong-triangular, 1–3 mm long, ⅓–½ as long as the corolla; stamens longer than the petals, the filaments pubescent. Fruit subglobose or pyriform, 6–16 mm long, whitish, yellowish, reddish, blue, or purple, glabrous or sparsely pubescent, glaucous; seeds ellipsoid, the surface reticulate.

Hammocks, flatwoods, and sandhills. Common; northern counties, central peninsula. Massachusetts and New York south to Florida, west to Ontario, Illinois, Kansas, Oklahoma, and Texas; Mexico. Spring.

EXCLUDED TAXA

Vaccinium crassifolium Andrews—Reported for Florida by Nuttall (1843, as *Metagonia myrtifolia* (Michaux) Nuttall). No Florida specimens known.

Vaccinium tenellum Aiton—Florida reports of this species such as those of Chapman (1860), Small (1903, 1913a, and 1933, all as *Cyanococcus tenellus* (Aiton) Small), Correll and Johnston (1970), Wunderlin (1998), and Wunderlin and Hansen (2003) are based on misidentification of material of *Gaylussacia dumosa*, fide Ward and Lyrene (2007). Also reported by Vander Kloet (2009).

EXCLUDED GENERA

Leiophyllum buxifolium (Bergius) Elliott—Reported for Florida by Small (1903, 1913a, as *Dendrium buxifolium* (Bergius) Desvaux; 1933). No Florida specimens known.

Zenobia pulverulenta (W. Bartram ex Willdenow) Pollard—Reported for Florida by Small (1903, 1913a, 1913e, 1933), probably based on Willdenow's (1799) erroneous citation of Florida as the type locality, followed by Pollard (1895). *Zenobia cassinifolia* (Ventenat) Pollard, a synonym of *Zenobia pulverulenta* was also reported for Florida by Small (1903, 1913a, 1913e, 1933), probably based on earlier erroneous citations, such as Ventenat (1801) in the original description and by Pollard (1895). Also *Andromeda speciosa* Michaux, a heterotypic synonym of *Zenobia pulverulenta*, was reported by Chapman (1860, 1883, 1897). *Zenobia* has never been found in Florida.

RUBIACEAE Juss., nom. cons. 1789. MADDER FAMILY

Herbs, shrubs, trees, or woody vines. Leaves opposite or whorled, simple, pinnate-veined, petiolate or epetiolate, stipulate. Flowers in terminal or axillary cymes, heads, panicles, or solitary, actinomorphic, bisexual or unisexual (plant monoecious), bracteate, bracteolate or ebracteolate; sepals 4–6, basally connate or free; petals 4–6, basally connate; stamens 4–6, epipetalous; ovary inferior or half-inferior, 2- to 3-carpellate and -loculate. Fruit a capsule, schizocarp, drupe, berry, or nutlet; seeds 1–many.

A family of about 600 genera and about 13,000 species; nearly cosmopolitan.

The circumscription of many herbaceous genera (e.g., *Diodia*, *Houstonia*, *Oldenlandia*, *Spermacoce*, *Sternaria*) are very likely to change in the coming years.

Selected reference: Rogers (2005).

1. Herb (sometimes suffrutescent).
  2. Leaves whorled.
    3. Flowers in involucrate heads.. **Sherardia**
    3. Flowers in cymes .... **Galium**
  2. Leaves opposite.
    4. Carpels few- to many-seeded.
      5. Fruit a berry.. **Coccocypselum**
      5. Fruit a capsule.
        6. Flowers primarily 5-merous.
          7. Plant glabrous; flowers axillary; calyx lobes equal; corolla tube 2–3 mm long...............
          .... **Pentodon**
          7. Plant pubescent; flowers terminal; calyx lobes unequal; corolla tube to ca. 2 cm long
          .... **Pentas**
        6. Flowers primarily 4-merous.
          8. Seeds crateriform (with a ventral depression and a linear hilar ridge within or a ventral subglobose cavity lacking a hilar ridge)... **Houstonia**
          8. Seeds noncrateriform.
            9. Seeds trigonous..**Oldenlandia**
            9. Seeds ellipsoid or subglobose.
              10. Leaves filiform, linear, narrowly lanceolate to oblanceolate; capsule longer than the hypanthium..**Stenaria**
              10. Leaves suborbicular to ovate; capsule not exceeding the hypanthium...............
              .... **Oldenlandiopsis**
    4. Carpels 1-seeded.
      11. Flowers with the ovaries of 2 of a pair fused ....**Mitchella**
      11. Flowers separate.
        12. Flowers in dense, terminal, involucrate heads .... **Richardia**
        12. Flowers in axillary or terminal clusters, but not involucrate.
          13. Fruit separating into 2 parts... **Diodia**
          13. Fruit not separating into 2 parts.
            14. Carpels opening longitudinally along the inner surface.....................**Spermacoce**
            14. Carpels opening transversely ....**Mitracarpus**
1. Tree, shrub (sometimes scrambling or reclining), or woody vine.
  15. Flowers or fruits in a capitate inflorescence.
    16. Fruit dry...**Cephalanthus**
    16. Fruit fleshy ....**Morinda**
  15. Flowers or fruits solitary or in an open inflorescence.
    17. Plant with axillary spines.
      18. Leaves 2–5 cm long; corolla lobes 5; fruit 6–13 mm long..............................**Randia**
      18. Leaves to 1 cm long; corolla lobes 4; fruit 2–4 mm long ....**Catesbaea**
    17. Plant lacking axillary spines.

19. Some flowers with 1–2 sepals greatly enlarged and petaloid (at least in the outer flowers of
  the inflorescence)..**Pinckneya**
19. Flowers not as above.
  20. Plant a vine or a scrambling or reclining shrub.
    21. Flower or fruit sessile or subsessile ....**Ernodea**
    21. Flower or fruit pedunculate or pedicellate.
      22. Corolla pale lilac, the outer surface pubescent; fruit yellowish orange....**Paederia**
      22. Corolla white or yellowish white, the outer surface glabrous; fruit white ....
      ....**Chiococca**
  20. Plant an erect shrub or tree.
    23. Flowers or fruits in axillary cymes on short shoots or solitary in the leaf axils.
      24. Leaves scabrous on the upper surface..**Guettarda**
      24. Leaves glabrous on the upper surface.
        25. Fruit a woody capsule..**Exostema**
        25. Fruit a berry.
          26. Fruit 5–8 cm long, green turning yellow and then dark before falling ....
          ....**Genipa**
          26. Fruit 6–13 mm long, white ....**Randia**
    23. Flowers or fruits in terminal compound cymes.
      27. Leaves with a single prominent midvein, the lateral veins inconspicuous.
        28. Leaves linear-lanceolate, pubescent, the margin revolute ....**Strumpfia**
        28. Leaves elliptic, ovate, or suborbicular, glabrous, the margin not revolute ....
        ....**Erithalis**
      27. Leaves with prominent lateral veins.
        29. Corolla finely pubescent on the outer surface......................................**Hamelia**
        29. Corolla glabrous on the outer surface.
          30. Corolla funnelform, the lobes valvate..**Psychotria**
          30. Corolla salverform, the lobes contorted...**Ixora**

Catesbaea L. 1753. LILYTHORN

Shrubs. Leaves opposite, simple, pinnate-veined, petiolate, stipulate. Flowers axillary, solitary, actinomorphic, bisexual, bracteate; sepals 4, basally connate; petals 4, basally connate; stamens 4, epipetalous; ovary inferior, 2-carpellate and -loculate, the style 1, the stigma 2-lobed. Fruit a berry; seeds few.

A genus of about 10 species; North America and West Indies. [Commemorates Mark Catesby (1683–1749), English traveler and naturalist.]

Selected reference: Rogers (1987).

Catesbaea parviflora Sw. [Small-flowered.] SMALLFLOWER LILYTHORN; DUNE LILYTHORN.

Catesbaea parviflora Swartz, Prodr. 30. 1788. *Scolosanthus parviflora* (Swartz) C. Wright, in Sauvalle,
  Anales Acad. Ci. Med. Habana 6: 125. 1869.
Catesbaea campanulata Sagra ex de Candolle, Prodr. 4: 401. 1830. *Echinodendrum campanulatum*

(Sagra ex de Candolle) A. Richard, in Sagra, Hist. Fis. Cuba, Bot. 11: 18. 1850. *Catesbaea parviflora* Swartz var. *septentrionalis* Krug & Urban, in Urban, Symb. Antill. 1: 429. 1899.

Shrub, to 2(3) m; branchlets slender, scaberulous or glabrous, green or grayish, with axillary spines 0.2–2 cm long. Leaves usually crowded on short lateral spurs, the blade obovate to suborbicular, 2–15 mm long, 3–8 mm wide, coriaceous, the apex rounded, often apiculate, the base cuneate to rounded, the margin entire, revolute, the upper surface glabrous, lustrous, the lower surface scaberulous or glabrous, paler, the lateral nerves obsolete, the petiole short; stipules triangular, short-acuminate. Flowers solitary, short-pedicellate or sessile; hypanthium obovoid, scaberulous or glabrous; sepal lobes linear-subulate to narrowly triangular-lanceolate, ca. 1 mm long, scaberulous or glabrous; corolla tubular or funnelform, 5–12 mm long, white, the lobes narrowly triangular, the apex subobtuse, ½ as long as the tube or shorter; stamens with the anthers exserted, slightly shorter than the filaments; style slightly exceeding the stamens. Fruit globose or subglobose, 3–5 mm long, white, the pericarp thin; seeds 4–8, oval, 1–2 mm long, reddish brown, reticulate.

Pinelands and hammocks. Rare; Monroe County keys. Florida; West Indies. All year.

Catesbaea parviflora is listed as endangered in Florida (Florida Administrative Code, Chapter 5B-40).

Cephalanthus L. 1753. BUTTONBUSH

Shrubs or trees. Leaves opposite or whorled, simple, pinnate-veined, petiolate, stipulate. Flowers in axillary or terminal spherical, pedunculate heads, actinomorphic, bisexual, bracteate; sepals 4, basally connate; petals 4, basally connate; stamens 4, epipetalous; ovary inferior, 4-carpellate and -loculate, the style 1, the stigma capitate. Fruit a schizocarp.

A genus of 6 species; North America, Mexico, Central America, South America, Africa, and Asia. [From the Greek *cephale*, head, and *anthos*, flower, in reference to the flowers occurring in a globose inflorescence.]

Selected references: Ridsdale (1976); Rogers (1987).

Cephalanthus occidentalis L. [Western.] COMMON BUTTONBUSH.

Cephalanthus occidentalis Linnaeus, Sp. Pl. 95. 1753. *Cephalanthus oppositifolius* Moench, Methodus 487. 1794, nom. illegit. *Cephalanthus acuminatus* Rafinesque, New Fl. 3: 25. 1838 ("1836"), nom. illegit.
Cephalanthus occidentalis Linnaeus var. *pubescens* Rafinesque, Med. Fl. 1: 101. 1828.

Shrub or small tree, to 3 m; bark rough, ridged, and furrowed, the branchlets reddish brown, glabrous or short-pilose, with raised, corky lenticels in age, the leaf scars U-shaped to nearly round, each with a broad, crescent-shaped vascular-bundle scar, the axillary buds sunken, obscure. Leaves opposite or in whorls of 3–4, the blade ovate, lanceolate, elliptic, or obovate, 7–15 cm long, 3–10 cm wide, the apex acute to acuminate, the base rounded to cuneate, the margin entire, the upper surface glabrous, the lower surface glabrous, pubescent, or short-pilose, at least on the main veins, the petiole 0.5–3 cm long, pubescent or glabrous; stipules deltate, 2–3 mm long. Flowers in a dense spherical head 3–4 cm long, sessile, these axillary or terminal, solitary or in a few-headed cyme, the peduncle 2–5 cm long; bracts several, subtending each flower, filiform, dilated distally, pubescent; hypanthium obovoid, short-pubescent; corolla tubular, 6–10 mm long, white, the lobes ovate or elliptic, the outer surface glabrous, the inner surface pubescent; style slender, exserted from the corolla, the stigma slightly 4-lobed. Fruit narrowly obconical, ca. 5 mm long, splitting from the base upward into 2–4 nutlets, these obconical, 5–6 mm long.

Swamps, cypress ponds, and lake, pond, and river margins. Common; nearly throughout. New Brunswick south to Florida, west to Ontario, Missouri, Nebraska, Kansas, Oklahoma, and Texas, also Arizona and California; West Indies and Mexico. Spring.

Chiococca P. Browne 1756. MILKBERRY

Shrubs or woody vines. Leaves opposite, simple, pinnate-veined, petiolate, stipulate. Flowers in axillary cymes, racemes, or panicles, actinomorphic, bisexual, bracteate; sepals 5, basally connate; petals 5, basally connate; stamens 5, epipetalous; ovary inferior, 2-carpellate and -loculate, the style 1, the stigma slightly clavate. Fruit a drupe; pyrenes 2.

A genus of about 20 species; North America, West Indies, Mexico, Central America, and South America. [From the Greek *chion*, snow, and *coccos*, berry, in reference to the white drupes.]

Selected reference: Rogers (2005).

Chiococca alba (L.) Hitchc. [White, in reference to the fruit.] SNOWBERRY; MILKBERRY.

> *Lonicera alba* Linnaeus, Sp. Pl. 175. 1753. *Chiococca racemosa* Linnaeus, Syst. Nat., ed. 10. 917. 1759, nom. illegit. *Chiococca racemosa* Linnaeus var. *jacquiniana* Grisebach, Fl. Brit. W.I. 337. 1861, nom. inadmiss. *Chiococca alba* (Linnaeus) Hitchcock, Rep. (Annual) Missouri Bot. Gard. 4: 94. 1893.
> *Chiococca racemosa* Linnaeus var. *floridana* de Candolle, Prodr. 4: 482. 1830. SYNTYPE: FLORIDA.
> *Chiococca parvifolia* Wullschlagel ex Grisebach, Fl. Brit. W.I. 337. 1861. *Chiococca alba* (Linnaeus) Hitchcock var. *parvifolia* (Wullschlagel ex Grisebach) Urban, Symb. Antill. 8: 675. 1921. *Chiococca alba* (Linnaeus) Hitchcock subsp. *parvifolia* (Wullschlagel ex Grisebach) Steyermark, Act. Bot. Venez. 6: 138. 1971.
> *Chiococca pinetorum* Britton, in Millspaugh, Publ. Field Columb. Mus., Bot. Ser. 2: 171. 1906.

Erect, reclining, or clambering shrub or woody vine, to 3 m; branchlets green when young, glabrous, the older ones pale brown to tan, with a raised band of stipular tissue around the node, this eventually sloughed. Leaves with the blade ovate to oblong or lanceolate, 2–6 cm long, 2–4 cm wide, the apex acuminate or acute, the base short-cuspidate, the margin entire, the upper and lower surfaces glabrous, the petiole to 1 cm long; stipules triangular, 2–3 mm long, the apex cuspidate, the base broad, connecting between the petiole bases. Flowers in a pedunculate, axillary cyme, raceme, or panicle shorter than to somewhat exceeding the leaves, the pedicel 1–5 mm long; bracts lanceolate, ca. 1 mm long; hypanthium campanulate, ca. 2 mm long; sepal-lobes minute; corolla funnelform, 8–10 mm long, greenish white or pale to golden-yellow, the tube ca. 5 mm long, the lobes bluntly triangular, shorter than the tube; stamens included, the filaments connate basally around the style. Fruit compressed-globose, 4–5 mm long, white, leathery; pyrenes 2.

Coastal hammocks, pinelands, and shell middens. Frequent; peninsula, Dixie County. Florida and Texas; West Indies, Mexico, Central America, and South America. All year.

Plants occurring in pine rocklands with smaller leaves and flowers have been recognized as *C. pinetorum* (e.g., Rogers, 2005), but the characters appear to intergrade with typical *C. alba*.

Coccocypselum P. Browne, nom. et orth. cons. 1756.

Herbs. Leaves opposite, simple, pinnate-veined, petiolate, stipulate. Flowers in terminal or axillary heads, actinomorphic, bisexual, bracteate, bracteolate; sepals 4, basally connate; petals 4, basally connate; stamens 4, epipetalous; ovary inferior, 2-carpellate and -loculate, the style 1, the stigma 2-lobed. Fruit a berry; seeds many.

A genus of about 20 species; North America, West Indies, Mexico, Central America, and South America. [From the Greek *coccos*, berry, and *cypsela*, an indehiscent, 1-seeded inferior ovary of the Asteraceae, in reference to its many-seeded fleshy fruit.]

Coccocypselum hirsutum Bartl. ex DC. [Rough hairy.] YERBA DE GUAVA.

Coccocypselum hirsutum Bartling ex de Candolle, Prodr. 4: 396. 1830. *Tontanea hirsuta* (Bartling ex de Candolle) Standley, in Britton, N. Amer. Fl. 32: 147. 1921.

Prostrate or creeping perennial herb; stem hirsute. Leaves with the blade ovate, ovate-rhomboid, oblong, or reniform-ovate, (2)3–4(6.5) cm long, 1.5–3 cm wide, the apex obtuse, apiculate, the base rounded to truncate, usually slightly inequilateral, the margin entire, the upper surface hirsute or pubescent, the lower surface hirsute, the petiole 0.5–1.5(2) cm long, hirsute; stipules subulate, 3–6 mm long, hirsute. Flowers (1)2–4(5) in a terminal or axillary head, the peduncle 0.3–2 cm long, slender, hirsute; bracts and bracteoles triangular, ca. 1 mm long, hirsute; hypanthium 1–2 mm long, hirsute; sepal lobes linear-lanceolate, 3–4 mm long, hirsute; corolla campanulate, 8–10 mm long, blue or white, the tube 5–7 mm long, the lobes triangular, 2–4 mm long; anthers 1–2 mm long. Fruit ovoid or ellipsoid, 1–2 cm long, blue or purplish blue, hirsute; seeds ca. 1 mm long, brown, rugose.

Disturbed mesic flatwoods. Rare; Manatee County. Florida; West Indies, Mexico, and Central America. Native to West Indies, Mexico, and Central America. All year.

Diodia L. 1753. BUTTONWEED

Herbs. Leaves opposite, simple, pinnate-veined, petiolate or epetiolate, stipulate. Flowers axillary, solitary or few in fascicles, actinomorphic, bisexual, bracteate; sepals 4, basally connate; petals 4, basally connate; stamens 4, epipetalous; ovary inferior, 2-carpellate and -loculate, the style 1, 2-fid or undivided. Fruit a schizocarp or tardily and irregularly dehiscent capsule.

A genus of 40–50 species; North America, West Indies, Mexico, Central America, South America, and Africa. [From the Greek *diodos*, thoroughfare, in reference to its occurrence in disturbed roadsides.]

Hexasepalum has been recognized as a segregate by some authors, but here *Diodia* s.l. is retained pending additional studies.

Diodella Small, 1913; *Hexasepalum* Bartl. ex DC., 1830.

1. Sepals 2 (rarely 4 and then dimorphic), 4–6 mm long; fruit 5–9 mm long; corolla lobes short-hirsute within ....**D. virginiana**

1. Sepals 4, subequal, 4 mm long; fruit 2–3(4.5) mm long; corolla lobes glabrous within..............**D. teres**

Diodia teres Walter [Terete, round in cross section, in reference to the stem.] POOR JOE; ROUGH BUTTONWEED.

Diodia teres Walter, Fl. carol. 87. 1788. *Diodella teres* (Walter) Small, in Small & J. J. Carter, Fl. Lancaster Co. 271, 319. 1913. *Diodia teres* Walter var. *typica* Fernald & Griscom, Rhodora 39: 307. 1937, nom. inadmiss. *Hexasepalum teres* Walter J. H. Kirkbride, J. Bot. Res. Inst. Texas 8: 17. 2014.

Diodia teres Walter var. *hirsutior* Fernald & Griscom, Rhodora 39: 307, t. 469(4). 1937. TYPE: FLORIDA: Duval Co.: near Jacksonville, Aug, *Curtiss 1116* (holotype: GH).

Spreading or ascending annual or short-lived perennial herb, to 8 dm; stem somewhat 4-angled above, terete below, puberulent to hirsute. Leaves with the blade linear to lanceolate, to 5 cm long, to 1 cm wide, the apex acute to acuminate, the base rounded to somewhat clasping, the margin entire, often revolute, the upper and lower surfaces scabrous, sessile or subsessile; stipules sheathing, with numerous, long filiform bristles along the margin, usually equaling the flowers and longer than the fruit, these often reddish brown. Flowers axillary, solitary or few in a fascicle, sessile; bracts minute; hypanthium 2–3 mm long, hirsute; sepals lanceolate, ca. 4 mm long, subequal; corolla funnelform, 4–6 mm long, whitish to pinkish purple, the lobes short-hirsute within; style undivided. Fruit a schizocarp, obovate-turbinate, ca. 4 mm long, hispid or hispidulous, crowned with the 4 subequal, ovate-lanceolate, ciliate calyx lobes, the carpels separating, often with 3 lobes on one carpel and the 1 on the other, the carpels indehiscent.

Sandhills, dunes, and flatwoods. Common; nearly throughout. New York and Massachusetts south to Florida, west to California; West Indies, Mexico, Central America, and South America. Spring–fall.

Diodia virginiana L. [Of Virginia.] VIRGINIA BUTTONWEED.

Diodia virginiana Linnaeus, Sp. Pl. 104. 1753. *Diodia virginiana* Linnaeus var. *linnaei* Torrey & A. Gray, Fl. N. Amer. 2: 29. 1841, nom. inadmiss.

Diodia tetragona Walter, Fl. Carol. 87. 1788. *Diodia virginiana* Linnaeus var. *latifolia* Torrey & A. Gray, Fl. N. Amer. 2: 29. 1841.

Diodia hirsuta Pursh, Fl. Amer. Sept. 106. 1814. *Diodia virginiana* Linnaeus var. *hirsuta* (Pursh) Torrey & A. Gray, Fl. N. Amer. 2: 29. 1841. *Diodia virginiana* Linnaeus forma *hirsuta* (Pursh) Fernald, Rhodora 44: 457. 1942.

Diodia harperi Small, Man. S.E. Fl. 1264. 1933.

Spreading or procumbent perennial herb, to 6 dm; stem somewhat 4-angled above, terete below, villous-hirsute to subglabrous. Leaves with the blade elliptic-oblong to elliptic-oblanceolate, to 9 cm long, to 2 cm wide, the apex acute to acuminate, the base cuneate, the margin slightly serrulate, the upper and lower surfaces glabrate to sparsely hirsute, sessile or subsessile; stipules sheathing, with a few strong, flat bristles along the margin, shorter than the flowers and fruit. Flowers axillary, solitary or few in a fascicle, sessile; bracts minute; hypanthium 2–3 mm long, hirsute; sepals lanceolate, 4–6 mm long, dimorphic (2 long and 2 short); corolla slender-filiform and abruptly expanding into large lobes, 7–10 mm long, white, the lobes glabrous within; style 2-fid. Fruit a capsule, 5–8 mm long, glabrous or villous, crowned with 2 lanceolate, glabrescent calyx lobes, the carpels suberose-crustaceous, 3-ribbed on the back, held together by a thin epicarp that tardily ruptures.

Wet flatwoods, swamps, and marshes. Common; nearly throughout. Connecticut south to Florida, west to Kansas, Oklahoma, and Texas. All year.

EXCLUDED TAXA

Diodia rigida Chamisso & Schlechtendal—Reported for Florida by Small (1903, 1913a, 1933, all as Diodella rigida (Chamisso & Schlechtendal) Small) and Long and Lakela (1971), probably based on a misidentification of D. teres.

Diodia ocymifolia (Willdenow ex Roemer & Schultes) Bremekamp—Reported for Florida by Liogier (1962, as Hemidiodia ocymifolia (Willdenow ex Roemer & Schultes) K. Schumacher). No Florida specimens known.

Erithalis P. Browne 1756.

Shrubs or trees. Leaves opposite, simple, pinnate-veined, petiolate, stipulate. Flowers in axillary or terminal cymes or corymbose panicles, actinomorphic, bisexual, bracteate; sepals 4–5, basally connate; petals 4–5, basally connate; stamens 4–5, epipetalous, the anthers basifixed, longitudinally dehiscent; ovary inferior, 5- to 10-carpellate and -loculate, the style 1, the stigmas 5- to 8-lobed. Fruit a drupe.

A genus of about 8 species; North America, West Indies, Mexico, Central America, and South America. [Adopted from a name used by Pliny for a plant of maritime cliffs.]

Selected reference: Negrón-Ortiz (2005).

Erithalis fruticosa L. [Shrubby, bushy.] BLACKTORCH.

Erithalis fruticosa Linnaeus, Syst. Nat., ed. 10. 930. 1759. Erithalis odorata Persoon, Syn. Pl. 1: 200. 1805, nom. illegit.

Shrub or small tree, to 4 m; stem glabrous, smooth. Leaves with the blade obovate, elliptic, ovate, or suborbicular, 3–7 cm long, 1.5–6 cm wide, subcoriaceous, the apex rounded to short acute, the base cuneate, the upper and lower surfaces glabrous, lustrous, the petiole 0.5–1.5 cm long; stipules connate, 1–2 mm long, mucronate. Flowers in an axillary or terminal, pedunculate cyme or corymbose panicle, 5–6 cm long, the pedicel 2–5 mm long, the peduncle 5–7 cm long; bracts lanceolate to 2 mm long; hypanthium globose to ovate, glabrous; sepal lobes 1–2 mm long, denticulate; corolla rotate, white, 4–7 mm long, the lobes linear-oblong, 3–5 mm long, spreading; stamens 3–5 mm long, the anthers subequaling the filaments; style slightly shorter than stamens. Fruit globose or depressed-globose, 3–5 mm long, black, 5- to 10-furrowed; pyrenes 5–10.

Coastal hammocks and dunes. Rare; Martin, Palm Beach, and Miami-Dade Counties, Monroe County keys. Florida; West Indies, Central America, and South America. All year.

Erithalis fruticosa is listed as threatened in Florida (Florida Administrative Code, Chapter 5B-40).

Ernodea Sw. 1788. BEACH CREEPER

Shrubs. Leaves opposite, simple, pinnate-veined, epetiolate, stipulate. Flowers axillary, solitary, actinomorphic, bisexual, bracteate; sepals 4–6, basally connate; petals 4–6, basally connate; stamens 4–6, epipetalous; ovary inferior, 2-carpellate and -loculate, the style 1, the stigma 2-lobed. Fruit a drupe.

A genus of 4 species; North America, West Indies, Mexico, and Central America. [From the Greek *ernos*, offshoot, in reference to its growth habit.]

Selected reference: Negrón-Ortiz and Hickey (1996).

1. Leaves 1-nerved, 1–3 mm wide .... **E. cokeri**
1. Leaves 3- to 5-nerved, 4–10 mm wide .... **E. littoralis**

Ernodea cokeri Britton ex Coker [Commemorates William Coker (1872–1953), American botanist.] COKER'S BEACH CREEPER; ONE-NERVED ERNODEA.

Ernodea cokeri Britton ex Coker, in Shattuck, Bahama Isl. 264. 1905.

Prostrate, trailing shrub; stem to 9 dm, glabrous to fine-pubescent. Leaves with the blade narrowly linear, 2–4 cm long, 2–3 mm wide, coriaceous, evergreen, 1-nerved, the apex acute, the base narrowly cuneate, the margin entire, the upper and lower surfaces rough-pubescent, sessile; stipules connate and forming a sheath with 1–4 apices, 1–2 mm long. Flowers axillary, solitary, sessile; bracts lanceolate, minute; sepals 4, the lobes subulate, 3–6 mm long; corolla tubular-funnelform, orange-red, pink, or white, the tube 6–7(9) mm long, the lobes 4, 4–6(7) mm long. Fruit subglobose, 4–6 mm long, orange-yellow, grooved; pyrenes 2, the endocarp cartilaginous.

Disturbed hammocks and pinelands. Rare; Miami-Dade County, Monroe County keys. Florida; West Indies. All year.

Ernodea cokeri is listed as endangered in Florida (Florida Administrative Code, Chapter 5B-40).

Ernodea littoralis Sw. [Pertaining to the sea-shore.] BEACH CREEPER; COUGHBUSH.

Ernodea littoralis Swartz, Prodr. 29. 1788.

Ernodea angusta Small, Bull. New York Bot. Gard. 3: 438. 1905. *Ernodea littoralis* Swartz var. *angusta* (Small) R. W. Long, Rhodora 72: 35. 1970. TYPE: FLORIDA: Miami-Dade Co.: between Cutler and Camp Longview, 9–12 Nov 1903, *Small & Carter 870* (holotype: NY).

Erect or decumbent shrub, to 1.5 m; stem glabrous or puberulent. Leaves with the blade elliptic, linear-oblong, or oblanceolate, 2–4 cm long, 3–9 mm wide, coriaceous, evergreen, the margin entire, 3- to 5-nerved, sessile; stipules connate and forming a sheath with 1–3 apices, 1–2 mm long. Flowers axillary, solitary, sessile; bracts minute; sepals 4–6, linear-lanceolate, 3–6 mm long; corolla tubular-funnelform, orange-red, pink, or yellow, the tube 4–10 mm long, the lobes 4–6, 8–10 mm long. Fruit subglobose, 4–6 mm long, yellow, grooved; pyrenes 2, the endocarp cartilaginous.

Coastal dunes. Frequent; central and southern peninsula. Florida; West Indies, Mexico, and Central America. All year.

Exostema (Pers.) Bonpl. 1807

Shrubs or trees. Leaves opposite, simple, pinnate-veined, petiolate, stipulate. Flowers axillary, solitary, actinomorphic, bisexual, bracteate; sepals 5, basally connate; petals 5, basally connate; stamens 4, epipetalous; ovary inferior, 2-carpellate and -loculate, the style 1, the stigma 2-lobed. Fruit a 2-valved capsule; seeds numerous, winged.

A genus of about 35 species; North America, West Indies, Mexico, Central America, and South America. [From the Greek *exo*, outside, and *stema*, stamen, in reference to the exserted stamens.]

Selected reference: Rogers (1987).

Exostema caribaeum (Jacq.) Schult. [Of the Caribbean.] CARIBBEAN PRINCEWOOD.

Cinchona caribaea Jacquin, Enum. Syst. Pl. 16. 1760. *Exostema caribaeum* (Jacquin) Roemer & Schultes, Syst. Veg. 5: 18. 1819.

Shrub or small tree, to 8 m; branches glabrous. Leaves with the blade oblong-lanceolate to narrowly ovate, 3–8 cm long, 1–3 cm wide, the apex acute to acuminate, the base cuneate, the margin entire, the upper and lower surfaces glabrous, the petiole 1–2 cm long; stipules broadly ovate, 1–2 mm long, the apex acuminate. Flowers axillary, solitary, the peduncle 4–5 mm long; bracts minute; hypanthium clavate-cylindric, 4–5 mm long, glabrous, the lobes linear, 1–2 mm long; corolla salverform, white or pinkish, the tube 2–3 cm long, the lobes slightly shorter than the tube; stamens long-exserted. Fruit oblong, 1–1.5 cm long, woody, smooth.

Pinelands and hammocks. Rare; Miami-Dade County, Monroe County keys. Florida; West Indies, Mexico, and Central America. Spring–summer.

Exostema caribaeum is listed as endangered in Florida (Florida Administrative Code, Chapter 5B-40).

Galium L. 1753. BEDSTRAW

Herbs. Leaves whorled, simple, pinnate-veined, petiolate or epetiolate, stipulate. Flowers in axillary or terminal simple or compound cymes, actinomorphic, bisexual, bracteate; sepals 3 or 4, basally connate; petals 3 or 4, basally connate; stamens (3)4; ovary inferior, 2-carpellate and -locular, the styles 2, the stigmas capitate. Fruit a dry or fleshy indehiscent schizocarp.

A genus of 300–400 species; nearly cosmopolitan. [From the Greek *gala*, milk, in reference to its ancient use for curdling milk.]

1. Fruit smooth or with a few scattered uncinate trichomes.
    2. Fruit dry.
        3. Edges of the stem angles acute, smooth or sometimes minutely retrorse-scabrid; leaves of the main stems 5 or 6 per node; corolla lobes 3 or sometimes 4, the apex obtuse .....**G. tinctorium**
        3. Edges of the stem angles rounded, always smooth; leaves of the main stem 4 per node; corolla lobes always 4, the apex acute ....**G. obtusum**
    2. Fruit fleshy.
        4. Leaves oblong or elliptic (2–3 times as long as wide); fruit blue...........................**G. hispidulum**
        4. Leaves linear-oblong (5–7 times as long as wide); fruit black.................................**G. uniflorum**
1. Fruit uncinate-hispid.
    5. Leaves (4)6–8 per node.
        6. Leaves usually 8 per node, linear to slightly oblanceolate; plant annual......................**G. aparine**
        6. Leaves usually 6 per node, narrowly elliptic; plant perennial .... **G. triflorum**
    5. Leaves usually 4 per node.
        7. Flowers borne in the upper 2 or 3 nodes...**G. circaezans**
        7. Flowers borne in the upper 6–12 nodes..**G. pilosum**

Galium aparine L. [From the Greek *apara*, to seize, in reference to the retrorse bristles.] GOOSEGRASS; SPRING CLEAVERS; STICKYWILLY.

Galium aparine Linnaeus, Sp. Pl. 108. 1753. *Galium lappaceum* Salisbury, Prodr. Stirp. Chap. Allerton 59. 1796, nom. illegit. *Galium uncinatum* Gray, Nat. Arr. Brit. Pl. 2: 484. 1821, nom. illegit. *Galium aparine* Linnaeus var. *verum* Wimmer & Grabowski, Fl. Siles. 1(1): 119. 1827, nom. inadmiss. *Galium aparine* Linnaeus var. *vulgare* Reichenbach, Icon. Fl. Germ. Helv. 17: 94. 1854, nom. inadmiss.

Reclining or scrambling annual herb, to 10(20) dm; stem 4-angled, retrorsely bristly on the angles, pubescent at the nodes. Leaves in whorls of (4–6)8, the blade linear-oblanceolate, 1–4(8) cm long, 2–4 mm wide, the apex cuspidate, the base narrowly cuneate, the margin retrorsely scabro-ciliate, the upper surface glabrous, the lower surface scabro-ciliate on the midrib, sessile. Flowers 1–3 on an axillary peduncle or a short branch, pedicellate; bracts minute; sepals 4; petals 4, the lobes 1–2 mm long, white. Fruit a pair of subglobose, indehiscent carpels, (2)3–4(5) mm wide, dry, uncinate-hispid.

Hammocks. Occasional; northern counties, central peninsula. Nearly throughout North America; Mexico, Central America, and South America; Europe, Africa, and Asia. Spring–summer.

Galium circaezans Michx. [Resembling *Circaea* (Onagraceae).] FOREST BEDSTRAW; LICORICE BEDSTRAW.

> *Galium circaezans* Michaux, Fl. Bor.-Amer. 1: 80. 1803. *Galium circaeoides* Roemer & Schultes, Syst. Veg. 3: 256. 1818, nom. illegit. *Galium rotundifolium* Linnaeus var. *circaezans* (Michaux) Kuntze, Revis. Gen. Pl. 1: 282. 1891. *Galium circaezans* Michaux var. *typicum* Fernald, Rhodora 39: 450. 1937, nom. inadmiss.

Erect or ascending perennial herb, to 4(6) dm; stem 4-angled, glabrate or pubescent on the angles. Leaves in whorls of 4(6), the blade ovate-lanceolate or elliptic, (1.5)2–5 cm long, 1–2.5 cm wide, the apex obtuse, the base cuneate, the margin entire, ciliate, the upper and lower surfaces glabrous or pubescent, sessile or short petiolate. Flowers 1–3 in a cyme in the terminal or upper 2–3 nodes, the cyme simple or with 1–2 divaricate forks, pedicellate, short pedunculate or sessile; bracts minute; sepals 4; petals 4, the lobes 1–2 mm long, greenish white, usually pubescent on the outer surface. Fruit a pair of subglobose, indehiscent carpels, 4–5 mm wide, dry, uncinate-hispid.

Hammocks. Occasional; central panhandle. Quebec south to Florida, west to Ontario, Minnesota, Nebraska, Kansas, Oklahoma, and Texas. Spring–summer.

Galium hispidulum Michx. [Covered with fine, coarse trichomes.] COASTAL BEDSTRAW.

> *Galium hispidulum* Michaux, Fl. Bor.-Amer. 1: 79. 1803. *Galium hispidum* Pursh, Fl. Amer. Sept. 104. 1814, nom. illegit. *Galium carolinianum* F. G. Dietrich, Nachtr. Vollst. Lex. Gaertn. 3: 429. 1817, nom. illegit. *Galium peregrinum* Britton et al., Prelim. Cat. 24. 1888, nom. illegit.; non Franchet, 1866. *Bataprine hispidula* (Michaux) Nieuwland, Amer. Midl. Naturalist 1: 264. 1910.

Erect or ascending perennial herb, to 6 dm; stem 4-angled, hirsute on the angles. Leaves in whorls of 4, the blade elliptic to oblong, 5–12 mm long, 2–3 times as long as wide, the apex apiculate, the base cuneate, the margin entire, the upper surface glabrous, the lower surface hirsute, sessile. Flowers 1–3 in an axillary cyme, pedicellate, the peduncle pubescent; bracts minute; sepals 4; petals 4, the lobes ca. 2 mm long, greenish white. Fruit a pair of subglobose, indehiscent carpels, 3–4 mm wide, fleshy, blue, smooth.

Sandhills, hammocks, and disturbed sites. Frequent; nearly throughout. New Jersey south to Florida, west to Texas. Spring–summer.

Galium obtusum Bigelow subsp. **filifolium** (Wiegand) Puff [Blunt, in reference to the leaf apex; with threadlike leaves.] BLUNTLEAF BEDSTRAW.

> *Galium tinctorium* Linnaeus var. *filifolium* Wiegand, Bull. Torrey Bot. Club 104: 307. 1977. *Galium filifolium* (Wiegand) Small, Man. S.E. Fl. 1268. 1933. *Galium obtusum* Bigelow var. *filifolium* (Wiegand) Fernald, Rhodora 37: 443. 1935. *Galium obtusum* Bigelow subsp. *filifolium* (Wiegand) Puff, Bull. Torrey Bot. Club 104: 207. 1977.

Erect or ascending perennial herb, to 6 dm; stem 4-angled, glabrous on the angles, with short, conic trichomes at the nodes or glabrous. Leaves in whorls of (4)5, the blade linear-filiform to narrowly oblanceolate, (0.8)1–2(2.5) cm long, 1–2 mm wide, the apex obtuse, apiculate, the base cuneate, the margin downward recurved, sometimes with a few conic trichomes, the upper and lower surfaces glabrous, sessile. Flowers (2)3–4(5) in a cyme, pedicellate, the peduncle (3)5–12(15) mm long; bracts minute; sepals 4; petals 4, the lobes 2–3 mm long, white. Fruit a pair of subglobose, indehiscent carpels, 2.5–4(5) mm wide, dry, glabrous, often with only 1 mericarp developed.

Wet hammocks. Rare; Wakulla County. New Jersey south to Florida, west to Alabama. Spring–summer.

Galium pilosum Aiton [With long, ascending trichomes.] HAIRY BEDSTRAW.

> *Galium pilosum* Aiton, Hort. Kew. 1: 145. 1789. *Galium puncticulosum* Michaux var. *pilosum* (Aiton) de Candolle, Prodr. 4: 601. 1830.
> *Galium puncticulosum* Michaux, Fl. Bor.-Amer. 1: 80. 1803. *Galium pilosum* Aiton var. *puncticulosum* (Michaux) Torrey & A. Gray, Fl. N. Amer. 2: 24. 1841. *Galium pilosum* Aiton subsp. *puncticulosum* (Michaux) W. Stone, Pl. S. New Jersey 705. 1912 ("1911").
> *Galium pilosum* Aiton var. *laevicaule* Weatherby & S. F. Blake, Rhodora 18: 194. 1916. *Galium orizabense* Hemsley subsp. *laevicaule* (Weatherby & S. F. Blake) Dempster, Allertonia 2: 411. 1981. TYPE: FLORIDA: Duval Co.: near Jacksonville, 18 Jun 1898, *Curtiss 6420* (holotype: GH).

Erect or ascending perennial herb, to 7(10) dm; stem 4-angled, spreading-pilose or glabrous. Leaves in whorls of 4, the blade oval to elliptic or oblong-elliptic, 0.8–1.2(2) cm long, 2–3 times as long as wide, the apex obtuse, apiculate, the base cuneate, the margin entire, ciliate, the upper and lower surfaces pilose, the lower surface usually sparsely glandular-punctate, sessile or short-petiolate. Flowers 3–10 in a paniculiform cyme in the terminal and upper axils, pedicellate, the peduncle 2- or 3-forked; bracts minute; sepals 4; petals 4, the lobes ovate to lanceolate, 1–2 mm long, greenish white to yellowish or purplish. Fruit a pair of subglobose, indehiscent carpels, 3–4 mm wide, dry, uncinate-hispid.

Sandhills and hammocks. Frequent; northern counties, central peninsula. New Hampshire south to Florida, west to Ontario, Michigan, Wisconsin, Illinois, Kansas, Oklahoma, and Texas. Summer.

Galium tinctorium L. [Used in dyeing.] STIFF MARSH BEDSTRAW.

> *Galium tinctorium* Linnaeus, Sp. Pl. 106. 1753. *Galium trifidum* Linnaeus var. *tinctorium* (Linnaeus) Torrey & A. Gray, Fl. N. Amer. 2: 22. 1841. *Galium trifidum* Linnaeus subsp. *tinctorium* (Linnaeus) H. Hara, Rhodora 41: 388. 1939.
> *Galium tinctorium* Linnaeus var. *floridanum* Wiegand, Bull. Torrey Bot. Club 24: 397. 1897. *Galium tinctorium* Linnaeus subsp. *floridanum* (Wiegand) Puff, Canad. J. Bot. 54: 1916. 1976. *Galium obtusum* Bigelow var. *floridanum* (Wiegand) Fernald ex R. W. Long & Lakela, Fl. Trop. Florida 803. 1971, nom. nud. TYPE: FLORIDA: Lake Co.: vicinity of Eustis, 22 Mar 1894, *Nash 152* (holotype: NY; isotypes: GH, MB, UC, WU).

Erect, ascending, or decumbent perennial herb, to 2.5 dm; stem 4-angled, retrorsely scabrous on the angles or glabrate. Leaves in whorls of 4–6, the blade linear-lanceolate or linear-spatulate, 1.5–2.5 cm long, 2–3 times as long as wide, the apex obtuse, apiculate, the base cuneate, the margin entire, ciliate, the upper and lower surfaces glabrous, sessile to short-petiolate. Flowers

2–3 in a terminal cyme, pedicellate, the peduncle to 1.5 cm long; bracts minute; sepals 3 or 4; petals 3 or 4, the lobes 1–2 mm long, greenish white. Fruit a pair of subglobose, indehiscent carpels, 2–3 mm wide, dry, smooth.

Swamps and bogs. Common; nearly throughout. Labrador south to Florida, west to Ontario, Minnesota, Nebraska, Kansas, Oklahoma, and Texas. Spring–summer.

Galium triflorum Michx. [Three-flowered.] FRAGRANT BEDSTRAW.

> *Galium triflorum* Michaux, Fl. Bor.-Amer. 1: 80. 1803. *Galium triflorum* Michaux forma *typicum* Leyendecker, Iowa State Coll. J. Sci. 15: 179. 1941, nom. inadmiss.

Scrambling perennial herb, to 8(12) dm; stem 4-angled, retrorsely scabrous or spreading pubescent on the angles below to glabrate. Leaves in whorls of (4–5)6(7–9) cm long, 5–10(15) mm wide, the blade elliptic-lanceolate to oblanceolate, 2–6 cm long, the apex cuspidate, the base narrowly cuneate, the margin entire, minutely ciliate, the upper and lower surfaces with minute cilia near the margin, the lower surface uncinate-scabrous on the midrib, sessile or short-petiolate. Flowers 1–3 in an axillary or sometimes terminal cyme, pedicellate, the peduncle to 1.5 mm long; sepals 4; petals 4, the lobes 1–2(4) mm long, whitish or greenish white. Fruit a pair of subglobose, indehiscent carpels, ca. 2 mm wide, dry, uncinate-hispid.

Hammocks. Rare; Leon, Liberty, and Jackson Counties. Throughout North America. Summer–fall.

Galium uniflorum Michx. [One-flowered.] ONEFLOWER BEDSTRAW.

> *Galium uniflorum* Michaux, Fl. Bor.-Amer. 1: 79. 1803. *Bataprine uniflora* (Michaux) Nieuwland, Amer. Midl. Naturalist 1: 264. 1910.

Herb, to 6 dm; stem 4-angled, glabrous. Leaves in whorls of 4, the blade linear, 0.5–2.5 cm long, 5–7 times as long as wide, the apex acute, the base narrowly cuneate, the margin entire, the upper and lower surfaces glabrous, sessile or short-petiolate. Flowers 1–2 in the axil of a leaf whorl, pedicellate; bracts minute; sepals 4; petals 4, the lobes broadly ovate, ca. 2 mm long, white. Fruit a pair of subglobose, indehiscent carpels, ca. 2 mm wide, fleshy, blue, smooth.

Hammocks. Occasional; northern and central peninsula west to central panhandle. Virginia south to Florida, west to Texas. Summer.

EXCLUDED TAXA

> *Galium bermudense* Linnaeus—Reported for Florida by Small (1913a; 1913b; 1913d; 1933), the name misapplied to our material of *G. hispidum*.
> *Galium hypocarpium* (Linnaeus) Hemsley—This Meso-American species was reported for Florida by Michaux (1803, as *Rubia brownei* Michaux, nom. illegit.).
> *Galium obtusum* Bigelow—Because infraspecific categories were not recognized, the typical subspecies was reported for Florida by implication by Correll and Johnston (1970) and Godfrey and Wooten (1981). Our material is all subsp. *filifolium*.
> *Galium trifidum* Linnaeus—This northern species was mistakenly attributed to Florida by Chapman (1860). No Florida specimens known.

Genipa L. 1754.

Shrubs or trees. Leaves opposite, simple, pinnate-veined, petiolate, stipulate. Flowers in axillary cymes, actinomorphic, unisexual (plants monoecious), bracteate; sepals 5–6, basally connate; petals 5–6, basally connate; stamens 5–6, epipetalous, the anthers sessile; nectariferous disk present; ovary inferior, 1-carpellate and -loculate, the style 1, the stigmas 2. Fruit a berry; seeds numerous.

A genus of 6 species; North America, West Indies, Central America, and South America. [From the Guayanan vernacular *genipago*.]

Genipa clusiifolia (Jacq.) Griseb. [With leaves like *Clusia* (Clusiaceae).]
SEVENYEAR APPLE.

Gardenia clusiifolia Jacquin, Collectanea 5: 37, t. 4(3). 1797 ("1796"). *Randia clusiifolia* (Jacquin) Chapman, Fl. South. U.S. 179. 1860. *Genipa clusiifolia* (Jacquin) Grisebach, Fl. Brit. W.I. 317. 1861. *Casasia clusiifolia* (Jacquin) Urban, Symb. Antill. 5: 505. 1908.

Shrub or small tree, to 3 m; stem glabrous. Leaves opposite, tending to cluster near the ends of the branches, the blade obovate, 5–15 cm long, coriaceous, the apex rounded to retuse, often mucronate, the base cuneate, the margin entire, the upper and lower surfaces glabrous, lustrous, the petiole to ca. 1 cm long. Flowers in a short-pedunculate cyme; bracts lanceolate; hypanthium 8–10 mm long; sepal-lobes linear-subulate; corolla salverform, white or yellow, fleshy, the tube 1.5–2 cm long, pubescent in the throat, the lobes lanceolate or oblong-lanceolate, shorter than the tube, spreading, contorted; stamens adnate near the mouth of the corolla tube, the anthers subsessile, with a subulate appendage; nectariferous disk cuplike; style stout, with 2 oblong lobes. Fruit ovoid to obovoid, 5–8 cm long, green, turning yellow and then dark before falling, the pericarp woody, the mesocarp pulpy; seeds numerous, flattened, angled, embedded in a mucilaginous material.

Coastal hammocks. Rare; Lee County, southern peninsula. Florida; West Indies. All year.

EXCLUDED TAXON

Genipa americana Linnaeus—Reported for Florida by Liogier (1997). No Florida specimens known.

Guettarda L. 1753. VELVETSEED

Shrubs or trees. Leaves opposite, simple, petiolate, stipulate. Flowers in axillary cymes, nearly actinomorphic, bisexual, bracteate, bracteolate; sepals 4–7, basally connate, 2-lobed; petals 4–7, basally connate; stamens 4–9, the filaments adnate to the corolla distally, the anthers dorsifixed; ovary inferior, 2- to 7-carpellate and -loculate, the style and stigma 1. Fruit drupelike; seeds in a hard endocarp.

A genus of about 130 species; North America, West Indies, Mexico, Central America, South America, Asia, and Pacific Islands. [Commemorates Jean-Étienne Guettard (1715–1786), French naturalist.]

1. Lower side of the leaf with sparse silky trichomes, the surface visible .... **G. elliptica**
1. Lower side of the leaf finely and closely pubescent, the surface not visible .... **G. scabra**

Guettarda elliptica Sw. [Shaped like an ellipse, in reference to the leaves.] HAMMOCK VELVETSEED.

> *Guettarda elliptica* Swartz, Prodr. 59. 1788. *Matthiola elliptica* (Swartz) Kuntze, Revis. Gen. Pl. 1: 288. 1891; non R. Brown ex de Candolle, 1821.
> *Guettarda blodgetii* Shuttleworth ex Chapman, Fl. South. U.S. 178. 1860. TYPE: FLORIDA: Monroe Co.: Key West.

Shrub or tree, to 8 m; branchlets grayish or rusty, appressed pubescent, the lenticels often numerous, whitish. Leaves with the blade ovate to broadly elliptic or obovate, 1–7.5 cm long, 1–4.5 cm wide, membranous or chartaceous, the apex rounded, rarely acute, apiculate, the base truncate or cuneate, the margin entire, usually flat, the upper surface sparsely appressed-pubescent, becoming glabrate, the lower surface paler, appressed-pubescent, becoming glabrate, the petiole 1–12 mm long; stipules deltate, 2–3 mm long, densely white pubescent, early caducous. Flowers 1–3 in an axillary cyme, the peduncle 0.5–3 cm long; bracteole shorter or subequaling the calyx; calyx tubular, 2–3 mm long; corolla salverform to funnelform, white, the tube 5–9 mm long, the lobes 4–5, obovate, 2–3 mm long, antrorsely pubescent externally. Fruit subglobose, 4–8 mm long, purple to black, 2- to 4-loculate.

Hammocks and pinelands. Rare; St. Lucie and Martin Counties, southern peninsula. Florida; West Indies, Mexico, Central America, and South America. All year.

Guettarda scabra (L.) Vent. [Rough to the touch, in reference to the leaves.] ROUGH VELVETSEED.

> *Matthiola scabra* Linnaeus, Sp. Pl. 1192. 1753. *Guettarda scabra* (Linnaeus) Ventenat, Choix Pl. 1: t. 1. 1803.
> *Guettarda ambigua* de Candolle, Prodr. 4: 455. 1830.

Shrub or tree, to 10 m; branchlets gray or reddish brown, pubescent, the lenticels brown to tawny. Leaves with the blade oblong, elliptic, ovate, obovate, or ovate-rhombic, 3–17 cm long, 2–11 cm wide, coriaceous, the apex rounded to acute, often mucronate, the base cordate to cuneate, the margin entire, often revolute, the upper surface densely scabrous, the lower surface densely pubescent, the petiole 0.5–2 cm long; stipules triangular, 4–8 mm long, tawny-pubescent, persistent or tardily dehiscent. Flowers 3–7 in an axillary cyme, the peduncle 2–12 cm long; bracteoles shorter than the calyx; calyx tubular, ca. 3 mm long; corolla salverform to funnelform, white to pinkish, red tinged externally, the tube 17–21 mm long, the lobes 5–7, oblong or obovate, 4–5 mm long, retrorsely pubescent externally. Fruit globose or depressed globose, 5–8 mm long, red, 5- to 7-loculate.

Hammocks and pinelands. Rare; Martin, Palm Beach, and Miami-Dade Counties, Monroe County keys. Florida; West Indies and South America. All year.

Hamelia Jacq. 1760.

Shrubs or trees. Leaves opposite or whorled, simple, pinnate-veined, petiolate, stipulate. Flowers in terminal dichasia, actinomorphic, bisexual, bracteate, ebracteolate; sepals 5, basally connate; petals 5, basally connate; stamens 5, epipetalous, the anthers basally sagittate, the connective extended; ovary inferior, 5-carpellate and -loculate, the style 1, the stigma fusiform. Fruit a berry; seeds numerous.

A genus of about 15 species; Florida, West Indies, Mexico, Central America, and South America; Asia. [Commemorates Henri-Louis Duhamel du Monceau (1700–1781), French botanist.]

Selected references: Elias (1976); Rogers (1987).

Hamelia patens Jacq. [Spreading, in reference to the inflorescence branches.] FIREBUSH.

Hamelia patens Jacquin, Enum. Syst. Pl. 16. 1760. *Hamelia coccinea* Swartz, Prodr. 46. 1788, nom. illegit. *Duhamelia patens* (Jacquin) Persoon, Syn. Pl. 1: 203. 1805.
Hamelia erecta Jacquin, Enum. Syst. Pl. 16. 1760.

Shrub or tree, to 7 m; branchlets strigose. Leaves opposite or in a whorl of 3–4, the blade ovate-elliptic to obovate-elliptic, 5.5–18 cm long, 3–8 cm wide, the apex acute to acuminate, the base rounded, the margin entire, the upper and lower surfaces villous, the petiole 1–3 cm long; stipules triangular to subulate, 2–6 mm long. Flowers 4–7 in a terminal, modified dichasia, secund, solitary, or fascicled, the peduncle 1–4 cm long, the floral branches usually outward curved, to 4.5 cm long, the flowers secund on the axis; bracts minute; calyx lobes ovate, ca. 1 mm long; corolla tube narrowly campanulate, 1–2 cm long, orange-red to red; anthers 8–12 mm long, the connective forming an ovate appendix. Fruit ellipsoidal to globose, 7–10 mm long, black; seeds 0.5–1 mm long, compressed, finely foveolate.

Coastal hammocks and shell middens, rarely inland. Frequent; central and southern peninsula, Leon County. Escaped from cultivation. Florida; West Indies, Mexico, Central America, and South America; Asia. Native to North America and tropical America. All year.

Houstonia L. 1753. BLUET

Annual or perennial herbs. Leaves opposite, simple, pinnate-veined, petiolate or epetiolate, stipulate. Flowers terminal or axillary, solitary or in cymes, actinomorphic, bisexual, bracteate; sepals 4, basally connate; petals 4, basally connate; stamens 4, epipetalous; ovary part-inferior, 2-carpellate and -loculate, the style and stigma 1. Fruit a capsule, extending beyond the hypanthium, dehiscing loculicidally across the top; seeds few to many.

A genus of 20 species in North America and Mexico. [Commemorates William Houstoun (1695–1733), Scottish physician and botanist who collected in the West Indies, Mexico, and South America.]

1. Flowers solitary, borne on pedicels extending above the leaves.
   2. Corolla 4–10(12) mm long, purple with a reddish center (yellowish center when white-flowered); calyx lobes shorter than or rarely subequaling the corolla tube .... **H. pusilla**
   2. Corolla 2–6 mm long, white; calyx lobes subequaling the corolla tube .... **H. micrantha**
1. Flowers mostly in cymes, or if solitary, then little if any extending above the leaves.
   3. Cymes axillary and terminal ....**H. procumbens**
   3. Cymes terminal.
      4. Leaves rounded to subcordate at the base.. **H. purpurea**
      4. Leaves tapering to narrowed at the base ....**H. longifolia**

Houstonia longifolia Gaertn. [With long leaves.] LONGLEAF SUMMER BLUET.

Houstonia longifolia Gaertner, Fruct. Sem. Pl. 1: 226, t. 49. 1788. *Hedyotis longifolia* (Gaertner) Hooker, Fl. Bor.-Amer. 1: 286. 1833. *Anotis longifolia* (Gaertner) G. Don, Gen. Hist. 3: 535. 1834. *Oldenlandia purpurea* (Linnaeus) A. Gray var. *longifolia* (Gaertner) A. Gray, Manual, ed. 2. 173. 1856. *Houstonia purpurea* Linnaeus var. *longifolia* (Gaertner) A. Gray, Manual, ed. 5. 212. 1867. *Chamisme longifolia* (Gaertner) Nieuwland, Amer. Midl. Naturalist 4: 92. 1915. *Hedyotis purpurea* (Linnaeus) Torrey & A. Gray var. *longifolia* (Gaertner) Fosberg, Castanea 19: 34. 1954.
Houstonia tenuifolia Nuttall, Gen. N. Amer. Pl. 1: 95. 1818. *Hedyotis longifolia* (Gaertner) Hooker var. *tenuifolia* (Nuttall) Torrey & A. Gray, Fl. N. Amer. 2: 40. 1841. *Oldenlandia purpurea* (Linnaeus) A. Gray var. *tenuifolia* (Nuttall) A. Gray ex Chapman, Fl. South. U.S. 181. 1860. *Houstonia longifolia* Gaertner var. *tenuifolia* (Nuttall) A. W. Wood, Class-Book Bot., ed. 1861. 403. 1861. *Houstonia purpurea* Linnaeus var. *tenuifolia* (Nuttall) A. Gray, Syn. Fl. N. Amer. 1(2): 26. 1884. *Chamisme tenuifolia* (Nuttall) Nieuwland, Amer. Midl. Naturalist 4: 92. 1915. *Hedyotis nuttalliana* Fosberg, Virginia J. Sci. 2: 111. 1941. *Hedyotis purpurea* (Linnaeus) Torrey & A. Gray var. *tenuifolia* (Nuttall) Fosberg, Castanea 19: 35. 1954.

Erect or ascending perennial herb, to 2.5 dm; stem glabrous or scaberulous, sometimes with short trichomes at the nodes. Leaves with the blade linear or linear-oblong, 1–3 cm long, 2–5 mm wide, 1-nerved, the apex acute to obtuse, the base cuneate, the margin entire, revolute, the upper and lower surfaces glabrous or minutely scaberulous, sometimes ciliate, sessile; stipules ca. 1 mm long, scarious, apiculate. Flowers in a terminal or axillary cyme, the pedicel 3–10 mm long; hypanthium ca. 1 mm long, glabrous; calyx lobes linear-lanceolate or subulate, ca. 2 mm long, glabrous; corolla salverform, purplish or pinkish to white, the tube 4–5(6) mm long, the lobes 2–3 mm long, pubescent within. Fruit subglobose, 2–3 mm long; seeds subglobose, ca. 1 mm long, minutely papillate, with an obscure ridge on the concave ventral face, black.

Sandhills. Rare; Washington and Walton Counties. Quebec south to Florida, west to Alberta, North Dakota, Kansas, and Oklahoma. Fall.

Houstonia micrantha (Shinners) Terrell [With small flowers.] SOUTHERN BLUET.

Hedyotis crassifolia Rafinesque var. *micrantha* Shinners, Field & Lab. 18: 100. 1950. *Hedyotis australis* W. H. Lewis & D. M. Moore, Southw. Naturalist 3: 208. 1958. *Houstonia micrantha* (Shinners) Terrell, Phytologia 31: 425. 1975.

Erect, ascending, or spreading annual herb, to 6(8) cm; stem glabrous. Leaves with the blade ovate or elliptic, 5–10 mm long, 3–4 mm wide, 1-nerved, the apex acute, the base cuneate, the margin entire, the petiole 2–5 mm long; stipules ca. 1 mm long, scarious, apiculate. Flowers solitary, terminal, borne on a pedicel extending above the leaves; hypanthium ca. 1 mm long; calyx lobes oblong-lanceolate, 2–3 mm long, subequaling the corolla tube; corolla salverform, 2–6 mm long, white. Fruit subglobose, 3–4 mm long, glabrous; seeds subglobose, ca. 1 mm long, with a deep pit on one side, the surface minutely papillate, brownish black.

Disturbed sites. Rare; Escambia County. Tennessee south to Florida, west to Oklahoma and Texas. Winter–spring.

Houstonia procumbens (J. F. Gmel.) Standl. [Prostrate, lying down.] INNOCENCE; ROUNDLEAF BLUET.

> *Poiretia procumbens* J. F. Gmelin, Syst. Nat. 2: 263. 1791. *Hedyotis veronicifolia* Steudel, Nomencl. Bot., ed. 2. 1: 729. 1840, nom. illegit. *Houstonia procumbens* (J. F. Gmelin) Standley, in Britton, N. Amer. Fl. 32: 26. 1918. *Hedyotis procumbens* (J. F. Gmelin) Fosberg, Castanea 19: 32. 1954.
>
> *Houstonia rotundifolia* Michaux, Fl. Bor.-Amer. 1: 85. 1803. *Anotis rotundifolia* (Michaux) de Candolle, Prodr. 4: 433. 1830. *Oldenlandia rotundifolia* (Michaux) A. Gray ex Chapman, Fl. South. U.S. 180. 1860. *Panetos rotundifolius* (Michaux) Nieuwland, Amer. Midl. Naturalist 4: 93. 1915. *Hedyotis rotundifolia* (Michaux) Torrey & A. Gray, Fl. N. Amer. 2: 39. 1841; non Sprengel, 1815. TYPE: "Floridae et Carolinae."
>
> *Hedyotis procumbens* (J. F. Gmelin) Fosberg var. *hirsuta* W. H. Lewis, Ann. Missouri Bot. Gard. 53: 378. 1966. *Houstonia procumbens* (J. F. Gmelin) Standley var. *hirsuta* (W. H. Lewis) D. B. Ward, Novon 14: 370. 2004. TYPE: FLORIDA: Walton Co.: Villa Tasso, 1 mi. W of Choctaw Bay, 28 May 1964, *McDaniel 4707* (holotype: FSU).

Prostrate or creeping perennial herb; stem to 4 dm long, sparsely pubescent or glabrate. Leaves with the blade elliptic, ovate, or suborbicular, 5–15 mm long, 3–12 mm wide, 3-nerved, the apex obtuse to rounded, the base rounded or cuneate, the margin entire, the upper and lower surfaces pubescent or glabrate, the petiole ca. 5 mm long; stipules ca. 1 mm long, scarious. Flowers solitary or few in a terminal or axillary cyme; sepal lobes elliptic or ovate, ca. 1–2 mm long; corolla salverform, 2–6 mm long; white. Fruit subglobose, 4–5 mm long, sparsely pubescent; seeds subglobose, concave on the ventral surface, minutely papillose.

Sandhills, dunes, flatwoods, hammocks, and disturbed sites. Common; nearly throughout. South Carolina south to Florida, west to Louisiana. Spring–fall.

Houstonia purpurea L. [Purple, in reference to the flower color.] LARGE SUMMER BLUET; VENUS' PRIDE.

> *Houstonia purpurea* Linnaeus, Sp. Pl. 105. 1753. *Knoxia purpurea* (Linnaeus) Poiret, in Lamarck, Encycl., Suppl. 3: 225. 1813. *Anotis purpurea* (Linnaeus) G. Don, Gen. Hist. 3: 535. 1834. *Hedyotis purpurea* (Linnaeus) Torrey & A. Gray, Fl. N. Amer. 2: 40. 1841. *Oldenlandia purpurea* (Linnaeus) A. Gray, Manual, ed. 2. 173. 1856. *Chamisme purpurea* (Linnaeus) Nieuwland, Amer. Midl. Naturalist 4: 92. 1915.

Erect or ascending perennial herb, to 4(5) dm; stem villous or glabrous. Leaves with the blade suborbicular, ovate, elliptic-ovate, to lanceolate-oblong, (0.5)2–4(5) cm long, 0.5–2(3) cm wide, 3-nerved, the apex acute to obtuse, the base rounded to subcordate, the margin entire, the upper and lower surfaces glabrate, sessile; stipules ca. 1 mm long, scarious, apiculate. Flowers in a terminal cyme, short-pedicellate; sepals lanceolate to elliptic or linear-lanceolate, 2–3 mm long, about as long as the hypanthium, glabrous or sometimes ciliate; corolla funnelform, purple, lilac, or white, the tube 6–7(11) mm long, the lobes ca. ½ as long as the tube, pubescent within. Fruit subglobose, 2–3 mm long; seed subglobose, concave on the ventral surface, minutely papillate.

Floodplain forests. Rare; Jackson County. Maine south to Florida, west to Michigan, Nebraska, Oklahoma, and Texas. Spring–summer.

Houstonia pusilla Schoepf [Very small.] TINY BLUET.

Houstonia pusilla Schoepf, Reise Nordamer. Staat. 2: 306. 1788.
Houstonia linnaei Michaux var. *minor* Michaux, Fl. Bor.-Amer. 1: 85. 1803. *Houstonia caerulea* Linnaeus var. *minor* (Michaux) Pursh, Fl. Amer. Sept. 106. 1814. *Hedyotis caerulea* (Linnaeus) Hooker var. *minor* (Michaux) Torrey & A. Gray, Fl. N. Amer. 2: 38. 1841. *Houstonia minor* (Michaux) Britton, Mem. Torrey Bot. Club 5: 302. 1894. TYPE: "Virginia ad Floridam," s.d., *Michaux s.n.* (holotype: P).
Houstonia patens Elliott, Sketch Bot. S. Carolina 1: 191. 1816. *Oldenlandia patens* (Elliott) Chapman, Fl. South. U.S., ed. 2. 625. 1883. *Oldenlandia caerulea* (Linnaeus) A. Gray var. *patens* (Elliott) M. Gómez de la Maza y Jiménez, Anales Soc. Esp. Hist. Nat. 23: 287. 1894.
Hedyotis crassifolia Rafinesque, Fl. Ludov. 77. 1817.

Ascending or spreading annual herb; stem to 1 dm long, glabrous. Leaves with the blade spatulate, ovate, or elliptic, 3–10 mm long, 2–8 mm wide, 1-nerved, the apex acute, the base cuneate, the margin entire, sessile or the petiole to 5 mm long; stipules 1–2 mm long, scarious, apiculate or cuspidate. Flowers terminal, solitary, the pedicel extending above the leaves; hypanthium ca. 1 mm long; sepal lobes oblong-lanceolate, 2–3 mm long, scabrous; corolla salverform, purple with a reddish center (yellow center when white-flowered), the tube 3–4 mm long, the lobes 3–4 mm long, oblong, glabrous. Fruit didymous-flattened, 3–4 mm wide; seeds subglobose, ca. 1 mm long, with a deep pit on one side, the surface minutely papillate, brownish black.

Disturbed sites. Occasional; Jefferson County, central and western panhandle. Maryland south to Florida, west to South Dakota, Nebraska, Kansas, Oklahoma, and Arizona. Winter–spring.

Ixora L. 1753. JUNGLEFLAME

Trees or shrubs. Leaves opposite, simple, pinnate-veined, petiolate or epetiolate, stipulate. Flowers in terminal, corymbose cymes or panicles, actinomorphic, bisexual, bracteate; sepals 4, basally connate; petals 4, basally connate; stamens 4, inserted at the corolla throat, the anthers dorsifixed; nectariferous disk present; ovary inferior, 2-carpellate and -loculate, the style 1, the stigmas 2. Fruit a drupe; pyrenes 2.

A genus of about 400 species; North America, West Indies, Mexico, Central America, South America, Africa, Asia, and Pacific Islands. [From the Sanskrit *Isvara*, the supreme lord, in reference to its beauty.]

1. Leaves short-petiolate; corolla tube to 0.7 cm long, white ....**I. pavetta**
1. Leaves sessile; corolla tube to nearly 5 cm long, red..**I. coccinea**

Ixora coccinea L. [Red, in reference to the flowers.] SCARLET JUNGLEFLAME.

Ixora coccinea Linnaeus, Sp. Pl. 110. 1753.

Shrub, to 1 m; branchlets glabrous. Leaves with the blade ovate, elliptic, or lanceolate, 4–8(16) cm long, 1.5–6.6 cm wide, the apex obtuse, apiculate, the base rounded or cordate, the margin entire, the upper and lower surfaces glabrous, sessile; stipules triangular, cuspidate. The flowers in a terminal corymbiform cyme, sessile, the peduncle short or absent; bracts ca. 8 mm long; hypanthium 1–2 mm long, glabrescent; sepal lobes ca. 1 mm long; corolla salverform, the tube filiform, 2.5–4.5 cm long, the lobes ovate, 8–10 mm long, red, glabrous; stamens with short filaments; style glabrous, exserted, the stigmas 2. Fruit globose, ca. 1 cm long, red; pyrenes 2, planoconvex.

Disturbed sites. Rare; Manatee, Palm Beach, and Lee Counties, southern peninsula. Escaped from cultivation. Florida; West Indies; Asia. Native to Asia. All year.

Ixora pavetta Andr. [From the Sri Lankan vernacular name *pawatta*.] SMALLFLOWER JUNGLEFLAME.

Ixora parviflora Vahl, Symb. Bot. 3: 11, t. 52. 1794; non Lamarck, 1791.
Ixora pavetta Andrews, Bot. Repos. 2: t. 78. 1799.
Ixora arborea Roxburgh ex Smith, in Rees, Cycl. 19: no. 5. 1811.

Shrub or small tree; branches glabrous. Leaves with the blade elliptic to oblong or oblanceolate, 7–15 cm long, 3.5–6.5 cm wide, coriaceous, the apex obtuse, acute, or short-acuminate, the base rounded, the margin entire, the upper and lower surfaces glabrous, the petiole 0.5–1 cm long; stipules triangular, cuspidate. Flowers numerous in a terminal corymbiform panicle, the pedicel 2–3 mm long, the peduncle 3–5 cm long; sepal lobes minute, obtuse; corolla salverform, the tube filiform, 6–7 mm long, white, the lobes reflexed, glabrous; anthers subequaling the corolla lobes; style pubescent, exserted, the stigmas 2. Fruit globose, ca. 1 cm long, black; pyrenes 2, planoconvex.

Disturbed sites. Rare; Palm Beach and Miami-Dade Counties. Escaped from cultivation. Florida; Asia. Native to Asia. All year.

Mitchella L. 1753.

Herbs. Leaves opposite, simple, pinnate-veined, petiolate, stipulate. Flowers terminal, in pairs, joined at their base, actinomorphic, bisexual, bracteate; sepals 4, basally connate; petals 4, basally connate; stamens 4, epipetalous; ovary inferior, 4-carpellate and -loculate, the style 1, the stigmas 4. Fruit a drupe.

A genus of 2 species; North America, Central America, and Asia. [Commemorates John Mitchell (1676–1768), Virginia botanist.]

Mitchella repens L. [Creeping, prostrate.] PARTRIDGEBERRY; TWINBERRY.

Mitchella repens Linnaeus, Sp. Pl. 111. 1753.

Creeping, suffrutescent, perennial herb; stem rooting at the nodes, irregularly short-pubescent in lines. Leaves with the blade ovate, 8–20 mm long and wide, persistent, the apex obtuse, the base rounded to cordate, the margin entire, the upper and lower surfaces glabrous, the petiole 5–10 mm long; stipules connate, membranous between the petiole bases, with triangular or ovate-triangular apex ca. 1 mm long. Flowers in pairs on a short stalk terminating the branchlets; hypanthium tube 3–4 mm long at the bases; sepal lobes minute; corolla narrow funnelform, white or pink-tinged, the tube 8–12 mm long, pubescent within and on the upper surface of the lanceolate to ovate, spreading or recurved lobes; stamens exserted and the style included or the stamens included and the style exserted. Fruits joined in pairs, each subglobose, 4–6 mm long, red or white, the minute calyx segments persistent; pyrenes 4.

Hammocks. Frequent; northern counties, central peninsula. Nova Scotia and New Brunswick south to Florida, west to Ontario, Minnesota, Iowa, Missouri, Oklahoma, and Texas; Mexico and Central America. Spring–fall.

Mitracarpus Zucc. 1827. GIRDLEPOD

Herbs. Leaves opposite, simple, pinnate-veined, petiolate, stipulate. Flowers in axillary clusters or terminal heads, actinomorphic, bisexual, bracteate; sepals 4, basally connate; petals 4, basally connate; stamens epipetalous; nectariferous disk present; ovary inferior, 2-carpellate and -loculate, the style 1, the stigmas 2. Fruit a circumscissile dehiscent capsule.

A genus of about 40 species; North America, West Indies, Mexico, Central America, South America, Africa, Asia, and Pacific Islands. [*Mitre*, turban, and *carpus*, fruit, in reference to the fruit shape.]

Mitracarpus hirtus (L.) DC. [With long, distinct trichomes.] TROPICAL GIRDLEPOD.

Spermacoce hirtus Linnaeus, Sp. Pl., ed. 2. 148. 1762. *Mitracarpus hirtus* (Linnaeus) de Candolle, Prodr. 4: 572. 1830.

Erect annual herb, to 5 dm; stem villous below the nodes. Leaves with the blade oblong or lanceolate, 3–5 cm long, 0.7–1.5 cm wide, the apex obtuse or acute, the base cuneate, the margin entire, the upper surface scabrous, the lower surface glabrous, the petiole short; stipules united with the petioles and forming a shallow sheath 1–4 mm long, with 8–10 bristles 1–5 mm long. Flowers in a terminal head and axillary clusters, short-stalked; hypanthium villous; sepal lobes subulate, to 1.5 mm long, basally connate, connate; corolla white, the tube ca. 2 mm long, the lobes ovate, ca. 0.5 mm long; stamens with the anthers short-oblong, glandular-apiculate, sessile; style included or exserted. Fruit globose, ca. 1 mm long, membranous, the upper part pubescent; seeds quadrangular, minutely pitted, the ventral surface with an X-shaped furrow, pale brown.

Disturbed sites. Rare; northern and central peninsula, central and western panhandle. Georgia south to Florida, west to Texas; West Indies, Mexico, Central America, and South America; Africa, Asia, Australia, and Pacific Islands. Native to tropical America. Summer.

Morinda L. 1753.

Shrubs or trees. Leaves opposite, simple, pinnate-veined, petiolate, stipulate. Flowers terminal or axillary, connate by the calyces and forming heads, actinomorphic, bisexual, bracteate; sepals 4–7, basally connate; petals 4–7, basally connate; stamens 5, epipetalous, with an evident connective; nectariferous disk present; ovary inferior, 2- to 4-carpellate and -loculate, the style 1, the stigmas 2. Fruit a fleshy syncarp, each segment a 2-loculate pyrene.

A genus of about 80 species; North America, West Indies, Central America, South America, Africa, Asia, Australia, and Pacific Islands. [Contraction of *Morus indica* L. (Moraceae).]

1. Larger leaf blades 25–30 cm long...**M. citrifolia**
1. Larger leaf blades 5–10 cm long ....**M. royoc**

Morinda citrifolia L. [With leaves like *Citrus* (Rutaceae).] INDIAN MULBERRY; NONI.

Morinda citrifolia Linnaeus, Sp. Pl. 176. 1753.

Shrub or tree, to 6 m; stems 4-angled, glabrous. Leaves with the blade broadly elliptic, 17–30 cm long, 8–18 cm wide, the apex acute to short acuminate, the base cuneate to rounded, the margin entire, the upper surface glabrous, the lower surface with tufts of axillary trichomes, the petiole stout, 1–2(3) cm long, slightly alate; stipules oblong or ovate, 1–2 cm long, the apex obtuse, connate with the petiole to form a sheath. Flowers in an axillary, pedunculate head; sepals forming a short tube, this truncate, the lobes minute; corolla infundibuliform, white, the tube 6–10 cm long, villous in the throat; stamens inserted in the corolla throat, the anthers 5–6 mm long, exserted; stigmas narrowly oblong, ca. 5 mm long. Fruit head oval to globose or ellipsoid-cylindric, 5–7(10) cm long, black.

Hammocks. Rare; Monroe County keys. Escaped from cultivation. Florida; West Indies, Mexico, Central America, and South America; Africa, Asia, Australia, and Pacific Islands. Native to Asia and Australia. All year.

Morinda royoc L. [From the Mayan *hoyoc*, the vernacular name for *M. yucatanensis*.] REDGAL.

Morinda royoc Linnaeus, Sp. Pl. 176. 1753.

Erect or clambering shrub, to 7 m; stem terete or slightly 4-angled when young, glabrous or glabrate. Leaves with the blade linear-lanceolate to oblanceolate or oblong-lanceolate, 5–10 cm long, 1–3 cm wide, the apex acute to acuminate, the base cuneate, the upper surface glabrous, the lower surface with tufts of axillary trichomes, the petiole stout, 5–10 mm long, slightly alate; stipules oblong to ovate, 1–2 mm long, the apex acute, connate with the petioles and forming a sheath. Flowers in a terminal or axillary, short-pedunculate head; sepals forming a short tube, this truncate, the lobes minute; corolla infundibuliform, white or pinkish, 6–8 cm long, the lobes shorter than the tube, villous in the throat; stamens inserted in the corolla throat, the anthers ca. 5–6 mm long, exserted; stigmas narrowly oblong, ca. 5 mm long. Fruiting head globose, 1–2.5 cm long, yellow.

Coastal hammocks, rarely inland. Occasional; central and southern peninsula. Florida; West Indies, Mexico, Central America, and South America. All year.

EXCLUDED TAXON

Morinda panamensis seemann—Dwyer (1980) cites this species as occurring in Florida. No Florida specimens known.

Oldenlandia L. 1753. MILLE GRAINES

Annual or perennial herbs. Leaves opposite, simple, pinnate-veined, petiolate or epetiolate, stipulate. Flowers axillary or terminal in subsessile or pedunculate cymes, actinomorphic, bisexual, bracteate; sepals 4, basally connate; petals 4, basally connate; stamens 4, epipetalous; ovary inferior or part-inferior, 2-carpellate and -loculate, the style 1, the stigmas 2. Fruit a capsule, dehiscing loculicidally across the top; seeds few to many.

A genus of about 100 species; nearly cosmopolitan. [Commemorates Henrik Bernard Oldenland (1663–1699), German-born physician and botanist employed by the Dutch East India Company to collect in South Africa.]

This genus will likely be broken up (e.g., *O. uniflora* appears to belong in *Edrastima*), however, additional species still require taxonomic consideration before revising the genus. Of the species here, only *O. corymbosa* appears to belong in *Oldenlandia* s.s.

Selected reference: Terrell (1990).

1. Flowers or fruits sessile or the pedicels to 3 mm long.
    2. Flowers all axillary; leaves linear to narrowly elliptic, 1–3(5) mm wide; stipules with conspicuous marginal teeth to 5 mm long; plant perennial ....**O. boscii**
    2. Flowers both axillary and terminal; leaves ovate to elliptic, 3–11 mm wide; stipules with inconspicuous marginal teeth to 2 mm long; plant annual..**O. uniflora**
1. Flowers or fruits distinctly pedicellate, the pedicels over 3 mm long.
    3. Plant an erect, spreading, or decumbent annual; corolla inconspicuous, somewhat obscured by the calyx lobes..**O. corymbosa**
    3. Plant a creeping perennial; corolla small, but not obscured by the calyx lobes...........**O. salzmannii**

Oldenlandia boscii (DC.) Chapm. [Commemorates Louis Augustin Guillaume Bosc (1759–1828), French botanist and invertebrate zoologist.] BOSC'S MILLE GRAINES.

Hedyotis boscii de Candolle, Prodr. 4: 420. 1830. *Oldenlandia boscii* (de Candolle) Chapman, Fl. South. U.S. 181. 1860.

Spreading, decumbent, or prostrate perennial herb; stem to 3 dm long, diffusely branching from the base, glabrous or short scabrid on the angles. Leaves with the blade linear-lanceolate or linear oblanceolate, 1–2.5(3) cm long, 1–3(5) mm wide, the apex acute, the base cuneate, the margin entire, the upper surface glabrous or minutely puberulent, the lower surface glabrous or scabrous on the midrib, sessile; stipules 1–2 mm long, connate between the leaf bases, with several marginal teeth or soft bristles ca. 2 mm long from the membranous part. Flowers 1–3 in an axillary and terminal sessile cyme; hypanthium short-cylindric, granular; sepal lobes triangular, 1–2 mm long, the outer surface glabrous or minutely papillose; corolla rotate, white, pink, or lavender, the lobes triangular to ovate, shorter than the calyx segments. Fruit subglobose, 2–3 mm long, fully or ⅞ inferior, papillose to glabrate; seeds numerous, ca. 0.5 mm long, trigonous, the surface smooth, dark purple.

Floodplain forests and cypress swamp margins. Occasional; Suwannee County, central and western panhandle. Virginia south to Florida, west to Oklahoma and Texas. Spring–fall.

Oldenlandia corymbosa L. [Flowers in a corymb.] FLATTOP MILLE GRAINES.

> *Oldenlandia corymbosa* Linnaeus, Sp. Pl. 119. 1753. *Hedyotis corymbosa* (Linnaeus) Lamarck, Tabl. Encycl. 1: 272. 1792.

Erect, spreading, decumbent, or prostrate annual herb; stem to 4 dm long, glabrous or puberulent. Leaves with the blade lanceolate or linear-lanceolate to narrowly elliptic, (5)1–2.5(4) cm long, 1–5(9) mm wide, the apex obtuse to acute, the base cuneate, the margin entire, glabrous or minutely scabrid, sessile or subsessile; stipules 1–2 mm long, connate, membranous between the leaf bases, usually with a few teeth or soft bristles from the membranous part, these to 3 mm long. Flowers (2)3(5) in an axillary cyme on a peduncle to ca. 1.5 cm long, pedicellate; hypanthium 2–3 mm long, glabrous, sepal lobes lanceolate, ca. 1 mm long, the margin minutely serrulate-pubescent; corolla rotate or short-funnelform, 1–2 mm long, white or pinkish, a little longer than the sepal lobes and somewhat obscured by them, the tube and lobes subequal, the lobes obovate, somewhat hooded, white bearded; ovary ⅘ to fully inferior. Fruit subglobose, 1–2 mm long, the apex extending slightly beyond the hypanthium top, glabrous; seeds numerous, trigonous, ca. 0.5 mm long, the surface alveolate-reticulate, dull brown.

Disturbed sites. Frequent; nearly throughout. North Carolina south to Florida, west to Texas; West Indies, Mexico, Central America, and South America; Africa, Asia, and Pacific Islands. Native to Africa, Asia, and Pacific Islands. Spring–summer.

Oldenlandia salzmannii (DC.) Benth. & Hook. f. ex B. D. Jacks. [Commemorates Philipp Salzmann (1781–1851), French botanist, entomologist, and physician.] SALZMANN'S MILLE GRAINES.

> *Anotis salzmannii* de Candolle, Prodr. 4: 433. 1830. *Hedyotis salzmannii* (de Candolle) Steudel, Nomencl. Bot., ed. 2. 1. 728. 1840. *Oldenlandia salzmannii* (de Candolle) Bentham & Hooker f. ex B. D. Jackson, Index Kew., Suppl. 2: 336. 1895.

Prostrate or creeping perennial herb; stem to 2 dm long, rooting at the nodes, glabrous. Leaves with the blade broadly elliptic to ovate, 2–5 mm long, 1–3 mm wide, sessile or subsessile, the apex obtuse, the base cuneate to rounded, the margin entire, the upper and lower surfaces glabrous or sparsely hirsute, sessile or subsessile; stipules ca. 0.5 mm long, connate, membranous between the leaf bases, entire or with 1–few marginal white trichomes to ca. 1 mm long. Flowers solitary, terminal or axillary, the pedicel 3–12 mm long; sepal lobes lanceolate, ovate, or oblong, 1–2 mm long, glabrate; corolla subsalverform, pink, lavender, purple, or white, 2–5 mm long, the tube 1–2 mm long, hirsutulous or pubescent within, the lobes ovate, equaling or longer than the tube. Fruit subglobose, 1–2 mm long, ⅞ to fully inferior, hirsute; seeds 4–14, ca. 0.5 mm long, trigonous.

Disturbed sites. Rare; Santa Rosa and Escambia Counties. Florida and Alabama; South America. Native to South America. Summer.

Oldenlandia uniflora L. [One-flowered.]. CLUSTERED MILLE GRAINES.

> *Oldenlandia uniflora* Linnaeus, Sp. Pl. 119. 1753. *Hedyotis uniflora* (Linnaeus) Lamarck, Tabl. Encycl. 1: 271. 1792. *Edrastima uniflora* (Linnaeus) Rafinesque, Actes Soc. Linn. Bordeaux 6: 269. 1834.
> *Oldenlandia glomerata* Michaux, Fl. Bor.-Amer. 1: 83. 1803. *Hedyotis glomerata* (Michaux) Elliott, Sketch Bot. S. Carolina 1: 188. 1816. *Stelmotis glomerata* (Michaux) Rafinesque, New Fl. 4: 101. 1838 ("1836"). *Stelmanis glomerata* (Michaux) Rafinesque, Autik. Bot. 13. 1840.
> *Hedyotis fasciculata* Bertoloni, Mem. Reale Accad. Sci. Ist. Bologna 2: 306, t. 17(2). 1850. *Oldenlandia fasciculata* (Bertoloni) Small, Fl. S.E. U.S. 1106, 1338. 1903. *Hedyotis uniflora* (Linnaeus) Lamarck var. *fasciculata* (Bertoloni) W. H. Lewis, Amer. J. Bot. 49: 865. 1962. *Oldenlandia uniflora* Linnaeus var. *fasciculata* (Bertoloni) D. B. Ward, Phytologia 94: 478. 2012.

Erect, ascending, or procumbent annual herb, to 6 dm; stem white-pilose or glabrous. Leaves with the blade lanceolate, ovate, or elliptic, 5–20 mm long, 4–10 mm wide, the apex obtuse, the base cuneate to subtruncate and clasping, the margin entire, the upper and lower surfaces glabrous or pubescent on the nerves, sessile or short-petiolate; stipules ca. 2 mm long, connate, membranous between the leaf bases, with 1–few linear or lanceolate marginal teeth, ciliate. Flowers 1–several, axillary or terminal, sessile or on a pedicel to 3 mm long; hypanthium pilose or glabrous; sepal lobes ovate or ovate-lanceolate, 1–2 mm long, pilose or glabrous; corolla rotate, white or pale blue, the tube short-cylindric, ca. 1 mm long, the lobes subequaling the tube. Fruit subglobose, 1–2 mm long, fully or ⅞ inferior; seeds numerous, trigonous, ca. 0.5 mm long, light to dark purple.

Flatwood ponds, pond margins, and coastal swales. Common; nearly throughout. New York south to Florida, west to Missouri, Oklahoma, and Texas; West Indies. Spring–fall.

Oldenlandiopsis Terrell & W. H. Lewis 1990.

Herbs. Leaves opposite, simple, pinnate-veined, petiolate, stipulate. Flowers axillary, solitary, actinomorphic, bisexual, bracteate; sepals 4, basally connate; petals 4, basally connate; stamens 4, epipetalous; ovary fully or partly inferior, 2-carpellate and -loculate, the style 1, the stigmas 2. Fruit a capsule; seeds numerous.

A monotypic genus; Florida and West Indies. [Resembling *Oldenlandia*.]

Selected reference: Terrell and Lewis (1990).

Oldenlandiopsis callitrichoides (Griseb.) Terrell & W. H. Lewis [Resembling *Callitriche* (Callitrichaceae).] CREEPING-BLUET.

Oldenlandia callitrichoides Grisebach, Mem. Amer. Acad. Arts, ser. 2. 8: 506. 1863. *Hedyotis callitrichoides* (Grisebach) W. H. Lewis, Rhodora 63: 222. 1961. *Oldenlandiopsis callitrichoides* (Grisebach) Terrell & W. H. Lewis, Brittonia 42: 185. 1990.

Creeping perennial herb; stem to 1 dm, rooting at the nodes, glabrous. Leaves with the blade ovate-orbiculate, 1–4 mm long, 1–5 mm wide, the apex obtuse to acute, the base truncate or rounded, the margin entire, the upper and lower surfaces glabrous or with a few scattered, stiff trichomes, the petiole 1–3 mm long; stipules triangular, ca. 1 mm long. Flowers solitary, the pedicel filiform, 2–3 times as long as the leaves; sepal lobes lanceolate to ovate, ca. 1 mm long, the margin ciliate; corolla subsalverform, white, the lobes sometimes purplish-tipped, 2–3 mm long, slightly longer than the tube; stamens inserted in the corolla throat, exserted; ovary fully or $^9/_{10}$ inferior, the stigma exserted. Fruit turbinate, 1–3 mm long, glabrous below the ciliate calyx lobes, dehiscent into 4 segments; seeds numerous, ellipsoid, ca. 0.5 mm long.

Wet, disturbed sites. Rare; Miami-Dade County. Florida; West Indies. Native to West Indies. All year.

Paederia L., nom. cons. 1767.

Vines. Leaves opposite, simple, pinnate-veined, petiolate, stipulate. Flowers axillary and/or terminal in thyrsiform, paniculate, cymose, or spiciform inflorescences, pedunculate, bracteate, sessile or pedicellate, actinomorphic, bisexual; sepals 5, basally connate; petals 5, basally connate; stamens 5, epipetalous, inserted in the corolla throat; ovary inferior, 2-carpellate and -loculate, the style 1, the stigmas 2. Fruit a drupe, dry and tardily schizocarpus, the exocarp fragmenting and exposing the 2 pyrenes.

A genus of about 30 species; North America, Mexico, South America, Africa, Asia, Australia, and Pacific Islands. [*Paedor*, odor, in reference to the fetid odor.]

Selected reference: Puff (1991).

1. Fruit laterally compressed; seeds conspicuously winged... P. cruddasiana
1. Fruit subglobose; seeds not winged.. P. foetida

Paederia cruddasiana Prain [Commemorates Lieutenant Cruddas, commandant of the battalion of the Frontier Police who assisted in the plant collections made in the Kachin Hills of British India (present day Myanmar).] SEWERVINE.

Paederia cruddasiana Prain, J. Asiat. Soc. Bengal 67: 295. 1898.

Vine, to 10 m; stem moderately to sparsely hirtellous or strigillose to glabrous. Leaves with the blade ovate to lanceolate, 5–16 cm long, 2–10.5 cm wide, the apex acute to acuminate, the base rounded to truncate or cordate, the margin entire, the upper surface glabrous or puberulent on the main veins, the lower surface sparsely hirtellous to glabrous on the blade, sparsely to moderately puberulent or hirtellous along the veins, the petiole 1–3 cm long, hirtellous or strigillose to glabrescent; stipules ovate-lanceolate to broadly triangular, 3–6 mm long, the apex acute or bifid. Flowers in an axillary and/or terminal pyramidal panicle 6–50 cm long, pedunculate, pedicellate, hirtellous to glabrescent; bracts triangular to linear, 0.5–2 mm long; hypanthium ellipsoid, ca. 1 mm long, puberulent to glabrescent; sepal lobes triangular, 1–3 mm long; corolla funnelform, purplish blue, lilac, or pink, the outer surface puberulent to tomentulose, tube 6–16 mm long, the lobes triangular, 2–4 mm long. Fruit ellipsoid to ovoid, 6–11 mm long, laterally compressed, glabrescent, yellowish gray; pyrenes ovate to elliptic, somewhat flattened, marginally winged.

Disturbed sites. Rare, but locally abundant; Broward and Miami-Dade Counties. Escaped from cultivation. Florida; Asia. Native to Asia. Spring–fall.

Paederia cruddasiana is listed as a Category I invasive species in Florida by the Florida Exotic Pest Plant Council (FLEPPC, 2017).

Paederia foetida L. [Bad-smelling.] SKUNKVINE.

Paederia foetida Linnaeus, Mant. Pl. 52. 1767.

Vine, to 5 m; stem densely puberulent, hirtellous, pilosulous, or glabrous. Leaves with the blade ovate, ovate-oblong, lanceolate, or elliptic, 5–9 cm long, 1–4(9) cm wide, the apex acute or acuminate, the base cuneate, rounded, truncate, or cordate, the margin entire, the upper and lower surfaces glabrous or puberulent, hirtellous, or strigillose at least on the principal veins, the petiole 0.5–9 mm long, glabrous, hirtellous, or pilosulous; stipules triangular to ovate, 2–6 mm long, the apex obtuse or acuminate, sometimes bifid. Flowers in an axillary and/or terminal pedunculate paniculate, thyrsiform, corymbiform, or cymose inflorescence 5–100 cm long, sessile or pedicellate, hirtellous, strigillose, or glabrous; bracts lanceolate to triangular, 1–3 mm long; hypanthium turbinate to ellipsoid, 1–2 mm long, puberulent or glabrous; sepal lobes triangular, ca. 1 mm long; corolla funnelform, pale purple, grayish pink, lilac, or grayish white, the outer surface mealy tomentulose, the tube 7–10 mm long, the lobes triangular to ovate, 1–2 mm long. Fruit globose, 4–7 mm long, glabrescent, yellow; pyrenes ovate to elliptic, concave-convex to plano-convex.

Hammocks and disturbed sites. Frequent; peninsula, central and western panhandle. Escaped from cultivation. North Carolina south to Florida, west to Texas; Africa, Asia, and Pacific Islands. Native to Asia. Spring–fall.

Paederia foetida is listed as a Category I invasive species in Florida by the Florida Exotic Pest Plant Council (FLEPPC, 2015).

Pentas Benth. 1844.

Perennial herbs or subshrubs. Leaves opposite, simple, pinnate-veined, petiolate, stipulate. Flowers in terminal cymes, bracteate, actinomorphic, bisexual, distylous; sepals 5, basally connate; petals 5, basally connate; stamens 5, epipetalous, inserted in the corolla throat; ovary inferior, 2-carpellate and -loculate, the style 1, the stigmas 2. Fruit a capsule, apically loculicidally dehiscent; seeds numerous.

About 15 species, North America, Africa, and Asia. [From the Greek *pente*, five, in reference to flower parts in fives.]

Pentas lanceolata (Forssk. Deflers [Lance-shaped, in reference to the leaves.] EGYPTIAN STARCLUSTER.

Ophiorrhiza lanceolata Forsskal, Fl. Aegypt.-Arab. 42. 1775. *Manettia lanceolata* (Forsskal) Vahl, Symb. Bot. 1: 12. 1790. *Pentas lanceolata* (Forsskal) Deflers, Voy. Yemen 142. 1889.

Erect perennial herb or subshrub, to 7 dm; stem 4-angled, moderately to densely pilosulous to villous, glabrescent in age. Leaves with the blade oblong-lanceolate to ovate, 5–14 cm long, 2–5.5 cm wide, the apex acute or short-acuminate, the base cuneate to obtuse, the margin entire, the upper surface scabrous or villous to glabrescent, the lower surface densely villous at least along the main veins, the petiole 0.5–3 cm long, pilosulous to villous; stipules truncate to broadly rounded, 1–2 mm long, villous, the margin with 1–5 bristles 1–4 mm long. Flowers in a congested terminal cyme 1.5–4 cm long and wide, sessile or subsessile, the peduncle 3–12 cm long; bracts narrowly triangular to linear, ca. 1 mm long; hypanthium subglobose to ovoid, densely hirtellous or villous, ca. 1 mm long; sepal lobes narrowly oblanceolate to spatulate, 2–8 mm long, usually unequal, the apex acute, densely hirtellous or villous; corolla salverform, purple, pink, red, white, or yellow, sparsely hirtellous to glabrescent, the tube to 2 cm long, slender except in the abruptly swollen throat around the stamens in the long-styled form, densely barbellate in the throat, the lobes elliptic to oblong-lanceolate, 3–4 mm long, the apex acute to obtuse. Fruit obovoid, 4–6 mm long, stiffly papery to woody, with a beak 1–2 mm long; seeds subglobose, ca. 1 mm long.

Disturbed sites. Rare; Miami-Dade County. Escaped from cultivation. Florida; Africa and Asia. Native to Africa and Asia. Summer.

Pentodon Hochst. 1844.

Herbs. Leaves opposite, simple, pinnate-veined, petiolate, stipulate. Flowers in axillary or terminal cymes; sepals 5, basally connate; petals 5, basally connate; stamens 5; epipetalous; ovary inferior, 1-carpellate and -loculate, the style 1, the stigmas 2. Fruit a capsule; seeds numerous.

A genus of 2 species; North America, West Indies, Mexico, Central America, South America, Africa, and Asia. [From the Greek *pente*, five, and *odontos* tooth, in reference to the calyx.]

Pentodon pentandrus (Schumach. & Thonn.) Vatke [From the Greek *pente*, five, and *andros*, in reference to the 5 stamens.] HALE'S PENTADON.

Hedyotis pentandra Schumacher & Thonning, in Schumacher, Beskr. Guin. Pl. 71. 1827. *Oldenlandia pentandra* (Schumacher & Thonning) de Candolle, Prodr. 4: 427. 1830. *Pentodon pentandrus* (Schumacher & Thonning) Vatke, Oesterr. Bot. Z. 25: 231. 1875.

Hedyotis halei Torrey & A. Gray, Fl. N. Amer. 2: 42. 1841. *Pentodon halei* (Torrey & A. Gray) A. Gray, Syn. Fl. N. Amer. 1(2): 28. 1884. *Oldenlandia halei* (Torrey & A. Gray) Chapman, Fl. South. U.S. 181. 1860.

Erect, weakly ascending, or creeping, annual, succulent herb, to 2 dm; stems 4-angled, usually much branched at the base, glabrous. Leaves with the blade ovate to elliptic or elliptic-lanceolate, 2–5 cm long, 0.5–2.5 cm wide, the apex obtuse to acute, the base cuneate, decurrent on the petiole, the margin with very small, thick, transparent trichomes that easily slough off leaving the edge rough or with minute cuplike depressions, the upper and lower surfaces glabrous, the petiole ca. 1 cm long; stipules connate, sheathing membranous between the petiole bases, irregularly few-toothed and -appendaged distally. Flowers in a short axillary or terminal cyme, the pedicel thick, 3–4 mm long; hypanthium clavate to obovate, glabrous; sepal lobes 5, deltoid to lanceolate, ca. 1 mm long, glabrous or with a few deciduous trichomes like those on the leaf margin; corolla funnelform, white, the tube 2–3 mm long, the lobes short, ovate to lanceolate, shorter than or subequaling the tube, pilose within and on the upper surface of the lobes; stamens with the filaments adnate above the middle of the corolla tube, the anthers included; ovary inferior. Fruit 2–4 mm long, 2-valved, glabrous; seeds angular, ca. 0.5 mm long, alveolate-reticulate, the reticules reddish brown, the pits pale.

Calcareous floodplain forests. Frequent; peninsula west to central panhandle. South Carolina south to Florida, west to Texas; West Indies, Mexico, Central America, and South America; Africa and Asia. Native to Africa and Asia. Spring–fall.

Pinckneya Michx. 1803.

Shrubs or trees. Leaves opposite, simple, pinnate-veined, petiolate, stipulate. Flowers in terminal and subterminal cymes; sepals 5, basally connate, 1 or more enlarged and petaloid; petals 5, basally connate; stamens 5, epipetalous; ovary partly inferior, 2-carpellate and -loculate, the style 1, the stigma 1. Fruit a 2-valved capsule; seeds numerous.

A monotypic genus; North America. [Commemorates Charles Coteswoth Pinckney (1746–1825), American statesman of South Carolina, Revolutionary War veteran, and delegate to the Constitutional Convention.]

Pinckneya bracteata (W. Bartram) Raf. [Bracteate, in reference to the large floral bracts.] FEVERTREE.

Bignonia bracteata W. Bartram, Travels Carolina 16, 468. 1791. Pinckneya bracteata (W. Bartram) Rafinesque, Casket (Philadelphia) 2: 194. 1827.
Pinckneya pubens Michaux, Fl. Bor.-Amer. 1: 103, t. 13. 1803. Cinchona caroliniana Poiret, in Lamarck, Encycl. 6: 40. 1804, nom. illegit. Pinckneya pubescens Persoon, Syn. Pl. 1: 197. 1805, nom. illegit. Cinchona pubens (Michaux) Hosack, Hort. Elgin. 15: 1811.

Shrub or small tree; branchlets pubescent, tawny, becoming glabrate and reddish brown, with pale, raised, corky lenticels, the leaf scars subcordate, each with a narrow, crescent-shaped vascular bundle scar. Leaves with the blade ovate or elliptic, 4–20 cm long, 2.5–12 cm wide, the apex acute to obtuse, the base broadly cuneate to rounded, the margin undulate, the upper surface with scattered short trichomes, the lower surface moderately pubescent, the petiole 1–3 cm long; stipules triangular, connate between the leaves, soon deciduous and leaving a scar between the petioles. Flowers few in a cyme on the branches of the season and often from 1–2 nodes below; sepals with at least 1 lobe of some flowers greatly enlarged, leaflike but petaloid, ovate, the largest ones 6–7 cm long, 4–5 cm wide, the apex obtuse to rounded, usually pink or yellow, the other lobes lance-subulate, 1–1.5 cm long; corolla greenish yellow, mottled with brown or purple, the tube narrowly funnelform, 1.5–2.5 cm long, the lobes triangular-lanceolate, much shorter than the tube, recurved; stamens inserted slightly above the corolla base, exserted well beyond the throat; ovary inferior or nearly so, the style longer than the stamens, the stigma capitate. Fruit subglobose to ovoid, 6–8 mm long, woody, the apex ringed by the perianth scar; seeds elliptic, 5–8 mm long, flat, with a thin membranous wing, the surface finely reticulate, tan.

Creek swamps and bogs. Occasional; northern peninsula west to central panhandle, Marion County. South Carolina, Georgia, and Florida. Spring–fall.

Pinckneya bracteata is listed as threatened in Florida (Florida Administrative Code, Chapter 5B-40).

Psychotria L., nom. cons. 1759. WILD COFFEE

Shrubs. Leaves opposite, simple, pinnate-veined, petiolate, stipulate. Flowers in terminal and subterminal, paniculate cymes, actinomorphic, bisexual, monomorphic or distylous, bracteate; sepals 5, basally connate, petals 5, basally connate; stamens 5, epipetalous; ovary inferior, 2-carpellate and -loculate, the style 1, the stigmas 2. Fruit a drupe; pyrenes 2.

A pantropical genus of about 2,000 species. [From the Greek *psyche*, mind or soul, in reference to the properties of some of its chemical compounds.]

Selected reference: Burch et al. (1975).

1. Leaf blade black-dotted .... P. punctata
1. Leaf blade not black-dotted.
    2. Calyx truncate, with 5 minute teeth on the rim; leaves glossy green when fresh .... P. nervosa
    2. Calyx with 5 conspicuous deltoid lobes; leaves dull blue-green when fresh.
        3. Domatia present in the axils of most secondary veins; leaves glabrous or nearly so, usually less than 6 cm long and 2 cm wide .... P. ligustrifolia
        3. Domatia absent; leaves densely and finely pilose on the lower surface, usually more than 10 cm long and 3 cm wide .... P. tenuifolia

Psychotria ligustrifolia (Northr.) Millsp. [With leaves like *Ligustrum* (Oleaceae).] BAHAMA WILD COFFEE.

Myrstiphyllum ligustrifolium Northrop, Mem. Torrey Bot. Club 12: 68, pl. 17. 1902. *Psychotria ligustrifolium* (Northrop) Millspaugh, Publ. Field Columb. Mus., Bot. Ser. 2: 172. 1906.
Psychotria bahamensis Millspaugh, Bull. New York Bot. Gard. 3: 451. 1905.

Shrub, to 2 m; branches smooth, glabrous. Leaves with the blade lanceolate to oblanceolate, 5–12 cm long, 1.5–3 cm wide, the apex acute to acuminate, the base cuneate, the margin entire, the upper surface glabrous, dull, the lower surface glabrous or sparsely pilose, especially on the midrib, the lateral veins 8–10 pairs, with domatia in the axils of most secondary veins, the petiole 0.5–1 cm long; stipules orbicular, sheathing, apiculate. Flowers in a paniculate cyme, the peduncle 1–2 cm long; calyx ca. 1 mm long, truncate, the lobes narrowly deltoid, the margin ciliate; corolla tubular, the tube 3–4 mm long, the outer surface glabrous, bearded within at the throat, the lobes elliptic, subequaling the tube, strongly reflexed, the apex acute. Fruit ellipsoid, red, ca. 5 mm long; pyrenes costate.

Rockland hammocks and pine rocklands. Rare; Miami-Dade County, Monroe County keys. Florida; West Indies. Spring–summer.

Psychotria ligustrifolia is listed as endangered in Florida (Florida Administrative Code, Chapter 5B-40).

Psychotria nervosa Sw. [In reference to the conspicuous leaf venation.] WILD COFFEE.

Psychotria nervosa Swartz, Prodr. 43. 1788.
Psychotria undata Jacquin, Pl. Hort. Schoenbr. 3: 5, pl. 260. 1798. *Uragoga undata* (Jacquin) Baillon, Hist. Pl. 7: 371. 1880. *Myrstiphyllum undatum* (Jacquin) Hitchcock, Rep. (Annual) Missouri Bot. Gard. 4: 95. 1893. *Psychotria undata* Jacquin var. *glabra* Urban, Symb. Antill. 4: 498. 1911, nom. inadmiss.
Psychotria lanceolata Nuttall, Amer. J. Sci. Arts 5: 290. 1822. *Psychotria nervosa* Swartz var. *lanceolata* (Nuttall) Sargent, Trees & Shrubs 2: 185. 1911. TYPE: FLORIDA: s.d., *Ware s.n.* (holotype: PH).

Shrub, to 3 m; branchlets smooth, glabrous or reddish brown-pubescent, becoming rough, brown, and glabrate. Leaves with the blade ovate to lanceolate or oblong, 6–12 cm long, 2–5 cm wide, the apex acuminate, the base cuneate, the margin entire or undulate, the upper surface glabrous, lustrous, the lateral veins in 12–14 pairs, deeply impressed along the veins, the lower surface glabrate or densely pubescent on the nerves, with domatia in the axils of most of the secondary veins, the petiole 0.5–1 cm long; stipules at first enclosing the terminal buds, later cleft, 1–1.5 cm long, caducous and leaving a rusty scar. Flowers in a paniculate cyme 3–5 cm long, sessile or subsessile, the branches reddish puberulent to glabrate; hypanthium narrowly ovoid, longer than the calyx; calyx ca. 1 mm long, truncate-undulate, with 5 minute lobes on the rim; corolla white, the tube ca. 4 mm long, glabrous externally, pilose within at the throat, the lobes ca. ½ as long as the tube. Fruit ellipsoid to oval, 5–7 mm long, red; pyrenes costate.

Hammocks. Common; peninsula. Florida; West Indies, Mexico, Central America, and South America. Spring–summer.

Psychotria punctata Vatke [In reference to the black nitrogenous dots on the leaves.] DOTTED WILD COFFEE.

Psychotria punctata Vatke, Oesterr. Bot. Z. 25: 230. 1875. *Uragoga punctata* (Vatke) Kuntze, Revis. Gen. Pl. 2: 962. 1891. *Apomuria punctata* (Vatke) Bremekamp, Verh. Kon. Ned. Akad. Wetensch., Afd. Natuuk., Tweede Sect., ser. 2. 54(5): 91. 1963.

Shrub, to 3 m; branches smooth, glabrous, glaucous. Leaves with the blade narrowly elliptic-obovate to ovate, 4–8 cm long, 2–4 cm wide, the apex rounded to obtuse, the base cuneate, the margin entire, narrowly revolute, the lateral veins usually in 5 pairs, inconspicuous, the upper surface glabrous, the lower surface paler, black-dotted with nitrogenous nodules, domatia absent, the petiole 1–2 cm long; stipules deltoid, bicuspidate. Flowers in a paniculate cyme, the peduncle 1–2 cm long; hypanthium narrowly ovoid, 1–2 mm long, glabrous; calyx truncate-undulate, ca. 1 mm long, glabrous; corolla white, the tube 3–4 mm long, glabrous externally, pilose at the throat within, the lobes subequaling the tube. Fruit ovoid, 5–6 mm long, red; pyrenes costate.

Disturbed sites. Rare; Monroe County keys. Escaped from cultivation. Florida; Africa. Native to Africa. All year.

Psychotria tenuifolia Sw. [*Tenuis*, slender, and *folium*, leaf, in reference to the slender leaves.] SHORTLEAF WILD COFFEE.

Psychotria tenuifolia Swartz, Prodr. 43. 1788. *Uragoga tenuifolia* (Swartz) Kuntze, Revis. Gen. Pl. 2: 963. 1891.
Psychotria sulzneri Small, Fl. Miami 176. 1913.
Psychotria pulverulenta Urban, Symb. Antill. 7: 456. 1913.

Shrub, to 2 m; branchlets pubescent. Leaves with the blade narrowly oblong to elliptic-lanceolate, 8–15 cm long, 3–10 cm wide, the apex acuminate or acute, the base cuneate, the margin entire, the lateral veins usually in 10–12 pairs, the upper surface pubescent on the veins, the lower surface densely and finely pilose, domatia absent, the petiole 3–6 mm long, pubescent; stipules suborbicular, the apex apiculate. Flowers in a paniculate cyme; hypanthium ca. 1 mm long; calyx ca. 1 mm long, with 5 deltoid lobes; corolla green, the tube 2–3 mm long, glabrous externally, pilose at the throat within, the lobes subequaling the tube. Fruit ca. 5 mm long, red, orange, or yellow; pyrenes costate.

Hammocks; frequent; central and southern peninsula. Florida; West Indies, Mexico, and Central America. Spring–summer.

Randia L. 1753. INDIGOBERRY

Shrubs or trees. Leaves opposite, simple, pinnate-veined, petiolate, stipulate. Flowers solitary or terminal or subterminal on short axillary branches, actinomorphic, bisexual; sepals 5, basally connate; petals 5, basally connate; stamens 5, epipetalous; ovary inferior, 2-carpellate and -loculate, the style 1, the stigma 2-lobed. Fruit a berry; seeds 5–10.

A genus of 90 species; North America, West Indies, Mexico, Central America, and South America. [Commemorates Isaac Rand (1674–1743), English botanist and apothecary, who was a lecturer and director at the Chelsea Physic Garden.]

Randia aculeata L. [With prickles or spines, in reference to the axillary spines.] WHITE INDIGOBERRY.

Randia aculeata Linnaeus, Sp. Pl. 1192. 1753. *Gardenia randia* Swartz, Prodr. 52. 1788, nom. illegit. *Gardenia aculeata* (Linnaeus) Aiton, Hort. Kew. 1: 295. 1789. *Randia latifolia* Lamarck, Encycl. 3: 24. 1789, nom. illegit.

Randia mitis Linnaeus, Sp. Pl. 1192. 1753. *Gardenia randia* Swartz var. *mitis* (Linnaeus) Swartz, Prodr. 52. 1788. *Randia latifolia* Lamarck var. *mitis* (Linnaeus) A. de Candolle, Prodr. 4: 385. 1830. *Randia aculeata* Linnaeus var. *mitis* (Linnaeus) Grisebach, Fl. Brit. W.I. 318. 1861. *Mussaenda mitis* (Linnaeus) Moçiño, in sessé y Lacasta & Moçiño, Fl. Mexic., ed. 2. 60. 1894. *Randia aculeata* Linnaeus forma *mitis* (Linnaeus) Steyermark, Mem. New York Bot. Gard. 23: 340. 1972.

Randia aculeata Linnaeus forma *minor* Steyermark, Mem. New York Bot. Gard. 23: 341. 1972.

Shrub or small tree, to 3 m; branchlets angular or terete, with paired woody, axillary spines 1–2 cm long or spineless, glabrous, scaberulous, or appressed pilose. Leaves usually clustered at the end of the branchlet or on a short lateral shoot, the blade obovate-oblong, oblanceolate, orbicular, or elliptic-oblong, 1–6 cm long, 0.5–3 cm wide, coriaceous, the apex rounded or obtuse, apiculate or slightly emarginate, the base cuneate or rounded, the margin entire, slightly revolute, the upper surface glabrous, lustrous, the lower surface glabrous or sparsely pilose along the midrib, sessile or the petiole to 2 mm long; stipules oblong-deltoid, ca. 2 mm long, apiculate, pilose within at the base. Flowers solitary or clustered at the end of short axillary branches, sessile or subsessile; hypanthium 2–3 mm long, glabrous or pilose; calyx lobes 5, linear to ovate, ca. 1 mm long, glabrous or ciliate; corolla funnelform, white, the tube 6–8 mm long, the outside glabrous, villous at the throat within, the lobes ovate to oblong, acute or acuminate, equaling or longer than the tube; styles connate or free. Fruit globose, 6–13 mm long, sessile, white, smooth, glabrous, the pericarp leathery; seeds 5–10, 4–5 mm long, brownish black, embedded in a mucilaginous material.

Coastal hammocks, Frequent; central and southern peninsula. Florida; West Indies, Mexico, Central America, and South America. All year.

Richardia L. 1753. MEXICAN CLOVER

Annual or perennial herbs. Leaves opposite, simple, pinnate-veined, petiolate or epetiolate, stipulate. Flowers sessile or short-pedicellate in pedunculate, terminal heads subtended by involucral leaves; sepals (3)4–6, basally connate; petals (3)4–6, basally connate; stamens (3)4—6, epipetalous; ovary inferior, 3–4(6)-carpellate and -loculate, the locules indehiscent, the style 1, the stigma 1- or 3–4(6)-lobed. Fruit a schizocarp, dehiscent into 3–4(6) dry, 1-seeded mericarps.

A genus of about 15 species; North America, Mexico, Central America, South America, Africa, Asia, Australia, and Pacific Islands. [Commemorates Richard Richardson (1663–1741), British bryologist.]

Selected reference: Lewis and Oliver (1974).

1. Corolla 4-merous; fruit smooth ....**R. humistrata**
1. Corolla usually 6-merous; fruit tuberculate or hirsute.
    2. Fruit hirsute, the inner surface of each mericarp broad and with a median keel ...... **R. brasiliensis**
    2. Fruit tuberculate, the inner surface of each mericarp with a narrow groove.
        3. Corolla 5–7 mm long ....**R. scabra**
        3. Corolla 12–20 mm long...**R. grandiflora**

Richardia brasiliensis Gomes [Of Brazil.] TROPICAL MEXICAN CLOVER.

Richardia brasiliensis Gomes, Mem. Ipecac. Bras. 31, t. 2. 1801. *Richardsonia brasiliensis* (Gomes) Hayne, Getreue Darstell. Gew. 8: t. 21. 1822.

Ascending or prostrate annual or perennial herb, stem to 4 dm long, sometimes rooting at the lower nodes, hirsute. Leaves with the blade elliptic to ovate, 1.5–4 cm long, 0.8–2 cm wide, the apex acute to obtuse, the base cuneate, the margin entire, the upper and lower surfaces scabrous, subsessile or the petiole to 1 cm long; stipules connate and forming a sheath ca. 2 mm long, the marginal setae 2–5 mm long. Flowers numerous in a terminal head, the involucrate leaves 1 or 2-paired, broadly ovate, the second pair smaller; calyx (4)6-lobed, the lobes lanceolate to linear-lanceolate, 2–4 mm long, glabrous or scabrous, the margin ciliate; corolla funnelform, white or rose, the tube 3–8 mm long, glabrous, the lobes (4)6, 1–3 mm long, with apical tufts of trichomes; stamens (4)6, the filaments short; ovary 3-carpellate, the stigma capitate, 3-lobed. Fruit with 3 mericarps, these broadly ovate, 3–4 mm long, papillose and strigose, the face with a prominent median keel, concave, glabrous.

Disturbed sites. Common; nearly throughout. Pennsylvania and New Jersey south to Florida, west to Texas; West Indies, Mexico, and South America; Africa, Asia, and Pacific Islands. Native to South America.

Richardia grandiflora (Cham. & Schltdl.) Steud. [Large-flowered.] LARGEFLOWER MEXICAN CLOVER.

Richardsonia grandiflora Chamisso & Schlechtendal, Linnaea 3: 351. 1828. *Richardia grandiflora* (Chamisso & Schlechtendal) Steudel, Nomencl. Bot., ed. 2. 2: 459. 1841.

Procumbent annual or perennial herb; stem to 7 dm long, rooting at the nodes, hirsute. Leaves with the blade linear-lanceolate or narrowly elliptic, 1.5–4.5 cm long, 2–8 cm wide, the apex acute to obtuse, the base cuneate, the margin entire, the upper and lower surfaces hirsute to strigose, sessile or the petiole 1–2 mm long; stipules connate and forming a sheath 2–3 mm long, the marginal setae 3–8 mm long. Flowers numerous in a terminal head, the involucral leaves 2-paired, broadly ovate, the second pair smaller; calyx 6-lobed, the lobes linear-lanceolate, 4–9 mm long, glabrous or scabrous, the margin ciliate; corolla funnelform to salverform, white, pink, or lilac, 12–20 mm long, the tube 3–8 mm long, the lobes 6, 3–5 mm long, often with apical tufts of trichomes; stamens 6; ovary 3-carpellate, the stigma clavate, 3-lobed. Fruit with 3 mericarps, these broadly ovate, 3-4 mm long, papillose and with blunt, rounded excrescences, the face with a narrow groove, glabrous.

Disturbed sites. Occasional; Alachua County, central and southern peninsula. Florida; South America. Native to South America. Summer–fall.

Richardia grandiflora is listed as a category II invasive species in Florida by the Florida Exotic Pest Plant Council (FLEPPC, 2017).

Richardia humistrata (Cham. & Schltdl.) Steud. [*Humus*, on the ground, and *stratosus*, in distinct layers, in reference to its growth habit.] SOUTH AMERICAN MEXICAN CLOVER.

Richardsonia humistrata Chamisso & Schlechtendal, Linnaea 3: 353. 1828. *Richardia humistrata* (Chamisso & Schlechtendal) Steudel, Nomencl. Bot., ed. 2. 1: 459. 1841.

Prostrate perennial herb; stem to 8 cm long, often rooting at the nodes, hispid. Leaves with the blade elliptic-ovate, 1–2 cm long, 6–8 mm wide, the apex acute to obtuse, the base cuneate, the margin entire, the upper and lower surfaces hispid, sessile or the petiole 1–2 mm long; stipules connate and forming a sheath ca. 1 mm long, the marginal setae ca. 3 mm long. Flowers numerous in a terminal head, the involucral leaves 2 pairs, ovate to broadly ovate; calyx 4-lobed, the lobes linear-lanceolate, ca. 2 mm long, hirsute, the margin ciliate; corolla funnelform or rotate, white, ca. 2 mm long, the lobes 4, ca. 1 mm long; stamens 4; ovary 4-carpellate, the stigma capitate, 4-lobed. Fruit with 4 mericarps, these ovate, 2 mm long, glabrous, the face flat, smooth, glabrous.

Disturbed sandhills and flatwoods. Occasional; Jackson County, western panhandle. New Jersey, Alabama, Mississippi, and Florida; Mexico and South America; Africa. Native to South America. Summer–fall.

Richardia scabra L. [Rough to the touch.] ROUGH MEXICAN CLOVER.

Richardia scabra Linnaeus, Sp. Pl. 330. 1753. *Richardsonia scabra* (Linnaeus) A. Saint-Hilaire, Pl. Usuel. Bras. 8: t. 8. 1822.

Erect or decumbent annual herb; stem to 8 dm long, sometimes rooting at the nodes, hirsute. Leaves with the blade ovate to elliptic-lanceolate, 1–6.5 cm long, 3–13 mm wide, the apex acute to obtuse, the base cuneate, the margin entire, the upper and lower surfaces scabrous, sessile or the petiole to 5 mm long; stipules connate and forming a sheath ca. 2 mm long, the marginal setae 2–5 mm long. Flowers numerous in a terminal head, the involucral leaves 1- to 2-paired, broadly ovate, the second pair smaller; calyx 3–4 mm long, 6-lobed, the lobes lanceolate, twice as long as the tube length, glabrous, the margin ciliate; corolla funnelform to salverform, white, occasionally pink-tinged, 6-lobed, 5–7 mm long, glabrous; stamens 6, the filaments short; ovary 3(6)-carpellate, the stigma capitate, 3-lobed. Fruit with 3(6) mericarps, these oblong or obovate, 2–4 mm long, papillose and strigillose, the face closed to a narrow groove.

Disturbed sites. Frequent; nearly throughout. New Jersey south to Florida, west to Texas; West Indies, Mexico, Central America, and South America; Africa. Native to North America and tropical America. All year.

Sherardia L. 1753.

Annual herbs. Leaves whorled, simple, petiolate, stipulate. Flowers in terminal and axillary, involucrate heads; sepals 4, basally connate; petals 4, basally connate; stamens 4, epipetalous; ovary 2-carpellate and -loculate, the style 1, the stigmas 2. Fruit a nutlet.

A monotypic genus; nearly cosmopolitan. [Commemorates William Sherard (1659–1728), English botanist.]

Sherardia arvensis L. [Of fields or cultivated land.] BLUE FIELDMADDER.

> *Sherardia arvensis* Linnaeus, Sp. Pl. 102. 1753. *Hexodontocarpus arvensis* (Linnaeus) Dulac, Fl. Hautes-Pyrénées 467. 1867.

Ascending, decumbent, or prostrate, tufted annual herb; stem to 2 dm long, 4-angled, hispidulous. Leaves in a whorl of 4(6), the blade of the lower ones usually obovate, the apex mucronate, that of the upper ones linear-lanceolate, 6–10 mm long, 2–4 mm wide, the apex acute, sharppointed, the base cuneate, the margin rough-ciliate, the upper and lower surfaces hispidulous; stipules foliaceous. Flowers 2–3 in a slender, pedunculate, involucrate head of 6–8 leaves, these lanceolate, the apex acute; calyx lobes obovoid or triangular; corolla funnelform or salverform, pink or bluish, (3)4–5 mm long, the lobes as long as the tube, spreading. Fruit obovoid, hispidulous, crowned with the persistent calyx lobes.

Disturbed sites. Rare; Sumter, Jefferson, and Leon Counties. Nearly cosmopolitan. Native to Europe and Asia. Spring.

Spermacoce L. 1753. FALSE BUTTONWEED

Annual or perennial herbs or subshrubs. Leaves opposite, simple, pinnate-veined, petiolate, stipulate. Flowers axillary and solitary or in axillary and terminal clusters or heads, actinomorphic, bisexual; sepals 4, basally connate; petals 4, basally connate; stamens 4, epipetalous; ovary inferior, 2-carpellate and -loculate, the style 1, the stigmas 2. Fruit a capsule, dehiscing into 2 mericarps, each 1-seeded, 1 carpel opening to release the seed, the other usually remaining closed by a fragile, easily removed septum.

A genus of about 150 species; nearly cosmopolitan. [From the Greek *sperma*, seed, and *acoce*, point, in reference to the calyx teeth on the fruits.]

Borreria is sometimes segregated to include species with fruits in which both cocci dehisce, however it is unclear if this distinction is tenable (Delprete, 2007).

Borreria G. Mey., nom. cons., 1818.

1. Calyx with 2 long lobes, the other 2 very short or vestigial.
    2. Terminal head (1)2–3 cm wide .... **S. densiflora**
    2. Terminal head 0.5–1 cm wide.
        3. Flowers in a dense terminal head, occasionally also in a penultimate leaf axil; corolla tube 1.5–2.5 mm long .... **S. neoterminalis**
        3. Flowers in a terminal and several upper leaf axils; corolla tube 0.5–1 mm long ..... **S. verticillata**
1. Calyx with 4 subequal lobes.
    4. Leaves and stem hispid or pilose.
        5. Leaves elliptic to ovate-oblong; corolla pilose-hirtellous on the outer surface .... **S. latifolia**
        5. Leaves linear-lanceolate to oblong-lanceolate; corolla glabrous on the outer surface ....
        .... **S. tetraquetra**
    4. Leaves and stem glabrous or scabrous.
        6. Calyx lobes with a green center and a conspicuous white margin .... **S. prostrata**
        6. Calyx lobes solid green.

7. Fruit 1–2 mm long..**S. keyensis**
7. Fruit 2.5–3 mm long.
   8. Fruit glabrous...**S. glabra**
   8. Fruit distinctly pubescent..**S. remota**

Spermacoce densiflora (DC.) Alain [With densely arranged flowers.] BOUQUET FALSE BUTTONWEED.

Borreria densiflora de Candolle, Prodr. 4: 542. 1830. *Spermacoce densiflora* (de Candolle) Alain, Phytologia 54: 113. 1983.

Erect annual herb, to 7 dm; stem 4-angled, glabrate. Leaves with the blade lanceolate, 6–10 cm long, 4–12 mm wide, the apex acuminate, the base cuneate, the margin entire, revolute, the upper and lower surfaces glabrous, sessile; stipules connate and forming a sheath, membranous, with marginal bristles ca. 4 mm long. Flowers many in a terminal or axillary cluster subtended by 4–8 small leaves, the terminal head (1)2–3 cm wide; hypanthium villous; sepals with 2 lobes 2–3 mm long, pubescent, with a broad-based, hyaline margin, the other lobes vestigial; corolla white, 2–3 mm long, the outer surface glabrous. Fruit subglobose, 3–4 mm long, villous; seeds narrowly oblong, ca. 2 mm long, brownish red, the ventral surface with a slightly sunken longitudinal groove, the dorsal surface convex, finely foveolate.

Disturbed sites. Occasional; central and western panhandle. Florida; West Indies, Mexico, Central America, and South America. Native to tropical America. Summer–fall.

Spermacoce glabra Michx. [Smooth.] SMOOTH FALSE BUTTONWEED.

Spermacoce glabra Michaux, Fl. Bor.-Amer. 1: 82. 1803. *Spermacoceodes glabra* (Michaux) Kuntze, Revis. Gen. Pl. 3(2): 123. 1898.
Spermacoce chapmanii Torrey & A. Gray, Fl. N. Amer. 2: 27. 1841. TYPE: FLORIDA: Gadsden Co.: Aspalaga, s.d., *Chapman s.n.* (holotype: NY).

Erect, spreading, or decumbent perennial herb; stem to 6 dm long, 4-angled, glabrous. Leaves with the blade lanceolate, elliptic, or oblanceolate, 2–4(7) cm long, 0.5–1.5 cm wide, the apex acute, the base cuneate or rounded, the margin entire, the upper surface glabrous, the lower surface glabrous or sparsely short-pubescent on the midrib, sessile or subsessile; stipules connate and forming a sheath, membranous, with a few long marginal bristles. Flowers in a terminal or axillary, sessile cluster; hypanthium glabrous; sepals lobes subequal, triangular, 2–3 mm long, the apex acute, the margin fine-toothed; corolla white, the tube short-cylindric, 2–3 mm long, equaling or slightly exceeding the sepal lobes, the outer surface glabrous, bearded within at the throat, the lobes oblong, spreading. Fruit turbinate, ca. 3 mm long, glabrous; seeds ellipsoid, ca. 2 mm long, brown, the ventral surface with a slightly sunken longitudinal groove, the dorsal surface convex, reticulate.

Floodplain forests. Rare; Madison County, central panhandle. Maryland south to Florida, west to Kansas, Oklahoma, and Texas. Summer–fall.

Spermacoce keyensis Small [Of the Florida Keys.] FLORIDA FALSE BUTTONWEED.

Spermacoce keyensis Small, Fl. Florida Keys 141, 155. 11 Aug 1913. TYPE: FLORIDA: Monroe Co.: Key
West, 15–16 Nov 1912, *Small 3748* (holotype: NY).

Spermacoce floridana Urban, Symb. Antill. 7: 550. 15 Aug 1913. *Spermacoce tenuior* Linnaeus var. *floridana* (Urban) R. W. Long, Rhodora 72: 36. 1970. TYPE: FLORIDA: Monroe Co.: Key West, Feb
1846, *Rugel 298* (lectotype: ?; isolectotype: GH). Lectotypified by Long (1970: 37).

Spermacoce floridana Gandoger, Bull. Soc. Bot. France 65: 35. 1918; non Urban, 1913. TYPE: FLORIDA.

Prostrate or low-spreading annual herb; stem to 5 dm long, 4-angled, glabrous. Leaves with
the blade elliptic to elliptic-lanceolate, 0.5–1.5 cm long, 3–5 cm wide, the apex acute to acuminate, the base cuneate to rounded, the margin entire, the upper and lower surfaces glabrous,
the petiole 1–3 mm long; stipules connate and forming a sheath, membranous, the margin
with several bristles. Flowers few in an axillary cluster; sepal lobes subequal, linear-lanceolate,
1–2 mm long, green, the outer surface glabrous; corolla campanulate, white, ca. 1 mm long,
glabrous, the lobes ovate-oblong. Fruit subglobose, 1–2 mm long, smooth, glabrous; seeds ellipsoid, ca. 2 mm long, brown, the ventral surface with a slightly sunken longitudinal groove,
the dorsal surface convex, finely papillate.

Pinelands. Rare; Miami-Dade County, Monroe County keys. Florida and Texas; West Indies. Spring–summer.

Spermacoce latifolia Aubl. WINGED FALSE BUTTONWEED.

Spermacoce latifolia Aublet, Hist. Pl. Guiane 55, t. 19(1). 1775. *Borreria latifolia* (Aublet) K. Schumann,
in Martius, Fl. Bras. 6(6): 61. 1888. *Tardavel latifolia* (Aublet) Standley, Contr. U.S. Natl. Herb. 18:
122. 1916.

Erect, spreading, or decumbent perennial herb. Stem to 1 mm long, 4-angled, the angles narrowly winged, hispidulous or pilosulous. Leaves with the blade elliptic to ovate-oblong, 1.2–7.5
cm long, 0.6–4 cm wide, the apex acute to obtuse, the base cuneate to obtuse, decurrent on
the petiole, the margin entire, the upper and lower surfaces sparsely to densely hispidulous to
pilosulous, sessile or the petiole to 4 mm long, pilosulous or hirtellous; stipules connate and
forming a sheath, the margin with 5–9 bristles or narrow triangular lobes, ciliate, the outer
surface hirtellous to hispidulous. Flowers few in a terminal or axillary cluster; bracts filiform,
1–4 mm long; hypanthium moderately to densely hirtellous or pilosulous; sepal lobes subequal,
lanceolate to elliptic or triangular, 1–2 mm long; corolla white tinged with blue to pale purple,
the tube funnelform, 2–3 mm long, the outer surface pilosulous to hirtellous, bearded within
at the throat, the lobes triangular, 1–2 mm long. Fruit ellipsoid to subglobose, 3–4 mm long,
densely hirtellous and sometimes also hirsute distally, densely puberulent to strigillose proximally; seeds elliptic to oblong, ca. 2 mm long, light to dark brown, the ventral surface with a
slightly sunken longitudinal groove, the dorsal surface convex, shallowly pitted.

Disturbed sites. Rare; Pasco County. Florida; West Indies, Mexico, Central America,
and South America; Africa, Asia, Australia, and Pacific Islands. Native to South America.
Spring–fall.

Spermacoce neoterminalis Govaerts [*Neo*, new, and *terminalis*, in reference to the inflorescence, a new name for what was previously known as *Spermacoce terminalis*.] EVERGLADES KEY FALSE BUTTONWEED.

Borreria terminalis Small, Bull. Torrey Bot. Club 51: 387. 1924. *Spermacoce terminalis* (Small) Kartesz & Gandhi, Brittonia 44: 370. 1992; non Vellozo, 1929. *Spermacoce neoterminalis* Govaerts, World Checkl. Seed Pl. 2(1): 18. 1996. TYPE: FLORIDA: Miami-Dade Co.: Everglade Keys, about Ross Hammock, 23 June 1915, *Small 6502* (lectotype: NY). Lectotypified by Kartesz and Gandhi (1992: 370).

Erect or ascending perennial herb, to 3 dm; stem 4-angled, glabrous. Leaves with the blade linear-subulate, 1–3 cm long, 1–4 mm wide, the apex acute, the base narrowly cuneate, the margin entire, the upper and lower surfaces glabrous, the margin entire, sessile or subsessile; stipules connate and forming a sheath, with a few marginal bristles. Flowers in a dense terminal head 0.5–1 cm wide, occasionally also in a penultimate axil, subtended by a few slightly reduced leaves; hypanthium glabrous; sepals with 2 lobes 1–2 mm long, the other 2 very short or vestigial; corolla white, ca. 3 mm long, the tube 0.5–1 mm long, the lobes ovate. Fruit subglobose, 2–3 mm long, glabrous; seeds oblong, ca. 2 mm long, brown, the ventral surface with a slightly sunken longitudinal groove, the dorsal surface convex, shallowly pitted.

Pinelands. Occasional; Hendry County, southern peninsula. Endemic. All year.

It is debatable whether *S. neoterminalis* is a good species or a synonym of *S. verticillata* s.l. Further study is needed.

Spermacoce neoterminalis is listed (as *S. terminalis*) as an endangered species in Florida (Florida Administrative Code, Chapter 5B-40).

Spermacoce prostrata Aubl. [Procumbent.] PROSTRATE FALSE BUTTONWEED.

Spermacoce prostrata Aublet, Hist. Pl. Guiane 58, t. 20(3). 1775. *Borreria prostrata* (Aublet) Miquel, Stirp. Surinam. Select. 177. 1851.

Borreria parviflora G. Meyer, Prim. Fl. Esseq. 83, t. 1(1–3). 1818. *Bigelovia parviflora* (G. Meyer) Sprengel, Syst. Veg. 1: 405. 1824. *Spermacoce parviflora* (G. Meyer) Hemsley, Biol. Cent.-Amer., Bot. 2: 59. 1881; non Salisbury, 1796.

Erect, ascending, or procumbent annual or perennial herb; stem to 6 dm long, 4-angled, smooth to sharp-angled, scabrous to glabrescent. Leaves with the blade elliptic-oblong or elliptic, 1–3 cm long, 1–7(10) mm wide, the apex acute to obtuse, the base cuneate, the upper and lower surfaces puberulent or scabrous, sessile; stipules connate and forming a sheath, puberulent, with 5–9 marginal bristles. Flowers few–many in a terminal head and in the uppermost leaf axils, subtended by numerous filiform bracts ca. 1 mm long; hypanthium obovoid, glabrescent; sepal lobes subequal, narrowly triangular, ca. 1 mm long, with a green center and a conspicuous white margin; corolla rotate to shortly tubular, white, ca. 1 mm long, the outer surface glabrous, pubescent within at the throat, the lobes triangular-spatulate, subequaling the tube. Fruit ellipsoid, ca. 1 mm long, slightly flattened laterally; seeds ellipsoid, 0.6–0.7 mm long, brownish yellow, the ventral surface with a slightly sunken longitudinal groove, the dorsal surface convex, coarsely pitted in definite lines.

Wet flatwoods and floodplain forests. Frequent; peninsula, central panhandle. Florida, Alabama, and Mississippi; West Indies, Mexico, Central America, and South America. Summer–fall.

Spermacoce remota Lam. [In reference to the widely separated flower clusters.] WOODLAND FALSE BUTTONWEED.

> *Spermacoce remota* Lamarck, Tabl. Encycl. 1: 273. 1792. *Borreria remota* (Lamarck) Bacigalupo & E. L. Cabral, Darwinia 37: 334. 1999.
>
> *Spermacoce assurgens* Ruiz López & Pavón, Fl. Peruv. 1: 69, t. 92(b). 1798. *Borreria assurgens* (Ruiz López & Pavón) Grisebach, Abh. Königl. Wis. Göttingen 19: 111. 1874.
>
> *Borreria micrantha* Torrey & A. Gray, Fl. N. Amer. 2: 28. 1841. TYPE: FLORIDA: Hillsborough Co.: Tampa Bay, s.d., *Leavenworth s.n.* (holotype: NY?).

Erect, decumbent, or procumbent annual or perennial herb; stem to 3 dm long, 4-angled, glabrous or sparsely pubescent. Leaves with the blade lanceolate, ovate-lanceolate, or oblong-lanceolate, 2–5 cm long, 1–2.5 cm wide, the apex acute, the base cuneate, the upper surface glabrous, the lower surface glabrous or pubescent, the petiole to 2 mm long; stipules connate and forming a sheath, subtruncate, with numerous marginal setae. Flowers several–many in a terminal or axillary head; hypanthium obovoid, pubescent distally; sepals triangular-subulate, 2 long and 2 short, glabrous, green; corolla white, often pink-tipped, ca. 4 mm long, the lobes subequaling the tube, the outer surface glabrous, pubescent within at the throat. Fruit ellipsoid, 2–3 mm long, somewhat flattened laterally, sparse-pubescent distally; seeds ellipsoid, ca. 2 mm long, brown, the ventral surface with a slightly sunken longitudinal groove, the dorsal surface convex, transversely striate.

Wet hammocks, wet flatwoods, marshes, and floodplain forests. Frequent; nearly throughout. Georgia, Alabama, and Florida; West Indies, Mexico, Central America, and South America. Summer–fall.

Spermacoce tetraquetra A. Rich. [From the Greek *tetra*, four, and the Latin *quetrus*, cornered, in reference to the four-sided stem.] PINELAND FALSE BUTTONWEED.

> *Spermacoce tetraquetra* A. Richard, in Sagra, Hist. Fis. Cuba, Bot. 11: 29. 1850.

Erect herbs, to 6 dm; stem 4-angled, hirsute or hispid. Leaves with the blade ovate or lanceolate to oblong-lanceolate, 2–8 cm long, 1–2.5 cm wide, the apex acute, the base cuneate, the margin entire, the upper and lower surfaces hirsute or hispid, subsessile to short-petiolate; stipules connate and forming a sheath, with numerous long, marginal bristles. Flowers few in an axillary head; hypanthium hispid; sepal subequal, lobes lanceolate, acuminate, hispid; corolla white, pink-tinged, twice as long as the calyx lobes, the outer surface glabrous, dense-barbellate within at the throat. Fruit subglobose, ca. 2 mm long, dense-hispidulous; seeds oblong, ca. 2 mm long, brown, the ventral surface with a slightly sunken longitudinal groove, the dorsal surface convex, minutely pitted.

Pinelands, hammocks, and disturbed sites. Rare; Collier and Miami-Dade Counties, Monroe County keys. Florida; West Indies. All year.

Spermacoce verticillata L. [Whorled, in reference to the flower clusters and leaves.] SHRUBBY FALSE BUTTONWEED.

Spermacoce verticillata Linnaeus, Sp. Pl. 102. 1753. *Borreria verticillata* (Linnaeus) G. Meyer, Prim. Fl. Esseq. 83. 1818. *Bigelovia verticillata* (Linnaeus) Sprengel, Syst. Veg. 1: 404. 1824.

Borreria podocephala de Candolle, Prodr. 4: 452. 1830. *Spermacoce podocephala* (de Candolle) A. Gray, Syn. Fl. N. Amer. 1(2): 34. 1884.

Borreria podocephala de Candolle var. *pumila* Chapman, Fl. South. U.S. 175. 1860. TYPE: FLORIDA: Monroe Co.: Big Pine Key, s.d., *Blodgett s.n.* (holotype: NY?).

Erect-ascending perennial herb or subshrub, to 1 m; stem slightly 4-angled, glabrous or sparsely scabrous. Leaves with the blade linear to linear-lanceolate, 2–6 cm long, 3–12 mm wide, with tufts of smaller ones in the axil, the apex acute, the base cuneate, the margin entire, the upper and lower surfaces glabrous, subsessile; stipules connate and forming a sheath, with marginal bristles. Flowers numerous in the terminal and upper 1–2 leaf nodes, the glomerule 0.5–1 cm wide; hypanthium turbinate; sepals with 2 lobes linear-lanceolate, ca. 1 mm long, the other 2 vestigial; corolla funnelform, white, ca. 2 mm long, the tube 0.5–1 mm long, the lobes ovate, acute, spreading. Fruit subglobose, ca. 1 mm long, glabrous; seeds oblong, ca. 2 mm long, dark brown, the ventral surface with a slightly sunken longitudinal groove, the dorsal surface convex, pitted.

Disturbed sites. Frequent; peninsula. Florida and Texas; West Indies, Mexico, Central America, and South America; Africa and Asia. Native to tropical America. All year.

Spermacoce verticillata is listed as a Category II invasive species in Florida by the Florida Exotic Pest Plant Council (FLEPPC, 2017).

EXCLUDED TAXA

Spermacoce confusa Rendle—Reported for Florida by Ward (2011), apparently based on a misidentification of material of *S. keyensis*.

Spermacoce laevis Lamarck—Reported for Florida by Small (1933, as *Borreria laevis* (Lamarck) Grisebach), Long and Lakela (1971, as *Borreria laevis* (Lamarck) Grisebach), Godfrey and Wooten (1981, as *Borreria laevis* (Lamarck) Grisebach), and Clewell (1985, as *Borreria laevis* (Lamarck) Grisebach). This is a misapplication of the name to material of *S. remota*.

Spermacoce ocymoides Burman f.—Reported for Florida by Small (1933, as *Borreria ocymoides* (Burman f.) de Candolle), Long and Lakela (1971, as *Borreria ocymoides* (Burman f.) de Candolle), and Clewell (1985, as *Borreria ocymoides* (Burman f.) de Candolle), based on a misapplication of the name to *S. prostrata*.

Spermacoce tenella Kunth—Reported for Florida by Small (1933), Wilhelm (1984), and Clewell (1985). No Florida specimens known.

Stenaria (Raf.) Terrell 2001. DIAMONDFLOWERS

Perennial herbs. Leaves opposite, simple, pinnate-veined, petiolate or epetiolate, stipulate. Flowers in terminal cymes, actinomorphic, bisexual, heterostylous; sepals 4, basally connate; petals 5, basally connate; stamens 4, epipetalous; ovary inferior, 2-carpellate and -loculate, the style 1, the stigmas 2. Fruit a capsule, dehiscing loculicidally, then septicidally through the septum.

A genus of 5 species; North America, West Indies, and Mexico. [From the Greek *stenos*, narrow, straight, apparently in reference to the leaves.]

Selected reference: Terrell (2001).

Stenaria nigricans (Lam.) Terrell [Blackish, in reference to the general appearance of the foliage.] DIAMONDFLOWERS.

Gentiana nigricans Lamarck, Encycl. 2: 645. 1788. *Houstonia nigricans* (Lamarck) Fernald, Rhodora 42: 299. 1940. *Hedyotis nigricans* (Lamarck) Fosberg, Lloydia 4: 298. 1941. *Stenaria nigricans* (Lamarck) Terrell, Sida 19: 600. 2001.

Houstonia angustifolia Michaux, Fl. Bor.-Amer. 1: 85. 1803. *Hedyotis stenophylla* Torrey & A. Gray, Fl. N. Amer. 2: 41. 1841. *Oldenlandia angustifolia* (Michaux) A. Gray, Smithsonian Contr. Knowl. 3(6): 68. 1853. *Chamisme angustifolia* (Michaux) Nieuwland, Amer. Midl. Naturalist 4: 92. 1915. TYPE: FLORIDA: s.d., *Michaux s.n.* (holotype: P).

Oldenlandia angustifolia (Michaux) A. Gray var. *filifolia* Chapman, Fl. South. U.S. 181. 1860. *Houstonia angustifolia* Michaux var. *filifolia* (Chapman) A. Gray, Syn. Fl. N. Amer. 1(2): 27. 1884. *Houstonia filifolia* (Chapman) Small, Fl. S.E. U.S. 1109, 1338. 1903. *Hedyotis nigricans* (Lamarck) Fosberg var. *filifolia* (Chapman) Shinners, Field & Lab. 17: 168. 1949. TYPE: FLORIDA: s.d., *Chapman s.n.* (lectotype: US). Lectotypified by Terrell (1986: 479).

Houstonia angustifolia Michaux var. *rigidiuscula* A. Gray, Syn. Fl. N. Amer. 1(2): 27. 1884. *Houstonia rigidiuscula* (A. Gray) Wooten & Standley, Contr. U.S. Natl. Herb. 16: 175. 1913. *Hedyotis nigricans* (Lamarck) Fosberg var. *rigidiuscula* (A.Gray) Shinners, Field & Lab. 17: 168. 1949. SYNTYPE: FLORIDA.

Houstonia pulvinata Small, Bull. New York Bot. Gard. 1: 289. 1899. *Hedyotis nigricans* (Lamarck) Fosberg var. *pulvinata* (Small) Fosberg, Castanea 19: 37. 1954. *Houstonia nigricans* (Lamarck) Fernald var. *pulvinata* (Small) Terrell, Sida 59: 79. 1985. TYPE: FLORIDA: St. Johns Co.: St. Augustine, Jul 1876, *Reynolds s.n.* (lectotype: NY; isolectotype: NA). Lectotypified by Terrell (1986: 479).

Erect, spreading, or decumbent perennial herb, to 6 dm; stem 4-angled, scabridulous to glabrate. Leaves with the blade linear-filiform, 1–4 cm long, 1–3 cm wide, the apex acute, the base narrowly cuneate, the margin entire, revolute, the upper and lower surfaces scabridulous to glabrate, sessile or short-petiolate; stipules deltate, 3–4 mm long, scarious. Flowers in a terminal cyme, the peduncle 5–10 mm long, sessile or the pedicels to 10 mm long; hypanthium glabrous to hirsute; calyx lobes lanceolate or triangular, 1–4 mm long; corolla salverform to funnelform, white, purple, or pink, 2–8 mm long, glabrous on the outer surface, the tube subequaling or longer than the lobes, the lobes lanceolate, puberulent on the inner surface. Fruit turbinate or subglobose, scabridulous to glabrate, 1–3 mm long; seeds ellipsoid, ca. 1 mm long, somewhat compressed, black or dark brown, the ventral face with a punctiform hilum center on the slightly concave or slightly ridged, reticulate surface.

1. Mature fruit turbinate..var. **nigricans**
1. Mature fruit subglobose..var. **floridana**

Stenaria nigricans var. **nigricans** DIAMONDFLOWERS.

Fruit turbinate. Sandhills, flatwoods, and shell beaches. Occasional; peninsula, central and western panhandle. New Hampshire south to Florida, west to Nebraska, Colorado, and New Mexico. Summer–fall.

Stenaria nigricans var. **floridana** (Standl.) Terrell [Of Florida.] FLORIDA DIAMONDFLOWERS.

> *Houstonia floridana* Standley, in Britton, N. Amer. Fl. 32: 36. 1918. *Hedyotis purpurea* (Linnaeus) Torrey & A. Gray var. *floridana* (Standley) Fosberg, Castanea 19: 36. 1954. *Houstonia nigricans* (Lamarck) Fernald var. *floridana* (Standley) Terrell, Phytologia 59: 79. 1985. *Hedyotis nigricans* (Lamarck) Fosberg var. *floridana* (Standley) Wunderlin, Sida 11: 400. 1986. *Stenaria nigricans* (Lamarck) Terrell var. *floridana* (Standley) Terrell, Sida 19: 605. 2001. TYPE: FLORIDA: Miami-Dade Co.: Coconut Grove, Biscayne Bay, Jul 1895, *Curtiss 5484* (holotype: US; isotypes: FLAS, ISC, NY, US, VT).

Fruit subglobose. Pine rocklands. Rare; Miami-Dade County, Monroe County keys. Florida; West Indies. Summer.

Strumpfia Jacq. 1760.

Shrubs or trees. Leaves whorled, simple, pinnate-veined, sessile or subsessile, stipulate. Flowers in axillary racemes, actinomorphic, bisexual, bracteate, bracteolate; sepals 4–5, basally connate; petals 4–5, basally connate; stamens 4–5, epipetalous; ovary inferior, 2-carpellate and -loculate, the style 1, the stigma 1. Fruit a drupe; seed 1.

A monotypic genus; Florida, West Indies, Mexico, Central America, and South America. [Commemorates Christopher Karl Strumpff (1711–1754), professor of botany at Halle, editor of Linnaeus's *Genera Plantarum* (1752).]

Strumpfia maritima Jacq. [Of the sea.] PRIDE-OF-BIG-PINE.

> *Strumpfia maritima* Jacquin, Enum. Syst. Pl. 28. 1760.

Shrub or small tree, to 2 m; branchlets densely tomentose, glabrescent in age, the branches covered with old stipule bases. Leaves in whorls of 3, the blade linear, 1–3 cm long, 1–2 mm wide, the apex acute, the base narrowly cuneate, the margin entire, strongly revolute, the upper surface glabrous, the lower surface gray-puberulent to glabrate, sessile or subsessile; stipules interpetiolar, the bases connate, subtriangular, 1–2 mm long, rigid, the apex acute, fimbriate or erose in the distal half, puberulous to glabrate. Flowers few in a short, axillary raceme to 1.5 cm long, the peduncle 3–9 mm long, gray-tomentose, the pedicel c. 1 mm long, puberulent; bract 1, subtriangular, ca. 1 mm long, acute, puberulent; bracteoles 2, similar to the bract; hypanthium ca. 1 mm long, puberulent; sepals ca. 1 mm long, the lobes triangular, ca. 0.5 mm long, obtuse; corolla 4–5 mm long, white or pink, the lobes linear-oblong, ca. 3 mm long, the apex obtuse or rounded, the outer surface puberulent, the inner surface tomentulose; stamens borne at the throat of the corolla tube, the anthers subsessile, connate by the connectives and forming a column style glabrous, surrounded at the base by a ring of trichomes. Fruit subglobose, 2–4 mm long, white, fleshy, the calyx persistent; seed oblong, the endosperm fleshy.

Wet pinelands and coastal strands. Rare; Monroe County keys. Florida; West Indies, Mexico, Central America, and South America. All year.

Strumpfia maritima is listed as endangered in Florida (Florida Administrative Code, Chapter 5B-40).

EXCLUDED GENERA

Hedyotis caerulea (Linnaeus) Hooker—This northern species was reported for Florida by Chapman (1860). No Florida specimens known.

Hedyotis rosea Rafinesque—Reported by Small (1903, as *Houstonia minor* (Michaux) Britton var. *pusilla* (A. Gray) Small) as from the Gulf States, which would presumably include Florida. No Florida specimens known.

GENTIANACEAE Juss., nom. cons. 1789. GENTIAN FAMILY

Herbs. Leaves opposite or alternate, simple, pinnipalmate-veined, petiolate or epetiolate, estipulate. Flowers in cymes or solitary, actinomorphic, bisexual, bracteate; sepals 4–14, basally connate or absent; petals 4, basally connate; stamens 6–14, epipetalous, the anthers 2-locular, longitudinally dehiscent; nectaries present or absent; ovary superior, 2-carpellate, 1-loculate, the style 1 or absent, the stigma 2-lobed. Fruit a capsule, septicidally or irregularly dehiscent; seeds numerous.

A family of about 100 genera and about 1,800 species; nearly cosmopolitan.

Selected reference: Wood and Weaver (1982).

1. Leaves scalelike (5 mm long or shorter).
    2. Plants achlorophyllous (stems white)..**Voyria**
    2. Plants chlorophyllous (stems green or purple).
        3. Calyx lobes 4; petals slightly connate at the base)..**Bartonia**
        3. Calyx lobes 2; petals connate about ½ their length...**Obolaria**
1. Leaves not scalelike.
    4. Corolla lobes shorter than the tube, the tube plaited, with toothlike or fringed appendages between the lobes ....**Gentiana**
    4. Corolla lobes longer than the tube, the tube not plaited, lacking appendages between the lobes.
        5. Stigmas shorter than the style...**Eustoma**
        5. Stigmas subequaling the style ....**Sabatia**

Bartonia Muhl. ex Willd., nom. cons. 1801. SCREWSTEM

Annual herbs. Leaves opposite or alternate, simple, reduced to scales, epetiolate, estipulate. Flowers in dichasial or racemiform cymes or solitary, bracteate; sepals 4, basally connate; petals basally connate; stamens 4, epipetalous, inserted at the sinus between the lobes; nectaries absent; ovary 2-carpellate, 1-loculate, the style 1, the stigmas 2. Fruit a capsule, apically or medially dehiscent; seeds numerous.

A genus of 3 species; North America. [Commemorates Benjamin Smith Barton (1766–1815), physician and botanist of Philadelphia.]

Bartonia is mycoheterotrophic, i.e., dependent on mycorrhizal fungi. It has a small root system with little branching and no root hairs and scalelike leaves.

Selected reference: Matthews et al. (2009).

1. Corolla lobes (2)5–10 mm long; style elongate; plants flowering in the spring ....**B. verna**
1. Corolla lobes 1–5 mm long; style short; plants flowering in the summer or fall.
    2. Flowers in a strongly ascending racemiform cyme; leaves usually opposite below the inflorescence; anthers with the apex mucronate...**B. virginica**
    2. Flowers in a spreading racemiform to compound cyme; leaves usually alternate below the inflorescence; anthers with the apex rounded...**B. paniculata**

Bartonia paniculata (Michx.) Muhl. [The inflorescence paniculate.] TWINING SCREWSTEM.

Centaurella paniculata Michaux, Fl. Bor.-Amer. 1: 98, t. 12(1). 1803. *Bartonia paniculata* (Michaux) Muhlenberg, Cat. Pl. Amer. Sept. 16. 1813. *Bartonia virginica* (Linnaeus) Britton et al. var. *paniculata* (Michaux) B. Boivin, Naturaliste Canad. 93: 1059. 1966. *Centaurium autumnale* Persoon, Syn. Pl. 1: 137. 1805, nom. illegit.

Bartonia lanceolata Small, Fl. S.E. U.S. 932, 1336. 1903. TYPE: FLORIDA: *s.d., Chapman s.n.* (holotype: NY).

Erect, decumbent, or twining annual herb, to 5 dm; stem yellowish green or purple, glabrous. Leaves alternate, opposite, or both, subulate and scalelike, 1–3 mm long. Flowers in a spreading simple or racemiform to compound cyme; sepal lobes lanceolate to ovate, 1–3 mm long, the lobe apex acute or acuminate; corolla campanulate, 3–5 mm long, white or pale yellow to green, purple tinged distally, the apex acute to acuminate, the margin entire; anthers triangular to oblong, ca. 1 mm long, yellow, the apex rounded; style ca. 1 mm long, the stigmas spreading, decurrent to the ovary. Fruit ovoid, elliptic, or oblong, 2–4 mm long, dehiscing apically; seeds compressed cylindric, minute, brown.

Wet flatwoods. Occasional; northern and central peninsula, central and western panhandle. Maine south to Florida, west to Wisconsin, Illinois, Missouri, Oklahoma, and Texas. Summer–fall.

Bartonia verna (Michx.) Raf. ex Barton [*Vernus*, spring, in reference to the flowering time.] WHITE SCREWSTEM.

Centaurella verna Michaux, Fl. Bor.-Amer. 1: 98, t. 12(2). 1803. *Centaurium vernum* (Michaux) Persoon, Syn. Pl. 1: 137. 1805. *Bartonia verna* (Michaux) Rafinesque ex Barton, Fl. Virgin. 51. 1812. *Centaurella vernalis* Pursh, Fl. Amer. Sept. 99. 1814, nom. illegit. *Andrewsia verna* (Michaux) Sprengel, Syst. Veg. 1: 428. 1824 ("1825").

Erect annual herb, to 2 dm; stem purplish or yellowish. Leaves alternate or subopposite, the blade subulate and scalelike, 1–3 mm long. Flowers solitary or in a racemiform cyme; sepal lobes lanceolate, 1–3 mm long, the apex obtuse or acute; corolla campanulate, white, 4–11 mm long, the lobes spatulate-obovate to elliptic, 3–6 mm long, the apex rounded to obtuse, the margin entire or undulate-erose distally; stamens 2–3 mm long, the filaments purple to yellow, the anthers oblong, ca. 1 mm long, yellow, the apex rounded; style 1–3 mm long, the stigmas connivent, slightly decurrent. Fruit ovoid, elliptic, or oblong, 3–5 mm long, dehiscing medially; seeds compressed-cylindric, minute, brown.

Wet flatwoods and bogs. Frequent; nearly throughout. North Carolina south to Florida, west to Texas. Winter–spring.

Bartonia virginica (L.) Britton et al. [Of Virginia.] YELLOW SCREWSTEM.

> *Sagina virginica* Linnaeus, Sp. Pl. 128. 1753. *Centaurella autumnalis* Pursh, Fl. Amer. Sept. 100. 1814, nom. illegit. *Andrewsia autumnalis* Sprengel, Syst. Veg. 1: 428. 1824 ("1825"), nom. illegit. *Bartonia virginica* (Linnaeus) Britton et al., Prelim. Cat. 36. 1888.
> *Bartonia tenella* Muhlenberg ex Willdenow, Ges Naturf. Freunde Berlin Neue Schriften 3: 445. 1803.

Erect annual herb, to 4.5 dm; stem yellowish green, usually with the base purplish. Leaves usually opposite below the inflorescence, subulate and scalelike, 1–4 mm long. Flowers in a strongly ascending racemiform cyme; sepal lobes subulate to lanceolate, 2–4 mm long, the apex acuminate; corolla campanulate, white to yellowish green, sometimes with the apex purple tinged, 2–4 mm long, the lobes oblong, 2–3 mm long, the apex rounded to acute, mucronate, the margin entire or erose-serrate distally; anthers 0.5–1 mm long, yellow or purple, the apex mucronate; style ca. 1 mm long, the stigmas connivent, decurrent to the ovary. Fruit ovoid, elliptic, or oblong, 3–4 mm long, dehiscing medially; seeds compressed-cylindric, minute, brown.

Swamps and wet prairies. Occasional; peninsula, central and western panhandle. Newfoundland south to Florida, west to Ontario, Minnesota, Missouri, and Louisiana. Fall.

Eustoma Salisb. 1806. PRAIRIE GENTIAN

Annual, biennial, or short-lived perennial herbs. Leaves opposite, simple, pinnipalmate-veined, epetiolate, estipulate. Flowers in monochasial cymes, bracteate; sepals 5, basally connate; petals 5, basally connate; stamens 5, epipetalous; ovary 2-carpellate, 1-loculate, the style 1, the stigmas 2. Fruit a capsule, septicidally dehiscent; seeds numerous.

A monotypic genus; North America, West Indies, Mexico, Central America, South America, and Pacific Islands. [From the Greek *eustomos*, beautiful mouth, or open mouth, in reference to the corolla throat.]

Eustoma exaltatum (L.) Salisb. ex G. Don [Raised high, lofty]. MARSH GENTIAN; CATCHFLY PRAIRIE GENTIAN.

> *Gentiana exaltata* Linnaeus, Sp. Pl., ed 2. 331. 1762. *Lisianthus exaltatus* (Linnaeus) Lamarck, Tabl. Encycl. 1: 478. 1793. *Erythraea plumieri* Kunth, in Humbolt et al., Nov. Gen. Sp. 3: 178. 1819, nom. illegit. *Eustoma exaltatum* (Linnaeus) Salisbury ex G. Don, Gen. Hist. 4: 211. 1837. *Chlora exaltata* (Linnaeus) Grisebach, Gen. Sp. Gent. 118. 1838.

Erect annual, biennial, or short-lived perennial herb, to 7(10) dm; stem glabrous, glaucous. Leaves with the blade 1.5–10(14) cm long, 0.4–3 cm wide, the basal ones spatulate-obovate to elliptic-oblong, 2–10 cm long, 0.5–2 cm wide, the upper cauline ones elliptic-oblong to lanceolate or ovate, 1.5–9(14) cm long, 0.4–3 cm wide, the apex obtuse, the uppermost usually with the apex acute to acuminate, the base somewhat clasping, the margin entire, the upper and lower surfaces glabrous, glaucous, sessile. Flowers solitary or few to many, when numerous, then essentially in an open paniculiform cyme, the pedicel 3–10 cm long; calyx 1–2.5 cm long, the lobes subulate to linear, overlapping below, keeled, the apex acuminate; corolla campanulate-funnelform, 1.8–4(4.5) cm long, blue, lavender, or white, the tube ca. 1 cm long, the lobes elliptic to narrowly obovate, about twice as long as the tube, the apex mucronate; stamens inserted at the sinus between the corolla lobes; style ca. 5 mm long, the stigma lobes ca. 2 mm long, oblong to rounded. Fruit compressed ovoid-ellipsoid, 2–3 cm long, glabrous, glaucous; seeds minute, subrotund, the surface pitted, lustrous, chestnut-brown.

Coastal marshes and dunes. Frequent; peninsula, Escambia County. Florida west to California; West Indies and Mexico; South America and Pacific Islands. Native to North America, West Indies, and Mexico. All year.

Gentiana L. 1753. GENTIAN

Perennial herbs. Leaves opposite, simple, pinnipalmate-veined, epetiolate, estipulate. Flowers solitary or in simple dichasia, these often congested into dense terminal and/or axillary clusters, bracteate; sepals 5, basally connate; petals 5, basally connate, the lobes connected with a thin tissue and pleated; stamens 5, epipetalous; ovary 2-carpellate, 1-loculate, the style 1, the stigmas 2. Fruit a capsule, septicidally dehiscent; seeds numerous.

A genus of about 360 species; North America, Mexico, Central America, South America, Europe, and Asia. [Commemorates the second century King of Illyria, who supposedly discovered medicinal properties in *G. lutea*.]

Dasystephana Adans., 1763.

Selected reference: Pringle (1967).

1. Flower solitary, not involucrate; corolla white spotted with blue-green on the inner surface ....
.... **G. pennelliana**
1. Flowers clustered, involucrate; corolla white suffused with blue or mostly blue, not spotted with blue green on the inner surface.
    2. Lower leaves obovate to spatulate, widest above the middle; corolla appendages obliquely deltoid; seeds not winged ....**G. villosa**
    2. Lower leaves linear to ovate or elliptic, widest below or at the middle; corolla appendages bifid, lacinate; seeds winged.
        3. Leaves ovate; calyx lobes longer than the tube; corolla lobes spreading .... **G. catesbaei**
        3. Leaves linear to elliptic; calyx lobes shorter than or subequaling the tube; corolla lobes usually incurved ....**G. saponaria**

Gentiana catesbaei Walter [Commemorates Mark Catesby (1683–1749), English naturalist.] CATESBY'S GENTIAN.

> *Gentiana catesbaei* Walter, Fl. Carol. 109. 1788. *Pneumonanthe catesbaei* (Walter) F. W. Schmidt, Arch. Bot. (Leipzig) 1: 10. 1796.
> *Gentiana elliottii* Chapman, Fl. South. U.S. 356. 1860; non *G. elliotea* Rafinesque, 1832.
> *Gentiana elliottii* Chapman var. *parvifolia* Chapman, Fl. South. U.S. 356. 1860. *Gentiana parvifolia* (Chapman) Britton, Man. Fl. N. States, ed. 2. 733. 1905; non Gilg, 1896. *Dasystephana parvifolia* (Chapman) Small, Fl. S.E. U.S. 930, 1336. 1903. *Pneumonanthe parvifolia* (Chapman) Greene, Leafl. Bot. Observ. Crit. 1: 71. 1904.

Erect or suberect perennial herb, to 7 dm; stem puberulent. Leaves with the blade ovate, lanceolate, or elliptic, usually widest below the middle, the larger ones (1)1.5–7.5 cm long, 0.5–3 cm wide, the apex acute or acuminate, the base rounded to subcordate, the margin entire, ciliate, the upper and lower surfaces glabrous, the primary nerves 1–3, sessile. Flowers solitary or in terminal clusters of 2–10, sometimes also in axillary clusters and/or clusters terminating the branches, the inflorescences subtended by 1–2 pairs of involucral leaves, each flower subtended by a pair of linear to obovate bracts; calyx tube 7–20 mm long, glabrous or puberulent in lines below the lobe margins, the lobes oblanceolate, usually somewhat foliaceous, 1–3.5 cm long, 2–7 mm wide, the margins ciliate; corolla funnelform, 3.5–5.5 cm long, the tube greenish white up to slightly below the point of stamen attachment, the veins outlined in green, above this level blue or occasionally with rose or rose-purple, the lobes and free part of the appendages nearly uniform in color, the lobes ovate-deltoid, 5–10 mm long, erect or recurved, entire, mucronate, the free part of the appendages 2–6 mm long, bifid, both segments erect, laciniate, the sinus on the outer side of each lobe slightly lower than that on the inner side; anthers connate at anthesis. Fruit ellipsoid, 2.5–3 cm long, glabrous; seeds obliquely ellipsoid to oblong, ca. 2 mm long, brown, winged.

Bogs and bluff seepages. Occasional, northern peninsula west to central panhandle. New Jersey and Pennsylvania south to Florida, west to Alabama. Fall.

Gentiana pennelliana Fernald [Commemorates Francis Whittier Pennell (1886–1952), American botanist.] WIREGRASS GENTIAN.

> *Diploma tenuifolia* Rafinesque, Fl. Tellur. 2: 27. 1837 ("1836"). *Dasystephana tenuifolia* (Rafinesque) Pennell, Bull. Torrey Bot. Club 46: 183. 1919. *Gentiana tenuifolia* (Rafinesque) Fernald, Rhodora 41: 557. 1939; non Petrie, 1913. *Gentiana penelliana* Fernald, Rhodora 42: 198. 1940. TYPE: FLORIDA: Gadsden Co.: s.d., *Croom s.n.* (lectotype: NY). Lectotypified by Pennell (1919: 183).
> *Gentiana angustifolia* Michaux var. *australis* Grisebach, in de Candolle, Prodr. 9: 114. 1845. TYPE: FLORIDA.
> *Gentiana angustifolia* Michaux var. *floridana* Grisebach, Gen. Sp. Gent. 291. 1838. *Gentiana frigida* Pallas var. *drummondii* Grisebach, in de Candolle, Prodr. 9: 111. 1945, nom. illegit. TYPE: FLORIDA: s.d., *Drummond 19* (holotype: ?).

Erect to decumbent perennial herb, to 3.5 dm; stem glabrous. Leaves with the blade linear to linear-spatulate or narrowly obovate, the larger ones 1.5–2.5 cm long, 2–7 mm wide, the apex acute to obtuse, the base cuneate, the margin entire, the upper and lower surfaces glabrous, the primary nerves 1–3, sessile. Flowers solitary, lacking involucral leaves; calyx tube 8–15 mm long, the lobes linear, 1–3 mm long, 1–3 mm wide, the margin entire; corolla campanulate, 3.5–6.5 cm long, the tube white with narrow green areas along the veins, with small purplish green spots on the inner surface in a row between the veins, suffused with greenish purple externally in the upper portion, darker along the veins, the lobes greenish blue, greenish purple, or nearly white toward the inner edges, the lobes ovate, 1.5–2.5 cm long, spreading, the apex mucronate, the margin minutely jagged, the free part of the appendages 5–10 mm long, somewhat equally bifid, both segments laciniate; anthers connate or free at anthesis. Fruit ellipsoid to lanceolate, laterally compressed, 2–4 cm long, glabrous; seeds ellipsoid to falcate, compressed, ca. 1 mm long, brown, winged.

Flatwoods. Rare; central panhandle, Walton County. Endemic. Winter–spring.

Gentiana pennelliana is listed as endangered in Florida (Florida Administrative Code, Chapter 5B-40).

Gentiana saponaria L. [Resembling *Saponaria* (Caryophyllaceae).] SOAPWORT GENTIAN; HARVESTBELLS.

Gentiana saponaria Linnaeus, Sp. Pl. 228. 1753. *Pneumonanthe saponaria* (Linnaeus) F. W. Schmidt, Arch. Bot. (Leipzig) 1: 10. 1796. *Ciminalis saponaria* (Linnaeus) Berchtold & J. Presl, Prir. Rostlin 1(Gentian.): 11. 1923. *Dasystephana saponaria* (Linnaeus) Small, Fl. S.E. U.S. 930, 1336. 1903.

Gentiana elliottii Chapman var. *latifolia* Chapman, Fl. South. U.S. 356. 1860. *Dasystephana latifolia* (Chapman) Small, Fl. S.E. U.S. 930, 1336. 1903. *Pneumonanthe latifolia* (Chapman) Greene, Leafl. Bot. Observ. Crit. 1: 71. 1903. *Gentiana latifolia* (Chapman) Britton, Man. Fl. N. States, ed. 2. 1075. 1905; non (Grenier & Godron) Jakow, 1899. TYPE: FLORIDA.

Erect or somewhat decumbent perennial herb, to 6.5 dm; stem glabrous or rarely puberulent. Leaves with the blade linear to broadly elliptic or obovate, usually widest near or above the middle, the larger ones 1.5–12 cm long, 0.5–3 cm wide, the apex acute, the base broadly cuneate, the margin entire, ciliate, the upper and lower surfaces glabrous, the primary nerves 1–3(5), sessile. Flowers solitary or in terminal clusters of 2–8, sometimes also in clusters terminating short, peduncle-like branches from the upper nodes, the inflorescence subtended by a pair of involucral leaves, each flower subtended by a pair of oblanceolate to linear bracts; calyx tube 5–15 mm long, glabrous, the lobes narrowly oblanceolate, 4–17 mm long, 1–4(6) mm wide, the margins ciliate; corolla funnelform, 3–5 cm long, the tube greenish white up to slightly below the point of stamen attachment, the veins outlined in green, above this level suffused in blue, the lobes 3–7(9) mm long, the apex rounded, mucronate, incurving, the free part of the appendages 2–6 mm long, bifid, both segments erect, laciniate, the sinus on the outer side of each lobe slightly lower than that on the inner side; anthers connate at anthesis. Fruit ellipsoid, ca. 2 cm long, glabrous; seeds obliquely ellipsoid to ovoid, ca. 2 mm long, brown, winged.

Creek swamps. Rare; Gadsden and Wakulla Counties. New York south to Florida, west to Michigan, Illinois, Oklahoma, and Texas. Fall.

Gentiana villosa L. [With long weak trichomes.] STRIPED GENTIAN.

> *Gentiana villosa* Linnaeus, Sp. Pl. 228. 1753. *Pneumonanthe villosa* (Linnaeus) F. W. Schmidt, Arch Bot. (Leipzig) 1: 10. 1796. *Dasystephana villosa* (Linnaeus) Small, Fl. S.E. U.S. 931, 1336. 1903.
>
> *Gentiana ochroleuca* Froelich, Gentiana 35. 1796. *Ciminalis ochroleuca* (Froelich) Berchtold & J. Presl, Prir. Rostlin 1(Gentian.): 11. 1823. *Pneumonanthe ochroleuca* (Froelich) G. Don, Gen. Hist. 4: 195. 1837.

Erect perennial herb, to 6 dm; stem glabrous. Leaves with the blade elliptic lanceolate to obovate, the lower ones obovate or patulate, 2.5–10 cm long, 1–4 cm wide, the apex acute to obtuse, the base broadly cuneate, the margin entire, the upper and lower surfaces glabrous, the primary nerves 5–12, sessile. Flowers in terminal clusters of 1–10, sometimes also in terminal clusters on short axillary branches, the inflorescence subtended by 2–3 pairs of involucral leaves, each flower subtended by a pair of small oblanceolate bracts; calyx tube 6–8 mm long, glabrous, the lobes linear to oblanceolate, 0.5–3.5 mm long, 1–2 mm wide, the margin entire; corolla funnelform, 3–3.5 cm long, the tube greenish to yellowish white, the veins outlined in green, sometimes suffused with blue-violet, the lobes ovate deltoid, 5–10 mm long, erect, entire, the apex mucronate, the free part of the appendages 1–2 mm long, shallowly bifid, both segments erect, laciniate, the sinus on the outer side of each lobe slightly lower than that on the inner side; anthers connate or free at anthesis. Fruit ellipsoid to oblong, laterally compressed, 2–2.5 cm long, glabrous; seeds ellipsoid to oblong, ca. 3 mm long, faintly reticulate, brown, wingless.

Sandhills and dry hammocks. Occasional; panhandle. New Jersey and Pennsylvania south to Florida, west to Indiana, Kentucky, Tennessee, Mississippi, and Louisiana. Fall.

EXCLUDED TAXON

> *Gentiana autumnalis* Linnaeus—Reported for Florida by Chapman (1860, 1883, 1897, all as *G. augustifolia* Michaux) and Small (1903, 1913a, 1933, all as *Dasystephena porphyio* J. F. Gmelin), the names misapplied to material of *G. pennelliana*.

Obolaria L. 1753.

Perennial herbs. Leaves opposite, simple, pinnate-veined, epetiolate, estipulate. Flowers axillary, bracteate; sepals 2, basally connate; petals 4, basally connate; stamens 4, epipetalous; ovary 2-carpellate, 1-loculate, the style 1, the stigma 2-lobed. Fruit a capsule, irregularly dehiscent; seeds numerous.

A monotypic genus; North America. [From the Greek *obolos*, nail, metal split, used for a small Greek coin (obolo), in reference to the fleshy, rounded leaves.]

Obolaria is mycoheterotrophic, i.e., dependent on mycorrhizal fungi. It has a coralloid root system with little branching and no root hairs.

Selected reference: Gillett (1959).

Obolaria virginica L. [Of Virginia.] VIRGINIA PENNYWORT.

Obolaria virginica Linnaeus, Sp. Pl. 632. 1753. *Schultzia virginica* (Linnaeus) Kuntze, Revis. Gen. Pl. 2: 430. 1891.

Fleshy perennial herb, to 2 cm; stem glabrous except for a few glandular trichomes in the leaf axil. Leaves with the blade of the upper ones spatulate to obdeltoid, 5–10 mm long, purple, glabrous, sessile, the base decurrent the entire length of the internode, the lower ones spatulate and scalelike, minute. Flowers 1–3, axillary, sessile or short pedicellate, subtended by 2 foliaceous bracts, these spatulate, with a few scales on the inner surface near the base; corolla campanulate, 7–10 mm long, white pinkish or purplish, divided ca. ½ to the base, the tube with fimbriate scales below the base of each stamen, the lobes spreading, oblanceolate, the apex erose; stamens inserted at the corolla throat; ovary sessile, glandular, the style short, the stigma lobes orbicular, erect. Fruit ovoid, 5–6 mm long, irregularly dehiscent; seeds ovoid, minute, translucent, striate, yellowish.

Mesic hammocks. Rare; Jefferson County. New Jersey and Pennsylvania south to Florida, west to Illinois, Missouri, Arkansas, and Texas. Winter–spring.

Sabatia Adans., 1763. ROSEGENTIAN

Annual, biennial, or perennial herbs. Leaves opposite, simple, pinnipalmate-veined, epetiolate, estipulate. Flowers in cymes, bracteate; sepals 5–14, basally connate; petals 5–14, basally connate; stamens 5–14, epipetalous; nectaries present at the ovary base; ovary 2-carpellate, 1-loculate, the style 1, the stigmas 2. Fruit a capsule, septicidally dehiscent; seeds numerous.

A genus of about 20 species; North America, West Indies, and Mexico. [Commemorates Liberato Sabbati (1714–1779), Italian botanist and surgeon.]

Lapithea Griseb., 1845.

Selected reference: Wilbur (1955).

1. Flowers sessile or subsessile; inflorescence appearing capitate.................................... **S. gentianoides**
1. Flowers pedicellate; inflorescence either of solitary flowers or loosely cymose.
    2. Flowers 6- to 14-merous.
        3. Corolla white and lacking a yellow eye in the throat; upper branches opposite throughout........ ....**S. difformis**
        3. Corolla white, pink, or rose with a yellow eye in the throat; upper branches usually alternate.
            4. Calyx and corolla lobes 5–7; calyx lobes more than half as long as the corolla lobes at anthesis; cauline leaves spatulate..**S. calycina**
            4. Calyx and corolla lobes 7–14; calyx lobes usually half as long or less than half as long as the corolla lobes at anthesis; cauline leaves elliptic, lanceolate, or linear.
                5. Leaves succulent, the cauline ones narrowly linear, the basal ones spatulate and rosulate; calyx lobes subulate, often succulent..**S. decandra**
                5. Leaves chartaceous, the cauline ones lanceolate, elliptic, or linear, the basal ones often absent, not rosulate; calyx lobes linear, chartaceous ....**S. dodecandra**
    2. Flowers 5-merous.
        6. Upper branches or peduncles opposite.

7. Corolla pink to rose (rarely white); pedicel 1–6 cm long..**S. angularis**
7. Corolla white; pedicel less than 1 cm long.
   8. Calyx lobes to 3 mm long; stem terete...**S. macrophylla**
   8. Calyx lobes 4–15 mm long; stem 4-angled.
      9. Stem 4-angled and winged proximally..**S. quadrangula**
      9. Stem subterete proximally, if 4-angled, then not winged ....**S. difformis**
  6. Upper branches or peduncles alternate.
   10. Calyx lobes foliaceous..**S. calycina**
   10. Calyx lobes linear to setaceous.
      11. Calyx lobes setaceous or subulate, less than 8 mm long; corolla white..........**S. brevifolia**
      11. Calyx lobes linear, usually more than 8 mm long; corolla pink.
         12. Sepals subequaling the petals; plants perennial ....**S. campanulata**
         12. Sepals ca. ¾ as long as the petals; plants annual.
            13. Uppermost leaves equal to or wider than the stem, thin and smooth, the venation prominent..**S. stellaris**
            13. Uppermost leaves not as wide as the stem, thick and rugose, the venation obscure...**S. grandiflora**

Sabatia angularis (L.) Pursh [*Angulus*, angled, in reference to the 4-angled and winged stem]. ROSEPINK.

Chironia angularis Linnaeus, Sp. Pl. 190. 1753. *Chironia angularis* Linnaeus var. *latifolia* Michaux, Fl. Bor.-Amer. 1: 147. 1803, nom. inadmiss. *Sabatia angularis* (Linnaeus) Pursh, Fl. Amer. Sept. 137. 1814.

Biennial herb, to 7.5(9) dm; stem 4-angled, glabrous, the branching opposite. Leaves cauline and sometimes basal, the blades of the basal ones oblong-spatulate to ovate-orbiculate, that of the cauline lanceolate to ovate, 1–4 cm long, 0.5–3 cm wide, the apex rounded to obtuse, the base cordate, the margin entire, the upper and lower surfaces glabrous, sessile. Flowers in an open cyme, the pedicel 1–6 cm long; sepals 5, the calyx tube shallowly campanulate, 1–2 mm long, the lobes linear to narrowly oblong-lanceolate, 0.5–1.5 cm long, somewhat foliaceous; petals 5, rotate, pink or rose, rarely white, with greenish yellow spots, these usually with a dark red border, the tube 4–7 mm long, the lobes narrowly spatulate, 0.5–2 cm long, the apex rounded to subacute; anthers yellow, 3–5 mm long; style 4–6 mm long, the stigma lobes slender, 3–6 mm long. Fruit cylindrical, 5–9 mm long, glabrous; seeds globose, pitted.

Calcareous hammocks. Rare; central and western panhandle. Massachusetts south to Florida, west to Ontario, Michigan, Illinois, Kansas, and New Mexico. Summer.

Sabatia brevifolia Raf. [With short leaves.] SHORTLEAF ROSEGENTIAN.

Sabatia brevifolia Rafinesque, Atl. J. 147. 1832. *Sabatia elliottii* Steudel, Nomencl. Bot., ed. 2. 2: 489. 1841, nom. illegit.

Annual herb, to 7 dm; stem terete, glabrous, the branching all or mostly alternate. Leaves all cauline or the basal sometimes persistent, the blade linear to oblong-lanceolate, 0.5–3 cm long, 1–5(7) mm wide, the apex obtuse to attenuate, the base cuneate, the margin entire, the upper and lower surfaces glabrous, sessile. Flowers in an open cyme or solitary at the end of the upper branches, the pedicel (1)2–4(5) cm long; sepals 5, the tube turbinate, 1–3 mm long, the lobes setaceous or subulate, 3–8 mm long; petals 5, the corolla rotate, white with triangular, greenish yellow basal spots, the tube 1–3 mm long, the lobes oblanceolate, 0.6–1.8 cm long, the apex obtuse to acute; anthers yellow, 1–2 mm long; style 1 mm long, the stigma lobes 3–5 mm long. Fruit cylindrical, 3–6 mm long, glabrous; seeds globose, pitted.

Flatwoods. Common; nearly throughout. South Carolina south to Florida, west to Alabama. Spring–summer.

Sabatia calycina (Lam.) A. Heller [In reference to the very evident calyx lobes.] COASTAL ROSEGENTIAN.

Gentiana calycina Lamarck, Encycl. 2: 638. 1788. *Sabatia calycina* (Lamarck) A. Heller, Bull. Torrey Bot. Club 21: 24. 1894.
Chironia calycosa Michaux, Fl. Bor.-Amer. 1: 147. 1803. *Sabatia calycosa* (Michaux) Pursh ex Sims, Bot. Mag. 39: pl. 1600. 1813.

Perennial herb, to 5 dm; stem terete or distally 4-angled, glabrous, the branching all or mostly alternate. Leaves all cauline, the blade elliptic to spatulate, 1–6(10) cm long, 0.5–3 cm wide, the apex rounded to acute or acute, the base cuneate, the margin entire, the upper and lower surfaces glabrous, sessile. Flowers few in an open cyme, the pedicel (1)3–6 cm long; sepals 5–7, the tube shallowly campanulate, 2–5 mm long, the lobes oblanceolate to spatulate, somewhat foliaceous, 0.8–2.5(3) cm long; petals 5–7, pink to pale pink or white, with triangular, yellow basal spots, the tube 3–6 mm long, the lobes oblanceolate to spatulate-obovate, 0.6–1.5 cm long, the apex rounded to obtuse; anthers yellow, 2–4 mm long; style 1–2 mm long, the stigma lobes 4–6 mm long. Fruit subglobose or broadly cylindrical, 6–10 mm long, glabrous; seeds globose, pitted.

Floodplain forests, wet flatwoods, and pond margins. Frequent; nearly throughout. Virginia south to Florida, west to Texas; West Indies. Summer.

Sabatia campanulata (L.) Torr. [Bell-shaped, in reference to the corolla.] SLENDER ROSEGENTIAN.

Chironia campanulata Linnaeus, Sp. Pl. 190. 1753. *Sabatia campanulata* (Linnaeus) Torrey, Fl. N. Middle United States 217. 1824.
Chironia gracilis Michaux, Fl. Bor.-Amer. 1: 146. 1803. *Chironia campanulata* Linnaeus var. *gracilis* (Michaux) Persoon, Syn. Pl. 1: 282. 1805. *Sabatia gracilis* (Michaux) Salisbury, Parad. Lond. t. 32. 1806. *Sabatia campanulata* (Linnaeus) Torrey var. *gracilis* (Michaux) Fernald, Rhodora 39: 444. 1937.

Perennial herb, to 6(9) dm; stem terete or distally 4-angled, the branching all or mostly alternate. Leaves all cauline, the blade narrowly lanceolate or oblong to linear, 1–4 cm long, 1–7(12) mm wide, the apex acute, the base rounded, the margin entire, flat or slightly revolute, the upper and lower surfaces glabrous, sessile. Flowers few in a cyme or solitary at the end of the branches, the pedicel (2)4–7(9) cm long; sepals 5, the tube turbinate to shallowly campanulate, the lobes linear, 7–15 mm long; petals 5, the corolla rotate, pink or white, with oblong, yellow basal spots, these with a red border, the tube 2–6 mm long, the lobes oblanceolate, 6–24 mm long, the apex obtuse; anthers yellow, 2–4 mm long; style 2–5 mm long, the stigma lobes linear, 3–7 mm long. Fruit cylindrical, 5–7 mm long, glabrous; seeds globose, pitted.

Wet flatwoods and pond margins. Occasional; northern counties. New York and Massachusetts south to Florida, west to Indiana, Arkansas, and Texas. Summer.

Sabatia decandra (Walter) R. M. Harper [From the Greek *deca*, ten, and *andra*, stamens]. BARTRAM'S ROSEGENTIAN.

Chironia decandra Walter, Fl. Carol. 95. 1788. *Sabatia chloroides* Pursh var. *erecta* Elliott, Sketch Bot. S. Carolina 1: 286. 1817. *Sabatia chloroides* Pursh var. *stricta* Grisebach, Gen. Sp. Gent. 125. 1838, nom. illegit. *Sabatia dodecandra* (Linnaeus) Britton et al. var. *stricta* A. Gray ex C. Mohr, Bull. Torrey Bot. Club 24: 26. 1897, nom. illegit. *Sabatia decandra* (Walter) R. M. Harper, Bull. Torrey Bot. Club 27: 432. 1900.
Sabatia chloroides Pursh var. *coriacea* Elliott, Sketch Bot. S. Carolina 1: 286. 1817. *Sabatia dodecandra* (Linnaeus) Britton et al. var. *coriacea* (Elliott) Ahles, J. Elisha Mitchell Sci. Soc. 80: 173. 1964.
Sabatia bartramii Wilbur, Rhodora 57: 91. 1955. TYPE: FLORIDA: Escambia Co.: 9 mi. W of Pensacola, s.d., *Wilbur & Webster 3577* (holotype: MICH).

Perennial herb, to 8(10) dm; stem terete, glabrous, the branching all or mostly alternate. Leaves basal and cauline, the blade of the basal leaves oblanceolate to spatulate, 4–10 cm long, 1–2.5 cm wide, the apex rounded, the base cuneate, the margin entire, the upper and lower surfaces glabrous, sessile, the cauline blades linear to lanceolate, 1.5–5(6.5) cm long, 1–8(15) mm wide, the apex acute. Flowers solitary or paired at the end of the upper branches, the pedicel (3)8–12 cm long; sepals 8–12(14), the tube shallowly campanulate, 2–8 mm long, the lobes linear to subulate, 0.4–2 cm long; petals 8–12(14), the corolla rotate, pink or white, with oblong, yellow, basal spots usually with red border, the tube 5–9 mm long, the lobes spatulate-obovate, 1.5–3.5 cm long, the apex rounded to obtuse; anthers yellow, 5–7 mm long, the stigma lobes 7–10 mm long. Fruit ovoid, 6–8 mm long, glabrous; seeds globose, pitted.

Wet flatwoods, marshes, and pond margins. Frequent; nearly throughout. South Carolina south to Florida, west to Mississippi. Summer–fall.

Sabatia difformis (L.) Druce [Irregularly formed]. LANCELEAF ROSEGENTIAN.

Swertia difformis Linnaeus, Sp. Pl. 226. 1753. *Sabatia difformis* (Linnaeus) Druce, Bot. Exch. Club Soc. Brit. Isles 3: 423. 1914.
Chironia lanceolata Walter, Fl. Carol. 95. 1788. *Sabatia paniculata* (Michaux) Pursh var. *latifolia* Pursh, Fl. Amer. Sept. 138. 1814. *Sabatia corymbosa* Baldwin ex Elliott, Sketch Bot. S. Carolina 1: 283. 1817, nom. illegit. *Sabatia lanceolata* (Walter) Torrey ex A. Gray, Manual 356. 1848.
Chironia paniculata Michaux, Fl. Bor.-Amer. 1: 146. 1803. *Sabatia paniculata* (Michaux) Pursh, Fl. Amer. Sept. 138. 1814.

Perennial herb, to 1 m; stem 4-angled distally, subterete proximally, glabrous, the branching opposite. Leaves all cauline, the blade linear-lanceolate to elliptic-ovate, 1–4(6) cm long, 3–14(20) mm wide, the apex acute, the base rounded to cordate, the margin entire, the upper and lower surfaces glabrous, sessile. Flowers in a corymbiform dichasium of compact cymules, the pedicel 1–8(15) mm long; sepals 5–6, the tube shallowly campanulate, 1–2(3) mm long, the lobes lanceolate-subulate, (2)4–9(14) mm long; petals 5–6, the corolla rotate, white, lacking a yellow eye in the throat, the tube 3–6 mm long, the lobes oblanceolate, (5)7–21 mm long, the apex rounded; anthers yellow, 2–3 mm long; style 2–5 mm long, the stigma lobes linear, 2–5 mm long. Fruit oblong-cylindrical, 4–8 mm long, glabrous; seeds globose, pitted.

Bogs and wet flatwoods. Frequent; northern and central peninsula, central and western panhandle. New Jersey south to Florida, west to Mississippi. Summer.

Sabatia dodecandra (L.) Britton et al. [From the Greek *dodec*, twelve, and *andra*, stamens.] MARSH ROSEGENTIAN.

Chironia dodecandra Linnaeus, Sp. Pl. 190. 1753. *Chlora dodecandra* (Linnaeus) Linnaeus, Syst. Nat., ed. 12. 2: 267. 1767. *Chironia chloroides* Michaux, Fl. Bor.-Amer. 1: 147. 1803, nom. illegit. *Sabatia chloroides* Pursh, Fl. Amer. Sept. 138. 1814, nom. illegit. *Sabatia chloroides* Pursh var. *flexuosa* Elliott, Sketch Bot. S. Carolina 1: 286. 1817, nom. inadmiss. *Sabatia dodecandra* (Linnaeus) Britton et al., Prelim. Cat. 36. 1888.
Pleienta leucantha Rafinesque, New Fl. 4: 92. 1838 ("1836"). TYPE: "south New Jersey to Florida."
Sabatia foliosa Fernald, Bot. Gaz. 33: 155. 1902. *Sabatia dodecandra* (Linnaeus) Britton et al. var. *foliosa* (Fernald) Wilbur, Rhodora 57: 87. 1955. TYPE: FLORIDA: Santa Rosa Co.: Blackwater River near Milton, 8 Jul 1897, *Curtiss 5928* (lectotype: GH). Lectotypified by Wilbur (1955: 87).
Sabatia harperi Small, Fl. S.E. U.S. 928, 1336. 1903.

Perennial herb, to 6 dm; stem terete or distally 4-angled, glabrous, the branching all or mostly alternate. Leaves all cauline, the blade elliptic- or oblong-lanceolate, 1.5–7 cm long, 4–12(16) mm wide, the apex rounded to obtuse or acute, the base rounded to subcordate or cuneate, the margin entire, the upper and lower surfaces glabrous, sessile. Flowers few in an open monochasium or solitary at the end of the distal branches, the pedicel 1–9(11) cm long; sepals 7–12(14), the tube turbinate to campanulate, 2–4 mm long, the lobes linear to oblong-lanceolate or spatulate, somewhat foliaceous, 4–20 mm long; petals 7–12(14), the corolla rotate, purplish pink or white, with an oblong, sometimes shallowly lobed, yellow basal spot with a red border, the tube (3)4–8 mm long, the lobes oblanceolate to spatulate-obovate, (1)1.2–2.5 cm long, the apex rounded to acute; anthers yellow, 3–5 mm long; style 3–5 mm long, the stigma lobes 5–9 mm long. Fruit cylindrical, 6–10 mm long, glabrous; seeds globose, pitted.

Cypress pond margins. Occasional; northern counties. New Jersey south to Florida, west to Texas. Summer–fall.

Sabatia gentianoides Elliott [Resembling some *Gentiana* species.] PINEWOODS ROSEGENTIAN.

Sabatia gentianoides Elliott, Sketch Bot. S. Carolina 1: 286. 1817. *Lapithea gentianoides* (Elliott) Grisebach, in de Candolle, Prodr. 9: 48. 1845.

Perennial herb, to 6.5 dm; stem terete or slightly 4-angled, the branching opposite or alternate. Leaves cauline and basal, the basal with the blade oblong-spatulate, the cauline linear, 1–10 cm long, 1–3 mm wide, the apex of the basal ones rounded to obtuse, that of the cauline acuminate, the base cuneate, the margin entire, the upper and lower surfaces glabrous, sessile. Flowers solitary or few in a dense cluster and appearing capitate, sessile or subsessile; sepals 7–12, the tube campanulate, 3–8 mm long, the lobes setaceous, 3–17 mm long; petals 7–12, the corolla pink, with oblong, greenish yellow basal spots, the tube 6–10 mm long, the lobes oblanceolate to narrowly spatulate-obovate, 1.2–3 cm long, 4–11 mm wide, the apex rounded to obtuse; anthers yellow, (3)4–6(7) mm long; style 5–8 mm long, the stigma lobes spatulate, 4–7 mm long. Fruit ovoid, 7–10 mm long, glabrous; seeds globose, pitted.

Wet flatwoods and bogs. Occasional; northern peninsula, Indian River County, central and western panhandle. North Carolina south to Florida, west to Arkansas and Texas. Summer.

Sabatia grandiflora (A. Gray) Small [With large flowers.] LARGEFLOWER ROSEGENTIAN.

Sabatia gracilis (Michaux) Salisbury var. grandiflora A. Gray, Syn. Fl. N. Amer. 2(1): 115. 1878. Sabatia grandiflora (A. Gray) Small, Fl. S.E. U.S. 928, 1336. 1903. Sabatia campanulata (Linnaeus) Torrey var. grandiflora (A. Gray) S. F. Blake, Rhodora 17: 52. 1915. TYPE: FLORIDA: Brevard Co.: Indian River, 1874, Palmer 430 (lectotype: GH). Lectotypified by Wilbur (1955: 65, 67).

Annual herb, to 9(11); stem terete, the branching alternate. Leaves all cauline, the blade linear to filiform, 1–5 cm long, 1–2(5) mm wide, the apex acuminate, the base cuneate, the margin entire, the upper and lower surfaces glabrous, sessile. Flowers few in a cyme or solitary, the pedicel (2)4–12 cm long; sepals 5, the tube campanulate, 2–6 mm long, the lobes subulate or linear, 0.6–2.5(3) cm long; petals 5, the corolla rotate, pink or white, with oblong, yellow basal spots, these usually with a red border, the tube 3–8 mm long, the lobes obovate, 1.5–3 cm long, the apex rounded to subacute; anthers dark yellow (3)5–7(8) mm long; style 2–5 mm long, the stigma lobes (4)6–8(9) mm long. Fruit cylindrical, (6)8–10(15) mm long, glabrous; seeds globose, pitted.

Wet flatwoods and marshes. Common; nearly throughout. Florida and Alabama; West Indies. All year.

Sabatia macrophylla Hook. [Large-leaved.] LARGELEAF ROSEGENTIAN.

Sabatia macrophylla Hooker, Campanion Bot. Mag. 1: 71. 1836.
Sabatia recurvans Small, Man. S.E. Fl. 1049. 1933. Sabatia macrophylla Hooker var. recurvans (Small) Wilbur, Rhodora 57: 17. 1955.

Perennial herb, to 1.4 m; stem terete, glabrous, the branching opposite. Leaves all cauline, the blade lanceolate to ovate-oblong or ovate, 2.5–6(8.5) cm long, 0.5–3(4.5) cm wide, the apex acute, the base rounded, the margin entire, the upper and lower surfaces glabrous, sessile. Flowers in a corymbiform dichasium of compact cymules, the pedicel 1–5 mm long; sepals 5, the tube campanulate to obconic, 1–2 mm long, the lobes triangular to linear subulate, 1–3 mm long; petals 5, the corolla rotate, white, lacking a yellow eye in the throat, the tube 2–4 mm long, the lobes oblong-lanceolate, 4–7(9) mm long, the apex rounded to obtuse; anthers white to pale yellow, 1–2 mm long; style 1–3 mm long, the stigma lobes slender, 1–3 mm long. Fruit globose to ovoid, 3–4 mm long; seeds globose, pitted.

Wet flatwoods, bogs, and cypress pond margins. Occasional; northern peninsula, central and western panhandle. Georgia and Florida, west to Louisiana. Summer.

Sabatia quadrangula Wilbur [Four-angled, in reference to the stem.] FOURANGLE ROSEGENTIAN.

Sabatia quadrangula Wilbur, Rhodora 57: 22. 1955.

Biennial herb, to 5 dm; stem 4-angled, glabrous, branching all opposite or the secondary and/ or tertiary sometimes alternate. Leaves basal and cauline or all cauline, the basal blades ob-ovate, the cauline linear-oblong to ovate-lanceolate, 0.8–2.5(6) cm long, 3–8(18) mm wide, the apex obtuse to acute, the base rounded to cordate, the margin entire, flat or somewhat revolute, the upper and lower surfaces glabrous, sessile. Flowers in a cyme of compact cymules, the pedicel 1–2(4) mm long; sepals 5, the tube turbinate to campanulate, 2–3 mm long, the lobes linear-subulate, 2–8(11) mm long; petals 5, the corolla rotate, white, lacking a yellow eye in the throat or occasionally with triangular, yellow basal spots, the tube 3–7 mm long, the lobes ob-lanceolate to narrowly spatulate-obovate, 5–15 mm long, the apex rounded to obtuse; anthers pale yellow, 2–3 mm long; style 1–2 mm long, the stigma lobes lanceolate to oblong, 2–6 mm long. Fruit cylindrical, 4–7 mm long, glabrous; seeds globose, pitted.

Flatwoods. Occasional; northern counties, Levy County. Virginia south to Florida, west to Alabama. Summer.

Sabatia stellaris Pursh [Star, in reference to the flower center being like a star.] ROSE-OF-PLYMOUTH.

Sabatia stellaris Pursh, Fl. Amer. Sept. 137. 1814. *Chironia stellaris* (Pursh) Eaton, Man. Bot., ed. 2. 204. 1818.
Sabatia gracilis (Michaux) Salisbury var. *stellaris* A. W. Wood, Amer. Bot. Fl. 267. 1870.

Annual or biennial herb, to 5(8) dm; stem terete or distally 4-angled, glabrous, the branch-ing alternate. Leaves all cauline, the blade linear to obovate, 0.5–6(9) cm long, (1)2–10(15) mm wide, the apex acute, the base cuneate, the margin entire, the upper and lower surfaces glabrous, sessile. Flowers few in a cyme or solitary, the pedicel (1)4–10 cm long; sepals 5, the tube turbinate, 2–6 mm long, the lobes setaceous to linear, (4)6–10(20) mm long; petals 5, the corolla rotate, pink or rarely white, with 3-lobed basal spots, these usually with a red border, the tube 3–8 mm long, the lobes oblanceolate, spatulate-obovate, or elliptic, 0.5–2 cm long, the apex rounded to obtuse; anthers yellow, 2–3 mm long; style 2–4 mm long, stigma lobes 3–8 mm long. Fruit subglobose to cylindrical, (4)6–8(14) mm long, glabrous; seeds globose, pitted.

Coastal swales and marshes. Frequent; nearly throughout. New Jersey south to Florida, west to Louisiana. Summer.

EXCLUDED TAXON

Sabatia brachiata Elliott—Reported for Florida by Small (1903, 1913a, both as *Sabatia angustifolia* (Michaux) Britton; 1933). This species is excluded from Florida by Wilbur (1955). No Florida speci-mens known.

Voyria Aubl. 1775. GHOSTPLANT

Perennial herb. Leaves opposite, simple, pinnipalmate-veined, epetiolate, estipulate. Flowers in dichasial cymes; sepals 5, basally connate; petals 5, basally connate; stamens 5, epipetalous; nectaries absent; ovary 2-carpellate, 1-loculate, the style 1, the stigmas 2. Fruit a capsule, septicidally dehiscent; seeds numerous.

A genus of about 20 species; Florida, West Indies, Mexico, Central America, South America, and Africa. [Aboriginal vernacular name for the plant in French Guiana, the name apparently applied to plants with large, edible roots.]

Voyria is mycoheterotrophic, i.e., dependent on mycorrhizal fungi. It has a coralloid root system with little branching and no root hairs.

Leiphaimos Schltdl. & Cham., 1831.

Selected reference: Maas & Ruyters (1986).

Voyria parasitica (Schltdl. & Cham.) Ruyters & Maas [Parasitic.] PARASITIC GHOSTPLANT.

Leiphaimos parasitica Schlechtendal & Chamisso, Linnaea 6: 387. 1831. *Voyria mexicana* Grisebach, Gen. Sp. Gent. 208. 1838, nom. illegit. *Voyria parasitica* Schlechtendal & Chamisso) Ruyters & Maas, Acta Bot. Neerl. 30: 143. 1981.

Perennial achlorophyllous herb, to 1.5 dm; stem white to pale tan, glabrous. Leaves with the blade scalelike, 3.5 mm long, 1 mm wide, the apex obtuse to acute, the base cuneate, the margin entire, the upper and lower surfaces glabrous, sessile. Flowers 2–30 in a dichasial cyme or solitary; calyx campanulate, 2–4 mm long, lobed to about the middle, the lobes lanceolate-triangular, the apex acute; corolla salverform, white to pale pink, 5–9 mm long, the lobes ovate, ca. 1 mm long, the apex obtuse; stamens inserted in the distal ¼ of the corolla tube, the anthers subsessile; ovary stipulate, the style 1, the stigma lobes peltate. Fruit ellipsoid, fenestrate (dehiscing in the middle and not to the base or apex); seeds spindle-shaped, with threadlike tails.

Hammocks. Rare; Miami-Dade County, Monroe County keys. Florida; West Indies, Mexico, and Central America. All year.

Voyria parasitica is listed as endangered in Florida (Florida Administrative Code, Chapter 5B-40).

EXCLUDED GENUS

Gentianella quinquefolia (Linnaeus) Small—Reported for Florida by Small (1903, 1913a, 1933). No Florida specimens known.

LOGANIACEAE R. Br. ex Mart., nom. cons. 1827. LOGANIA FAMILY

Shrubs or herbs. Leaves opposite or whorled, simple, pinnipalmate-veined, petiolate or epetiolate, stipulate. Flowers in terminal or axillary cymose spikes, spiciform racemes, or solitary, actinomorphic, bisexual, bracteate, bracteolate or ebracteolate; sepals 5, basally connate, petals 5, basally connate, stamens 5, epipetalous, the anthers dorsifixed, introrse, longitudinally dehiscent; ovary superior, or partly inferior, 2-carpellate and -loculate, the style 1, the stigmas 1- or 2-lobed or unlobed. Fruit a capsule or berry; seeds numerous.

A family of 15 genera and about 400 species; North America, West Indies, Mexico, Central America, South America, Africa, Asia, Australia, and Pacific Islands.

Spigeliaceae Bercht. & J. Presl, 1823; *Strychnaceae* DC. ex Perb., 1818.

Selected reference: Rogers (1986).

1. Shrubs or trees; fruit a berry with a hard exocarp..**Strychnos**
1. Herbs; fruit a capsule.
    2. Fruit conspicuously cleft into 2 rounded lobes at the apex, leaving a woody cuplike remnant after dehiscing; style articulated ....**Spigelia**
    2. Fruit shallowly emarginate or acuminate at the apex, separating into 2 hornlike structures, not leaving a woody cuplike remnant after dehiscing; style not articulated..............................**Mitreola**

Mitreola L. 1758. HORNPOD

Annual herbs. Leaves opposite, simple, pinnipalmate-veined, petiolate or epetiolate, stipulate. Flowers in terminal or axillary dichasial cymes, the branches of secund spiciform racemes, bracteate, bracteolate; sepals 5, basally connate, petals 5, basally connate; stamens 5, epipetalous; ovary partly inferior, the style 1, the stigma 1. Fruit a capsule, apically 2-horned when the style splits, dehiscing longitudinally along the adaxial face; seeds numerous.

A genus of about 8 species; North America, West Indies, Mexico, Central America, South America, Africa, Asia, Australia, and Pacific Islands. [Diminitive of *mitra*, in reference to the mitriform capsule.]

Cynoctonum J. F. Gmel., 1791.

1. Leaves narrowed to a short-petiolate base or the petiole to 3 cm long, the distal leaves divaricate or ascending...**M. petiolata**
1. Leaves sessile or subsessile, the distal leaves ascending or appressed to the stem.
    2. Leaves ovate to elliptic; seeds smooth, shiny, but not iridescent ....**M. sessilifolia**
    2. Leaves lance-elliptic to linear; seeds alveolate-reticulate, iridescent.........................**M. angustifolia**

Mitreola angustifolia (Torr. & A. Gray) J. B. Nelson [With narrow leaves].
NARROWLEAF HORNPOD.

Mitreola sessilifolia (J. F. Gmelin) G. Don var. *angustifolia* Torrey & A. Gray, Fl. N. Amer. 2: 45. 1841. *Cynoctonum angustifolium* (Torrey & A. Gray) Small, Bull. Torrey Bot. Club 23: 129. 1896. *Mitreola angustifolia* (Torrey & A. Gray) J. B. Nelson, Phytologia 46: 339. 1980. TYPE: FLORIDA: *s.d.*, *Chapman s.n.* (holotype: GH; isotypes K, OXF, P).

Erect annual herb, to 5 dm; stem 4-angled, glabrous. Leaves all cauline or sometimes also forming a basal rosette, the blade lance-elliptic to linear, 2–8 cm long, 1–3.5 cm wide, the apex acute to obtuse, the base rounded to truncate, the margin minutely toothed, the upper and lower surfaces glabrous, sessile or subsessile; stipules interpetiolar, truncate or deltoid, ca. 1 mm long, membranous. Flowers in a terminal spiciform cyme, pedunculate; bracts subequaling or longer than the calyx, the bracteoles minute; calyx 1–1.5 mm long; corolla globose-funnelform, 2–3 mm long, white or pink, the lobes oblong-elliptic, about 1 mm long, glabrous on the outer surface, pilose in a ring at the throat on the inner surface. Fruit 2–4 mm long, apically 2-horned, glabrous, the horns smooth or slightly tuberculate; seeds obliquely ellipsoid, somewhat plano-convex, ca. 0.5 mm long, dark yellow, brown, or black, alveolate-reticulate, iridescent.

Cypress-gum swamps and wet flatwoods. Occasional; northern and central peninsula, central and western panhandle. South Carolina south to Florida, west to Mississippi. Summer–fall.

Mitreola petiolata (J. F. Gmel.) Torr. & A. Gray [Petiolate.] LAXHORN HORNPOD.

> *Ophiorrhiza mitreola* Linnaeus, Sp. Pl. 150. 1753. *Mitreola ophiorrhizoides* Richard, Mém. Soc. Hist. Nat. Paris 1: 63. 1823. *Cynoctonum mitreola* (Linnaeus) Britton, Mem. Torrey Bot. Club 5: 258. 1894. *Cynoctonum mitreola* (Linnaeus) Britton var. *intermedia* Hochreutiner, Bull. New York Bot. Gard. 6: 284. 1910, nom. inadmiss.
>
> *Cynoctonum petiolatum* J. F. Gmelin, Syst. Nat. 2: 443. 1791. *Ophiorrhiza lanceolata* Elliott, Sketch Bot. S. Carolina 1: 238. 1817, nom. illegit.; non Forsskal, 1775. *Mitreola lanceolata* Torrey ex Croom, Cat. Pl. New Bern 23, 45. 1837, nom. illegit. *Mitreola petiolata* (J. F. Gmelin) Torrey & A. Gray, Fl. N. Amer. 2: 45. 1841.
>
> *Cynoctonum succulentum* R. W. Long, Rhodora 72: 29. 1970. TYPE: FLORIDA: Manatee Co.: near Manatee, 11 Jun 1890, *Simpson s.n.* (holotype: GH).

Erect annual herb, to 8 dm; stem 4-angled to narrowly winged glabrous. Leaves with the blade ovate-elliptic to elliptic-lanceolate, 2–8 cm long, 1–3.5 cm wide, the apex acuminate to obtuse, the base cuneate, the margin minutely toothed, the upper and lower surfaces glabrous, the petiole to 3 cm long; stipules interpetiolar, triangular, ca. 1 mm long, membranaceous. Flowers in a terminal spiciform cyme, pedunculate; bracts triangular, 2–3 mm long, the bracteoles minute; calyx obconical, ca. 1 mm long; corolla globose-funnelform, 2–3 mm long, white or pink, the lobes oblong-elliptic, about 1 mm long, glabrous on the outer surface, pilose at the throat on the inner surface. Fruit 3–4 mm long, apically 2-horned, smooth or with a few scattered papillae, glabrous; seeds ovate to suborbicular, somewhat plano-convex, ca. 0.5 mm long, dark brown to black, alveolate-reticulate, iridescent.

Marshes, pond margins, wet flatwoods, and swamps. Common; nearly throughout. Virginia south to Florida, west to Texas; West Indies, Mexico, Central America, and South America; Africa, Asia, and Australia. Summer–fall.

Mitreola sessilifolia (J. F. Gmel.) G. Don [With sessile leaves.] SWAMP HORNPOD.

> *Cynoctonum sessilifolium* J. F. Gmelin, Syst. Nat. 2: 443. 1791. *Ophiorrhiza croomii* M. A. Curtis, Boston J. Nat. Hist. 1: 105. 128. 1835., nom. illegit. *Mitreola sessilifolia* (J. F. Gmelin) G. Don, Gen. Hist. 4: 171. 1837.
>
> *Cynoctonum sessilifolium* J. F. Gmelin var. *microphyllum* R. W. Long, Rhodora 72: 27. 1970. TYPE: FLORIDA: Brevard Co.: 4 mi. W of Melbourne, 22 Aug 1958, *Kral 7961* (holotype: GH; isotypes NY, USF.)

Erect annual herb, to 5 dm; stem 4-angled, glabrous. Leaves all cauline or sometimes also forming a basal rosette, the blade ovate to elliptic, 1–2 cm long, 1–2 cm wide, the apex obtuse to acute, the base rounded, the margin minutely toothed, the upper and lower surfaces glabrous, sessile; stipules interpetiolar, deltate, ca. 1 mm long, membranaceous. Flowers in a terminal spiciform cyme, pedunculate; bracts narrowly triangular to elliptic, 2–3 mm long, the bracteoles minute; calyx 1–2 mm long, prominently keeled; corolla globose-funnelform, 2–3 mm long, white, the lobes oblong-elliptic, ca. 1 mm long, glabrous on the outer surface, pilose at the throat on the inner surface. Fruit 2–3 mm long, apically 2-horned, glabrous, papillose or warty; seeds elliptic, 1 face flattened, the other convex, dark brown or black, smooth, not iridescent.

Wet flatwoods, bogs, and pond margins. Frequent; nearly throughout. Virginia south to Florida, west to Missouri, Oklahoma, and Texas; West Indies. Summer–fall.

Spigelia L. 1753. PINKROOT

Annual or perennial herbs. Leaves opposite, simple, pinnipalmate-veined, petiolate or epetiolate, stipulate. Flowers in terminal or axillary cymes, subtended by 1 bract, each flower subtended by 2–3 bracteoles or ebracteolate; sepals 5, basally connate; petals 5, basally connate; stamens epipetalous; ovary superior, 2-carpellate and -loculate, the style 1, articulated above the base, the stigmas unlobed or 2-lobed. Fruit a capsule, septicidally dehiscent; seeds numerous.

A genus of about 50 species; North America, West Indies, Mexico, Central America, South America, Africa, and Asia. [Commemorates Adrian van den Spieghel (1578–1625), Dutch-born physician and botanist of Padua, Italy.]

Coelostylis Torr. & A. Gray ex Endl. & Fenzl, 1839.

1. Inflorescence subtended by a pseudowhorl of bracteate leaves .... **S. anthelmia**
1. Inflorescence not subtended by a pseudowhorl of bracteate leaves.
  2. Corolla scarlet on the outer surface, yellow on the inner surface, 3–5 cm long; anthers exserted from the corolla tube...**S. marilandica**
  2. Corolla white or pink, 1.5–3 cm long; anthers included within the corolla tube.
    3. Flowers in long terminal cymes, the branches secund, scorpioid; style articulated near the base .... **S. gentianoides**
    3. Flowers in short terminal cymes; style articulated near the middle......................**S. loganioides**

Spigelia anthelmia L. [In reference to its use as an anthelmintic, an intestinal parasite medication.] WEST INDIAN PINKROOT.

Spigelia anthelmia Linnaeus, Sp. Pl. 149. 1753. *Spigelia quadrifolia* Stokes, Bot. Mat. Med. 1: 307. 1812, nom. illegit.

Erect annual herb, to 4 dm; stem glabrous, the upper leaves in 2 pairs at the stem apex. Leaves subtending the inflorescence in 2 unequal pairs and forming a bracteate pseudowhorl, the lower ones 1 pair or absent, smaller than the bracteate ones, the blade ovate-lanceolate, 4–16(18) cm long, 2–5(7) cm wide, the apex acuminate, the base cuneate, the margin entire, minutely ciliolate, the upper surface scabridulous, the lower surface puberulous-setulose on the veins, sessile or the petiole 1–3 mm long; stipules interpetiolar, membranous. Flowers in a terminal or axillary cyme, the branches secund, scorpioid, 5–10(12) cm long, the inflorescence subtended by a pseudowhorl of bracteate leaves; sepals deeply cleft, linear-lanceolate, 1–2 mm long; corolla funnelform, 5–10 mm long, purplish white or pinkish, with 2 magenta submedian stripes, the lobes deltoid or ovate-deltoid; stamens included in the floral tube; styles articulate near the middle. Fruit globose, 2-lobed, 5–6 mm long, tuberculate; seeds subglobose, dorsiventrally compressed, 2–3 mm long, verrucose.

Rocky pinelands. Rare; Palm Beach County, southern peninsula. Florida; West Indies, Mexico, Central America, and South America; Asia and Africa. Native to Florida and tropical America. All year.

Spigelia gentianoides Chapm. ex A. DC. [Resembling *Gentiana* (Gentianaceae).] PURPLEFLOWER PINKROOT; GENTIAN PINKROOT.

> *Spigelia gentianoides* Chapman ex A. de Candolle, in de Candolle, Prodr. 9: 5. 1845. TYPE: FLORIDA: *s.d., Chapman s.n.* (holotype: G).

Erect or ascending perennial herb, to 3 dm; stem glabrous, scabrous at the nodes and in lines just below them. Leaves ovate to elliptic or lanceolate, 1.5–5 cm long, 0.5–2 cm wide, the apex acute to acuminate, the base rounded to cordate, the margin entire, the upper surface slightly scabrous, the lower surface scabrous on the margins and veins, sessile; stipules interpetiolar, membranous. Flowers 3–8 in a terminal cyme; calyx deeply cleft, the lobes subulate, 4–6 mm long; corolla funnelform, 2.5–3 mm long, light pink on the outer surface, with 2 dark pink vertical stripes on each lobe, lighter pink to white on the inner surface, the lobes deltoid or ovoid-deltoid; stamens included; style articulate near the base. Fruit subglobose, 2-lobed, 6–8 mm long, tuberculate; seeds subglobose, dorsiventrally compressed, 2–3 mm long, verrucose.

Sandhills. Rare; Jackson, Calhoun, and Washington Counties. Alabama and Florida. Spring.

Spigelia gentianoides is listed as endangered in Florida (Florida Administrative Code, Chapter 5B-40) and in the United States (U.S. Fish and Wildlife Service, 50 CFR 23).

Spigelia loganioides (Torr. & A. Gray ex Endl. & Fenzl) A. DC. [Resembling *Logania*.] FLORIDA PINKROOT; LEVY PINKROOT.

> *Coelostylis loganioides* Torrey & A. Gray ex Endlicher & Fenzl, Nov. Stirp. Dec. 33. 1839. *Spigelia loganioides* (Torrey & A. Gray ex Endlicher & Fenzl) A. de Candolle, in de Candolle, Prodr. 9: 4. 1845. TYPE: FLORIDA: Marion Co.: near Fort King, *s.d., Burrows s.n.* (holotype: ?).

Erect perennial herb; to 3 dm; stem glabrous. Leaves with the blade lanceolate, elliptic, or oblanceolate, 1–4 cm long, 0.5–2 cm wide, the apex acute to obtuse, the base cuneate, the margin entire, the upper and lower surfaces glabrous; stipules interpetiolar, membranous, triangular-subulate. Flowers few in a terminal cyme; sepals deeply cleft, the lobes linear-subulate, 3–5 mm long; corolla funnelform, 12–15 mm long, white, sometimes with pale lavender lines, the lobes triangular, 2–3 mm long; stamens included in the floral tube; style articulate near the middle. Fruit subglobose, 2-lobed, 5–6 mm long, tuberculate; seeds subglobose, dorsiventrally compressed, 2–3 mm long, verrucose.

Wet calcareous hammocks. Rare; Volusia, Levy, and Marion Counties, south to Sumter and Hernando Counties. Endemic. Spring.

Spigelia loganioides is listed as endangered in Florida (Florida Administrative Code, Chapter 5B-40).

Spigelia marilandica (L.) L. [Of Maryland.] WOODLAND PINKROOT.

Lonicera marilandica Linnaeus, Sp. Pl. 175. 1753. *Spigelia marilandica* (Linnaeus) Linnaeus, Syst. Nat., ed. 12. 2: 734. 1767. *Spigelia lonicera* Miller, Gard. Dict., ed. 8. 1768, nom. illegit. *Spigelia oppositifolia* Stokes, Bot. Mat. Med. 1: 307. 1812, nom. illegit.

Erect perennial herb, to 7 dm; stem glabrous. Leaves with the blade ovate to lanceolate or elliptic, 3–11 cm long, the apex acute to acuminate, the base obtuse to truncate, the margin entire, the upper and lower surfaces glabrous, sessile; stipules interpetiolar, membranous. Flowers few in a solitary, terminal, racemiform, pedunculate cyme; calyx deeply divided, the lobes linear, 10–11 mm long; corolla funnelform, 3–5 cm long, scarlet on the outer surface, yellow on the inner surface, the lobes lanceolate, ca. 1 mm long. Fruit subglobose, 2-lobed, 5–9 mm long, tuberculate; seeds subglobose, dorsiventrally compressed, 2–3 mm long, verrucose.

Calcareous hammocks. Occasional; panhandle. Virginia south to Florida, west to Illinois, Missouri, Oklahoma, and Texas. Spring.

Strychnos L. 1753.

Shrubs or trees. Leaves opposite, simple, pinnipalmate-veined, petiolate, stipulate. Flowers in terminal simple or compound cymes, bracteate; sepals 5, basally connate; petals 5, basally connate; stamens 5, epipetalous; ovary superior, 1-carpellate and -loculate, the style and stigma 1. Fruit a berry with a hard exocarp; seeds numerous.

A genus of about 190 species; nearly cosmopolitan. [From the Greek *strychnon*, deadly, originally applied to various Solanaceae species.]

Strychnos spinosa Lam. [With spines.] WOOD ORANGE; NATAL ORANGE.

Strychnos spinosa Lamarck, Tabl. Encycl. 2: 38. 1794. *Brehmia spinosa* (Lamarck) Harvey ex A. de Candolle, in de Candolle, Prodr. 9: 18. 1845.

Shrub or small tree, to 3 m; bark gray or brown, shallowly fissured, tending to flake in rectangular segments, the branchlets glabrous or puberulent, with axillary or terminal spines ca. 1 cm long. Leaves with the blade elliptic-obovate or suborbicular, 1.5–9(13) cm long, 1.2–7.5 cm wide, the apex rounded to obtuse or acute, mucronate, the base cuneate or rounded, the margin entire, usually undulate, the upper surface glabrous, the lower surface glabrous or with a tuft of silky trichomes in the angles between the main nerves and the midrib, the petiole 2–10 mm long, glabrous; stipules narrowly triangular, 1–2 mm long, caducous. Flowers in a simple or branched umbelliform dichasium terminating a short, young branchlet, puberulent; bracts linear, 3–4 mm long; calyx campanulate, 2–6 mm long, the tube very short, the lobes linear-subulate, glabrous or puberulent; corolla campanulate, 2–6 mm long, greenish white, the lobes triangular, ca. 2 mm long, the apex acute, the outer surface glabrous or sparsely puberulent, the inner surface with a ring of trichomes in the throat; stamens adnate near the tube base, bearded, included. Fruit globose, 5–12 cm long, the exocarp hard, greenish, yellow, or yellow-brown, smooth to slightly roughened, glabrous; seeds obliquely ovate to elliptic, ca. 2 cm long, laterally compressed, brown, smooth, slightly puberulent, embedded in the pulp.

Disturbed sites. Rare; Hillsborough, Highlands, and Miami-Dade Counties. Escaped from cultivation. Florida; Africa. Native to Africa. Spring.

GELSEMIACEAE Struwe & V. A. Albert 1995. JESSAMINE FAMILY

Woody vines. Leaves opposite, simple, pinnate-veined, petiolate, stipulate. Flowers axillary, solitary or in cymes, actinomorphic or slightly zygomorphic, bisexual, bracteolate; sepals 5, free; petals 5, basally connate; stamens 5, epipetalous, the anthers dorsifixed, extrorse, longitudinally dehiscent; ovary superior, 2-carpellate and -loculate, the style 1, the stigmas 4-lobed. Fruit a capsule; seeds several.

A family of 2 genera and 11 species; North America, Mexico, Central America, Africa, and Asia.

Selected reference: Rogers (1986).

Gelsemium Juss. 1789. JESSAMINE; TRUMPETFLOWER

Woody vines. Leaves opposite, simple, pinnate-veined, petiolate, stipulate. Flowers axillary, solitary or few in cymes, actinomorphic or slightly zygomorphic, bisexual, heterostylous, bracteolate; sepals 5, free; petals 5, basally connate; stamens 5, epipetalous; ovary superior, 2-carpellate and -loculate, the style 1, the stigmas 4-lobed. Fruit a capsule, septicidally and loculicidally dehiscent; seeds several.

A genus of 3 species; North America, Mexico, Central America, and Asia. [From the Italian *gelsomino*, a name for the Old World jasmine or Jessamine (*Jasminum* (Oleaceae)).]

Gelsemium has been placed in the Loganiaceae or Spigeliaceae by some workers.

Selected reference: Ornduff (1970).

1. Sepals not persistent on the fruit, the apex obtuse to rounded; seeds with a prominent membranaceous wing; flowers fragrant ....**G. sempervirens**
1. Sepals persistent on the fruit, the apex acuminate; seeds not winged; flower without fragance.............
....**G. rankinii**

Gelsemium rankinii Small [Commemorates Henry Ashby Rankin (1872–1947), North Carolina plant collector.] RANKIN'S JESSAMINE; SWAMP JESSAMINE.

Gelsemium rankinii Small, Addisonia 13: 37, pl. 435. 1928.

Twining, climbing, or creeping woody vine; stem glabrous or puberulent, the rhizomes with adventitious roots. Leaves with the blade lanceolate to narrowly ovate, 3–7.5(9) cm long, 1–1.5 cm wide, the apex acuminate to obtuse, the base rounded, the margin entire, the upper surface glabrous, the lower surface glabrous or puberulent, the petiole 3–6 mm long, puberulent on the lower surface; stipules interpetiolar, minute, membranous, caducous. Flowers solitary or 2–8 in a cyme, bracteolate in the proximal ½, the pedicel 2–10 mm long; sepals lanceolate, 3–6 mm long, the apex acuminate; corolla funnelform, 2–3 cm long, the lobes 8–15 mm long, yellow, the throat darker yellow, odorless. Fruit oblong, 1–1.6 cm long, flattened contrary to the septum, woody, the apex with the persistent stylar beak, the calyx persistent at the base, tardily dehiscent or partly so into 4 segments; seeds 3–4 mm long, flattened, pale brown, not winged.

Bogs, acid swamps, and floodplain forests. Occasional; northern counties. North Carolina south to Florida, west to Louisiana. Winter–spring.

Gelsemium sempervirens (L.) W. T. Aiton [Evergreen.] YELLOW JESSAMINE; CAROLINA JESSAMINE; EVENING TRUMPETFLOWER.

> *Bignonia sempervirens* Linnaeus, Sp. Pl. 623. 1753. *Jeffersonia sempervirens* (Linnaeus) Brickell, Med. Repos. 1: 555. 1899. *Gelsemium sempervirens* (Linnaeus) W. T. Aiton, Hortus Kew. 1: 64. 1811. *Gelsemium lucidum* Poiret, in Lamarck, Encycl., Suppl. 2: 714. 1811, nom. illegit.
>
> *Gelsemium nitidum* Michaux, Fl. Bor.-Amer. 1: 120. 1803. TYPE: FLORIDA.

Twining, climbing, or creeping woody vine. Stem glabrous or puberulent, the rhizomes with adventitious roots. Leaves with the blade lanceolate to narrowly ovate, sometimes suborbicular, 3–7(9) cm long, 1–1.5 cm wide, the apex acuminate to obtuse, the base cuneate to rounded, the margin entire, the upper surface glabrous, the lower surface glabrous or puberulent, the petiole 3–5 mm long, puberulent on the lower surface; stipules interpetiolar, minute, membranous, caducous. Flowers solitary or 2–8 in a cyme, usually densely bracteolate throughout, the pedicel 2–12 mm long; sepals elliptic, the apex obtuse; corolla funnelform, 2.5–3.5 cm long, the lobes 8–12 cm long, yellow, darker yellow in the throat, strongly fragrant. Fruit broadly oblong, 1.5–2.5 cm long, flattened perpendicular to the septum, woody, the apex with the persistent stylar beak, tardily dehiscent or partly so into 4 segments; seeds 7–10 mm long, flattened, pale brown, asymmetrically thin-winged apically.

Flatwoods and hammocks. Frequent; northern counties, central peninsula. Virginia south to Florida, west to Arkansas and Texas; Mexico and Central America. Winter–spring.

APOCYNACEAE Juss., nom. cons. 1797. DOGBANE FAMILY

Trees, shrubs, vines, or herbs, with white or colorless latex. Leaves opposite, alternate, or whorled, simple, pinnate-veined, petiolate or epetiolate, estipulate. Flowers in terminal or axillary cymes or solitary, actinomorphic, bisexual, bracteate or ebracteate, bracteolate or ebracteolate; sepals 5, basally connate; petals 5, basally connate, the tube or throat sometimes with squamallae or a corona; stamens 5, epipetalous, the filaments free or connate, the anthers free, connate, or adnate to the stigma head, the pollen free or in pollinia; nectaries present or absent; ovaries superior or partly inferior, 1- or 2-carpellate and -loculate. Fruit a capsule, follicle, drupe, or berry; seeds 1-many.

A family of about 366 genera and about 4,550 species; nearly cosmopolitan.

The Asclepiadaceae and Apocynaceae have long been recognized as separate families until recently.

Asclepiadaceae R. Br., nom. cons. 1811.

Selected references: Endress and Bruyns (2000); Endress et al. (2014); Rosatti (1989).

1. Filaments connate around the gynostegium, the anthers adnate into a head around the style-stigma; petals with 1 or 2 coronas.
  2. Stems erect or ascending, not twining or trailing.
    3. Corolla lobes each with an upcurved dorsal spur at the base...**Calotropis**
    3. Corolla lobes without a dorsal spur at the base..**Asclepias**
  2. Stems twining or trailing.
    4. Stout woody vines; corolla funnelform .... **Cryptostegia**
    4. Slender herbaceous or only partly woody vines; corolla rotate.
      5. Leaves hastate or triangular .... **Araujia**
      5. Leaves linear, elliptic, or cordate.
        6. Peduncles usually twice as long as the subtending leaves or longer ....**Funastrum**
        6. Peduncles subequaling or shorter than the leaves.
          7. Leaves linear or elliptic.
            8. Petals united to about the middle.. **Cynanchum**
            8. Petals united only slightly at the base.
              9. Corolla lobes pubescent on the inner surface; corona lobes narrow ....
              .... **Metastelma**
              9. Corolla lobes glabrous on the inner surface; corona lobes broad.
                10. Peduncles 1–2 cm long.. **Seutera**
                10. Peduncles less than 0.5 cm long...**Orthosia**
          7. Leaves cordate or cordate-lanceolate.
            11. Anthers lacking an appendage; foliar pubescence a mixture of glandular and simple trichomes; fruit muricate-tuberculate, not winged .... **Matelea**
            11. Each anther with an abaxial appendage; foliar pubescence usually of simple trichomes only; fruit smooth, 5-angled and -winged................................ **Gonolobus**
1. Filaments and anthers not united into a common structure (the anthers sometimes adnate to the stigma head, but the filaments free).
  12. Leaves alternate.
    13. Fruit a pair of follicles.
      14. Herbs; corolla blue ....**Amsonia**
      14. Shrubs or trees with the branches thick and fleshy; corolla various, but never blue.............
      .... **Plumeria**
    13. Fruit a drupe.
      15. Corolla salverform, less than 1 cm long; leaves lanceolate, 4–7 cm long .... **Vallesia**
      15. Corolla funnelform, ca. 5 cm long; leaves linear, 9–12 cm long .... **Thevetia**
  12. Leaves opposite or whorled.
    16. Vines.
      17. Carpels united, the fruit a 1-celled, spiny capsule; leaves mostly in whorls of 3–4.................
      ....**Allamanda**
      17. Carpels free, the fruit a pair of smooth follicles; leaves opposite.
        18. Corolla salverform.

19. Leaves 4.5–13 cm long, 2–8 cm wide; corolla tube 4–6 cm long ....**Echites**

19. Leaves 2–4.5 cm long, 1–1.5 cm wide; corolla tube 0.8–1.2 cm long .... **Vinca**

18. Corolla campanulate above the short cylindric base.

20. Corolla tube to 1 cm long.

21. Leaves chartaceous; corolla lobes acute, pale yellow......................... **Thyrsanthella**

21. Leaves coriaceous; corolla lobes obtuse to rounded, white ...... **Trachelospermum**

20. Corolla tube 2–6 cm long.

22. Corolla white to light pink; calyx without squamellae within.......... **Rhabdadenia**

22. Corolla yellow; calyx with squamellae within.

23. Calyx lobes narrowly lanceolate; anthers with linear apical appendages ....
....**Pentalinon**

23. Calyx lobes ovate; anthers lacking apical appendages.....................**Angadenia**

16. Erect shrubs to small trees or perennial herbs.

24. Plants with thorns..**Carissa**

24. Plants unarmed.

25. Herbs (sometimes suffrutescent).

26. Fruit a drupe, shiny purple or brown when ripe; leaves whorled ....**Rauvolfia**

26. Fruit a pair of dry follicles, greenish or brown when ripe; leaves opposite.

27. Flowers cream or yellow ....**Angadenia**

27. Flowers white or pink.

28. Calyx lobes linear-subulate; follicles 2–3 cm long .... **Catharanthus**

28. Calyx lobes lanceolate; follicles 12–21 cm long................................. **Apocynum**

25. Shrubs or small trees.

29. Corolla tube ca. 2 cm long or longer.

30. Leaves opposite, oblong to elliptic-lanceolate, the secondary veins ca. 1 cm apart; fruit fleshy, 3–4 cm long; corolla always white........................... **Tabernaemontana**

30. Leaves whorled, narrowly lanceolate, the secondary veins ca. 1 mm apart; fruit dry, cylindric, 12–18 cm long; corolla purple, pink, or white, rarely yellowish.... **Nerium**

29. Corolla tube less than 2 cm long.

31. Leaves opposite.. **Tabernaemontana**

31. Leaves whorled.

32. Flowers yellow, more than 2 cm long .... **Allamanda**

32. Flowers white, less than 2 cm long.

33. Leaves oblanceolate; pedicels about as long as the calyx, glabrous; fruit fleshy, 4–5 cm long, bright red when ripe.. **Ochrosia**

33. Leaves lanceolate to elliptic; pedicels several times longer than the calyx, puberulent; fruit dry, 30–45 cm long, greenish to brown when ripe.............
....**Alstonia**

Allamanda L. 1771.

Woody vines or subshrubs, with white latex. Leaves whorled or opposite, simple, pinnate-veined, petiolate, estipulate. Flowers in terminal or axillary cymes, bracteate; sepals 5, basally connate; petals 5, basally connate, the throat with squamallae; stamens 5, epipetalous, the filaments and anthers free; nectaries present; ovary superior, 2-carpellate, 1-loculate, the style and stigma 1. Fruit a capsule; seeds numerous.

A genus of about 14 species; North America, West Indies, Mexico, Central America, South America, Africa, Australia, and Pacific Islands. [Commemorates Frédéric-Louis Allamand (1735–1803), Swiss botanist.]

1. Leaves with the lateral veins usually flat or a few slightly raised near the midvein on the lower surface; corolla tube below the stamen insertion 2–4 cm long; collectors between the calyx base and the corolla tube base absent; seeds winged..**A. cathartica**
1. Leaves with lateral veins usually prominently raised on the lower surface; corolla tube below the stamen insertion 1–1.8 cm long; collectors present between the calyx base and the corolla base; seeds not winged...**A. schottii**

Allamanda cathartica L. [Containing a substance that accelerates defecation.] GOLDEN TRUMPET; BROWNBUD ALLAMANDA.

> *Allamanda cathartica* Linnaeus, Mant. Pl. 214. 1771. *Orelia grandiflora* Aublet, Hist. Pl. Guiane 271. 1775, nom. illegit. *Allamanda grandiflora* Poiret, in Lamarck, Encycl. 4: 601. 1797, nom. illegit. *Allamanda linnei* Pohl, Pl. Bras. Icon. Descr. 1: 74. 1827, nom. illegit. *Allamanda cathartica* Linnaeus var. *grandiflora* L. H. Bailey & Raffill, in L. H. Bailey, Stand. Cycl. Hort. 1: 247. 1914, nom. inadmiss.

Woody vine or subshrub, to 3 m; stem glabrous. Leaves in whorls of 3–4 distally, opposite below, the blade obovate to oblong-lanceolate, 5–12 cm long, 2.5–6 cm wide, the apex acuminate, the base cuneate, the margin entire, the upper and lower surfaces glabrous, the lateral veins usually flat or a few slightly raised near the midvein on the lower surface, the petiole 2–3 mm long. Flowers in a terminal or axillary cyme; sepal lobes elliptic, 5–12 mm long; corolla tubular-campanulate, yellow, 7–9 cm long, with ciliate squamallae closing the throat, the tube below the stamen insertion 2–4 cm long, the lobes spreading; collectors between the calyx base and the corolla tube base absent; stamens inserted near the top of the corolla tube, the filaments and anthers free, included; nectaries 2, forming a ring around the ovary base with shallow notches opposite the 2 lines of carpel fusion; style filiform and slightly widened distally, the stigma head cylindrical, the base umbraculiform, with 2 deltate lobes apically and a tuft of trichomes subapically. Fruit globose, 4–6 cm long, densely covered with soft spines 7–15 mm long, dehiscent by 2 valves; seeds elliptic, laterally compressed, tan, the margin winged.

Dry disturbed sites. Occasional; central and southern peninsula. Escaped from cultivation. Florida; West Indies, Mexico, Central America, and South America; Africa, Asia, Australia, and Pacific Islands. Native to South America. Spring–fall.

Allamanda schottii Pohl [Commemorates Heinrich Wilhelm Schott (1794–1855), Austrian botanist and participant in the Austrian Brazil Expedition (1817–1821).] BUSH ALLAMANDA.

> *Allamanda schottii* Pohl, Pl. Bras. Icon. Descr. 1: pl. 58. 1827. *Allamanda cathartica* Linnaeus var. *schottii* (Pohl) L. H. Bailey & Raffill, in L. H. Bailey, Stand. Cycl. Hort. 1: 247. 1: 247. 1914.

Shrub, to 2 m; stem glabrous. Leaves in whorls of 3–5 distally, opposite below, the blade elliptic or narrowly obovate, 5–14 cm long, 2–4 cm wide, the apex acuminate, the base cuneate, the margin entire, the upper and lower surfaces glabrous, the lateral veins usually prominently raised on the lower surface, the petiole 2–3 mm long. Flowers in a terminal or axillary cyme; sepal lobes elliptic, 5–8 mm long; corolla tubular-campanulate, yellow, 6–8 cm long, with ciliate squamalae closing the throat, the tube below the stamen insertion 1–1.8 cm long, the lobes spreading; collectors present between the calyx base and the corolla tube base; stamens inserted near the top of the corolla tube, the filaments and anthers free, included; nectaries 2, forming a ring around the ovary base with shallow notches opposite the 2 lines of carpel fusion; style filiform and slightly widened distally, the stigma head cylindric, the base umbraculiform, with 2 delicate lobes apically and a tuft of trichomes subapically. Fruit globose, 4–6 cm long, densely covered with soft spines 7–15 mm long, laterally compressed, tan, the margin not winged.

Disturbed sites. Rare; Broward County. Escaped from cultivation. Florida; Mexico, Central America, and South America; Pacific Islands. All year.

Alstonia R. Br., nom. cons. 1810.

Trees, with white latex. Leaves whorled, simple, pinnate-veined, petiolate, estipulate. Flowers in terminal cymes, bracteate; sepals 5, basally connate; petals 5, basally connate; stamens 5, epipetalous, the filaments and anthers free; ovary superior, 2-carpellate and -loculate, the style and stigma 1. Fruit a follicle; seeds numerous.

A genus of about 60 species; North America, Central America, South America, Asia, Australia, and Pacific Islands. [Commemorates Charles Alston (1683–1760), Scottish botanist and physician.]

1. Corolla less than 10 mm long, the tube glabrous or with a few trichomes; leaves mostly 4 or fewer per node..**A. macrophylla**
1. Corolla more than 10 mm long, the tube densely pubescent apically; leaves mostly 5 or more per node..**A. scholaris**

Alstonia macrophylla Wall. ex G. Don [With large leaves.] DEVILTREE.

Alstonia macrophylla Wallich ex G. Don, Gen. Hist. 4: 87. 1837.

Tree, to 20 m; bark smooth, branchlets somewhat 4-angled. Leaves in whorls of 3–4 distally, the blade narrowly obovate or elliptic, 1–5 dm long, 4–15 cm wide, the apex acuminate, the base cuneate, the margin entire, the upper surface glabrous, the lower surface pubescent, the petiole 1–4 cm long. Flowers in a terminal cyme, pubescent, the peduncle 4–6 cm long, the pedicel 4–5 mm long; corolla salverform, white, the tube 4–6 mm long, slightly longer than the lobes, the lobes ciliate; style filiform, the stigma conic, with 2 erect apical lobes. Fruit linear, to 60 cm long, 2–5 mm wide; seeds pubescent, the ends with deltoid wings with stiff trichomes on the margin.

Disturbed sites. Rare; Miami-Dade County. Escaped from cultivation. Florida; Africa, Asia, and Pacific Islands. Native to Asia. All year.

Alstonia macrophylla is listed as a category II invasive species in Florida by the Florida Exotic Pest Plant Council (FLEPPC, 2015).

Alstonia scholaris (L.) R. Br. [In Theravada Buddhism, the tree is said to have been used as the tree for achieved enlightenment, or Bodhi by the first Lord of Buddha called "Thanhankara."] DITA; WHITE CHEESEWOOD.

> *Echites scholaris* Linnaeus, Mant. Pl. 53. 1767. *Alstonia scholaris* (Linnaeus) R. Brown, Asclepiadeae 75. 1810. *Pala scholaris* (Linnaeus) Roberty, Bull. Inst. Franc. Afrique Noire 15: 1426. 1953.

Tree, to 40 m; bark gray, the branchlets glabrous, copiously lenticellate. Leaves in whorls of 3–10, the blade obovate to narrowly spatulate, 7–28 cm long, 1–11 cm wide, the petiole 1–3 cm long, the apex obtuse to rounded, the base cuneate, the margin entire, the upper and lower surfaces glabrous. Flowers in a terminal cyme, the peduncle usually as long as or shorter than the calyx; corolla salverform, white, the tube 6–10 mm long, the lobes broadly ovate or broadly obovate, 2–4.5 mm long; the style filiform, the stigma conic, with 2 erect lobes apically. Fruit linear, to 55 cm long, 2–5 mm wide; seeds oblong, the margin ciliate, the ends with tufts of trichomes.

Wet, disturbed sites. Rare; Palm Beach and Broward Counties. Escaped from cultivation. Florida; Asia and Australia. Native to Asia and Australia. Summer.

Amsonia Walter 1788. BLUESTAR.

Perennial herbs, with white latex. Leaves alternate, simple, pinnate-veined, petiolate or epetiolate, estipulate. Flowers in terminal or axillary cymes; sepals 5, basally connate; petals 5, basally connate; stamens 5, epipetalous, the filaments and anthers free; nectaries present or absent; ovary superior, 2-carpellate and -loculate, the style and stigma 1. Fruit a follicle; seeds numerous.

A genus of about 20 species; North America and Asia. [Commemorates John Amson (1698–1763), American physician, mayor of Williamsburg, Virginia, and associate of botanist John Clayton.]

1. Leaves linear to narrowly lanceolate, the uppermost sessile...**A. ciliata**
1. Leaves lanceolate to elliptic, the uppermost distinctly short-petiolate.................. **A. tabernaemontana**

Amsonia ciliata Walter [In reference to the ciliate leaves.] FRINGED BLUESTAR.

> *Amsonia ciliata* Walter, Fl. Carol. 98. 1768. *Amsonia angustifolia* Michaux, Fl. Bor.-Amer. 1: 121. 1803, nom. illegit.
> *Amsonia tenuifolia* Rafinesque, New Fl. N. Amer. 4: 58. 1838 ("1836"). *Amsonia ciliata* Walter var. *tenuifolia* (Rafinesque) Woodson, Ann. Missouri Bot. Gard. 15: 400. 1928.

Erect perennial herb, to 3.5 dm; stem pubescent. Leaves with the blade linear to narrowly lanceolate, 3–8 cm long, 1–5 mm wide, the apex acute, the base cuneate, the margin entire, long-ciliate, the upper and lower surfaces glabrous or glabrate, the upper ones sessile, the lower ones subsessile or with the petiole to 3 mm long. Flowers in a terminal cyme; sepal lobes narrow triangular to ovate, to 3 mm long, glabrous or villous on the margin; corolla salverform, light blue, the tube often darker than the lobes or sometimes green-tinged, 9–10 mm long, glabrous externally, lanose internally, the lobes oblong to lanceolate, 3.5–11 mm long, glabrous; style filiform, the stigma conic, the base umbraculiform. Fruit cylindric, 1–2 cm long, 3–4 mm wide, glabrous; seeds cylindric, 5–10 mm long, with corky ridges, glabrous, brown.

Sandhills. Occasional; northern counties, central peninsula. Virginia south to Florida, west to Missouri, Oklahoma, and Texas. Spring–summer.

Amsonia tabernaemontana Walter [Originally placed in the genus *Tabernaemontana*.] EASTERN BLUESTAR.

> *Tabernaemontana amsonia* Linnaeus, Sp. Pl., ed. 2. 301. 1762. *Tabernaemontana humilis* Salisbury, Prodr. Stirp. Chap. Allerton 148. 1796, nom. illegit. *Amsonia latifolia* Michaux, Fl. Bor.-Amer. 1: 121. 1803, nom. illegit. *Tabernaemontana latifolia* J. Parmentier, Cat. Arbr. Parm. 77. 1818, nom. illegit. *Amsonia amsonia* (Linnaeus) Britton, Mem. Torrey Bot. Club 5: 262. 1894, nom. inadmiss. *Amsonia tabernaemontana* Walter, Fl Carol. 98. 1788.
>
> *Amsonia rigida* Shuttleworth ex Small, Fl. S.E. U.S. 935. 1903. TYPE: FLORIDA: Wakulla Co.: near St. Marks, Jun 1843, *Rugel s.n.* (holotype: BM?, NY?; isotype: MO).

Erect perennial herb, to 1 m; stem glabrous. Leaves with the blade lanceolate to elliptic, (1.5)3–12 cm long, 1–6 cm wide, the apex acute to acuminate, the base cuneate, the margin entire, the upper surface glabrous to slightly pubescent, the lower surface glabrous to moderately pubescent, the petiole 2–10 mm long. Flowers in a terminal or axillary, dense to loose cyme; calyx lobes triangular, 0.5–1.5 mm long, glabrous, the margin scabrous; corolla salverform, light blue, the tube often darker than the lobes or green-tinged, the tube 6.5–8 mm long, the upper half pubescent on the outer surface, lanose on the inner, the lobes spatulate to broadly lanceolate, 6–9 mm wide, glabrous or pubescent on the median area on the outer surface; style filiform, the stigma conic, the base umbraculiform. Fruit linear-cylindric, 8–13 cm long, glabrous; seeds cylindric, 6–9 mm long, with low, irregular corky ridges, glabrous, dark brown.

Floodplain forests and wet hammocks. Occasional; northern counties, Levy County. Massachusetts and New York south to Florida, west to Kansas, Oklahoma, and Texas. Spring–summer.

Angadenia Miers 1878. PINELAND GOLDEN TRUMPET.

Subshrubs or woody vines, with white latex. Leaves opposite, simple, pinnate-veined, petiolate or epetiolate, estipulate. Flowers in axillary cymes, bracteate; sepals 5, basally connate; petals 5, basally connate; stamens 5, epipetalous, the filaments free, the anthers agglutinated to the stigma-style head; nectariferous disk present; ovary superior, 2-carpellate and -loculate. Fruit a follicle; seeds numerous.

A genus of 2 species; North America and West Indies. [From the Greek *angeion*, diminutive of jar, and *adin*, gland, in reference to the urceolate nectariferous disk surrounding the ovaries.]

Angadenia berteroi (A. DC.) Miers [Commemorates Carlo Luigi Guiseppe Bertero (1789–1831), Italian physician and botanist who collected in the West Indies and South America.] PINELAND GOLDEN TRUMPET.

> *Echites berteroi* A. de Candolle, in de Candolle, Prod. 8: 447. 1844. *Rhabdadenia berteroi* (A. de Candolle) Müller Argoviensis, Linnaea 30: 435. 1860. *Angadenia berteroi* (A. de Candolle) Miers, Apocyn. S. Amer. 180. 1878.

Echites sagrae A. de Candolle, in de Candolle, Prodr. 8: 450. 1844. *Rhabdadenia sagrae* (A. de Candolle) Müller Argovensis, Linnaea 30: 435. 1860. *Angadenia sagrae* (A. de Candolle) Miers, Apocyn. S. Amer. 181. 1878.

Rhabdadenia corallicola Small, Bull. New York Bot. Gard. 3: 434. 1905. TYPE: FLORIDA: Miami-Dade Co.: pinelands between Coconut Grove and Cutler, Nov 1903, *Small & Carter 714* (Holotype: NY).

Erect subshrub or vine, to 1 m; stem scabrous. Leaves with the blade elliptic, ovate, to oblong, 1–3 cm long, the apex obtuse, apiculate, the base rounded, the margin revolute, the upper surface glabrous, the lower surface scabrous, the petiole to 3 mm long. Flowers in a terminal cyme; calyx lobes deltoid-ovate, 2–3 mm long; corolla funnelform, yellow or cream-colored, 2–3 cm long, the tube 5–6 mm long, the lobes 1–1.5 cm wide; style filiform, the stigma conic, the base umbraculiform. Fruit linear, 8–11 cm long, 2–3 mm wide; seeds elliptic, the apex comose.

Pine rocklands. Rare; Miami-Dade County, Monroe County keys. North Carolina and Florida; West Indies. Native to Florida and West Indies. All year.

Angadenia berteroi is listed as threatened in Florida (Florida Administrative Code, Chapter 5B-40).

Apocynum L. 1753. DOGBANE

Perennial herbs, with white latex. Leaves opposite or alternate, simple, pinnate-veined, petiolate or epetiolate, estipulate. Flowers in terminal cymes, bracteate; sepals 5, basally connate; petals 5, basally connate; stamens 5, epipetalous, the filaments and anthers free; nectaries present; ovary partly inferior, 2-carpellate and -loculate, the style and stigma 1. Fruit a follicle; seeds numerous, comose.

A genus of 7 species; North America, Mexico, Europe, and Asia. [From the Greek *apo*, away from, and *cynos*, dog, in reference to its use as a dog poison.]

Apocynum cannabinum L. [Hemp-like, in reference to the strong cordage that was made by weaving together the long stem fibers.] INDIANHEMP.

Apocynum cannabinum Linnaeus, Sp. Pl. 213. 1753. *Apocynum cannabinum* Linnaeus var. *typicum* Béguinot & Belosersky, Atti. Reale Accad. Lincei, Rendiconti Cl. Sci. Fis., ser. 5. 9: 690. 1913, nom. inadmiss. *Cynopaema cannabinum* (Linnaeus) Lunell, Amer. Midl. Naturalist 4: 509. 1916.

Apocynum sibiricum Jacquin, Hort. Bot. Vindob. 3: 37, pl. 66. 1777. *Apocynum hypericifolium* Aiton, Hort. Kew. 1: 304. 1789, nom. illegit. *Apocynum cannabinum* Linnaeus var. *hypericifolium* A. Gray, Manual 365. 1848. *Apocynum hypericifolium* Aiton var. *typicum* Béguinot & Belosersky, Atti. Reale Accad. Lincei, Rendiconti Cl. Sci. Fis., ser. 5. 9: 703. 1913, nom. inadmiss. *Cynopaema hypericifolium* (A. Gray) Lunell, Amer. Midl. Naturalist 4: 509. 1916, nom. illegit.

Apocynum dimidiatum Rafinesque, Autik. Bot. 181. 1840. TYPE: FLORIDA.

Erect perennial herb, to 1 m; stem glabrous to villose, sometimes glaucous. Leaves opposite or alternate, the blade ovate to oblong or lanceolate, 2–14 cm long, 0.3–4.5(7) cm wide, the apex acute to rounded, the base cuneate to cordate, the margin entire, the upper surface glabrous, the lower surface glabrous or villous, sessile or the petiole to 5 mm long. Flowers in a terminal cyme; bracts linear to lanceolate, foliaceous or scalelike and inconspicuous; calyx lobes linear to lanceolate, 1–3 mm long, glabrous; corolla narrowly campanulate to urceolate or short-cylindric, white or greenish, 3–5 mm long, the lobes ca. ½ the tube length, erect or slightly spreading; style clavate, the stigma 2-lobed. Fruit linear, 7–20 cm long, 2–3 mm wide, glabrous; seeds narrowly fusiform, 3–6 mm long, with a white or tawny apical coma.

Open hammocks. Occasional; northern counties, central peninsula. Nearly throughout North America; Mexico; Asia. Native to North America and Mexico. Summer.

EXCLUDED TAXON

Apocynum ×*floribundum* Greene (*A. androsaemifolium* L. ×*A. cannabinum*)—Reported for Florida by Woodson (1930, as *A. medium* Greene), based on a misidentification of *Rolfs 720* (MO) from Columbia County, which is *A. cannabinum*.

Araujia Brot. 1817.

Perennial herbaceous vines, with white latex. Leaves opposite, simple, pinnate-veined, petiolate, estipulate. Flowers in axillary umbelliform cymes, bracteate; sepals 5, basally connate; petals 5, basally connate, with a 5-lobed tubular corona; stamens, style, and stigma adnate and forming a gynostegium, the pollen in pollinia, the ovary superior, 2-carpellate and -loculate. Fruit a follicle; seeds comose.

A genus of about 12 species; North America, South America, Europe, Africa, Australia, and Pacific Islands. [Commemorates António de Araújo e Azevedo (1754–1817), Portuguese statesman, author, and amateur botanist.]

Selected references: Goyder (2003); Rapini et al. (2011); Spellman and Gunn (1976).

Araujia odorata (Hook. & Arn.) Fontella & Goyder [With an odor, in reference to the fragrant flowers.] LATEXPLANT.

Cynanchum odoratum Hooker & Arnott, J. Bot. (Hooker) 1: 294. 1834. *Morrenia odorata* (Hooker & Arnott) Lindley, Edward's Bot. Reg. 24(Misc.): 71. 1838. *Araujia odorata* (Hooker & Arnott) Fontella & Goyder, in Rapani et al., Phytotaxa 26: 11. 2011.

Perennial vine; stem to 3 m, finely white pubescent when young, glabrous in age. Leaves with the blade hastate to narrowly triangular, 5–8 cm long, 3–5 cm wide, the apex acute, the base truncate to cuneate or shallowly cordate, the margin entire, the upper and lower surfaces minutely appressed puberulent, the petiole 2–4 cm long. Flowers 4–10 in an axillary umbelliform cyme, the peduncle 3–5 mm long, finely pubescent, the pedicel 8–12 mm long, pubescent; bracts lanceolate, minute; sepal lobes narrowly lanceolate, 6–8 mm long, spreading, the outer surface pubescent; corolla white to greenish white, rotate, the lobes narrowly lanceolate, 8–12 mm long, spreading or slightly reflexed, the corona a 5-lobed tube, 5–8 mm long, longer than the gynostegium, each lobe apically 2-lobed; gynostegium short stipitate, 3–4 mm long. Fruit narrowly ovoid, 8–10 cm long, 2–4 cm wide; seeds narrowly elliptical to narrowly ovate, 5–6 mm long, the surface irregularly tuberculate and minutely reticulate, the apex with an apical coma.

Hammocks and disturbed sites. Frequent; northern and central peninsula. Escaped from cultivation. Native to South America. Summer.

Asclepias L. 1753. MILKWEED

Perennial herbs, with white or colorless latex. Leaves opposite, whorled, or alternate, simple, pinnate-veined, petiolate or epetiolate, estipulate. Flowers in terminal and axillary umbelli-form cymes or solitary, bracteate; sepals 5, basally connate; petals 5, basally connate, with a corona of 5 hoods adnate at the base of the staminal column, each with or without a basal horn on the inner side; stamens, styles, and stigma adnate to form a gynostegium, the pollen in pollinia; ovary 2-carpellate and -loculate. Fruit a follicle; seeds numerous, with or without an apical coma.

A genus of about 140 species; nearly cosmopolitan. [Named for Asclepius, Greek god of medicine because of the many medicinal uses of the plants.]

Acerates Elliott, 1817; *Anantherix* Nutt., 1818; *Asclepiodora* A. Gray, 1876; *Asclepiodorella* Small, 1933; *Biventraria* Small, 1933; *Oxypteryx* Greene, 1897; *Podostigma* Elliott, 1817.

Selected reference: Woodson (1954).

1. Leaves all alternate.
    2. Corolla orange-red; sap clear..**A. tuberosa**
    2. Corolla various, but not orange-red; sap white.
        3. Leaves ovate to lance-oblong ....**A. viridis**
        3. Leaves linear or linear-lanceolate.
            4. Corona hoods subequaling the gynostegium in length, with horns..................**A. michauxii**
            4. Corona hoods shorter than gynostegium, lacking horns....................................**A. longifolia**
1. Leaves opposite, subopposite, or whorled, sometimes also with a few alternate.
    5. Corolla lobes erect at anthesis.
        6. Corolla greenish yellow; leaves linear, less than 1 cm wide......................................**A. pedicellata**
        6. Corolla greenish white; leaves 1.5–3 cm wide..**A. viridis**
    5. Corolla lobes reflexed at anthesis.
        7. Corolla lobes orange-red, scarlet, or bright yellow.
            8. Leaves linear to linear-lanceolate, less than 1 cm wide ....**A. lanceolata**
            8. Leaves lanceolate, more than 1 cm wide...**A. curassavica**
        7. Corolla lobes various, but not as above.
            9. Leaves oblong to ovate.
                10. Leaves sessile or subsessile; floral umbels pedunculate, the peduncle as long as or longer than the pedicels.
                    11. Leaves broadly cuneate at the base; corolla greenish yellow...................**A. connivens**
                    11. Leaves auriculate-clasping at the base; corolla purple or green suffused with purple.
                        12. Leaves ovate-deltoid; corolla lobes 5–6 mm long..............................**A. humistrata**
                        12. Leaves elliptic; corolla lobes 9–11 mm long ....**A. amplexicaulis**
                10. Leaves petiolate, or if subsessile, then the umbels sessile.

13. Floral umbels long-pedunculate, the peduncles as long as or longer than the petiole.

    14. Larger leaves usually more than 5 cm long; inflorescence terminal; hoods broadly rounded ....**A. variegata**

    14. Larger leaves usually less than 4 cm long; inflorescences lateral; hoods narrowly lanceolate...**A. curtissii**

13. Floral umbels or the peduncle much shorter than the pedicels.

    15. Horns absent; corolla lobes 6–7 mm long ....**A. viridiflora**

    15. Horns present; corolla lobes 9–10 mm long.

        16. Longest petioles usually more than 4 mm long; horns longer than the hoods..
.... **A. tomentosa**

        16. Longest petioles usually less than 4 mm long; horns shorter than the hoods....
....**A. obovata**

9. Leaves linear to lanceolate.

    17. Leaves whorled..**A. verticillata**

    17. Leaves opposite or subopposite.

        18. Hoods shorter than the gynostegium.

            19. Horns present .... **A. cinerea**

            19. Horns absent.

                20. Larger leaves less than 1.5 mm wide; corolla lobes spreading, 6–8 mm long ....
....**A. feayi**

                20. Larger leaves more than 1.5 mm wide; corolla lobes reflexed, to 5 mm long.....
....**A. longifolia**

        18. Hoods longer than the gynostegium.

            21. Floral umbels terminal..**A. michauxii**

            21. Floral umbels terminal and from the upper leaf axils.

                22. Leaves 1–2 mm wide...**A. viridula**

                22. Leaves 5 mm wide or wider.

                    23. Leaves sessile or subsessile .... **A. rubra**

                    23. Leaves petiolate.

                        24. Corolla white or pale pink; leaf base long tapering...............**A. perennis**

                        24. Corolla dull rose-purple; leaf base rounded or short-tapering................
.... **A. incarnata**

Asclepias amplexicaulis Sm. [With leaves clasping the stem.] CLASPING MILKWEED.

Asclepias amplexicaulis Smith, Nat. Hist. Lepidopt. Georgia 1: 13, pl. 7. 1797.

Erect perennial herb, to 8 dm; stem glabrous, glaucous. Leaves opposite, the blade broadly ovate or elliptic to oblong, 4–12 cm long, 1.8–8 cm wide, the apex obtuse to rounded, usually mucronate, the base cordate-clasping, the margin entire, undulate, usually sparsely puberulent, the upper and lower surfaces glabrous, sessile. Flowers numerous in a terminal umbelliform cyme, the peduncle (3)6–20(30) cm long, the pedicel (2)3–4.5 cm long; bracts lanceolate, minute; sepal lobes green or purple-tinged, lanceolate to ovate-lanceolate, 3–5.5 cm long, reflexed; corolla green or purple-tinged, lanceolate, 9–11 mm long, reflexed, the corona hoods oblong, erect or spreading, 5–6 mm long, slightly longer than the gynostegium, involute, open above, the horns subulate and arching over the gynostegium; gynostegium pale purple to rose. Fruit fusiform, 9–15 cm long, 8–18 mm wide, smooth, puberulent or glabrous, glaucous; seeds ovate, laterally compressed, the margin narrowly winged, apically comose.

Sandhills. Occasional; northern counties, central peninsula. New Hampshire and Vermont south to Florida, west to Minnesota, Nebraska, Kansas, Oklahoma, and Texas. Spring–summer.

Asclepias cinerea Walter [Grayish, in reference to the general appearance.] CAROLINA MILKWEED.

Asclepias cinerea Walter, Fl. Carol. 105. 1788.

Erect perennial herb, to 7 dm; stem glabrous. Leaves opposite, the blade narrowly linear, 4–10 cm long, 2–3 mm wide, the apex acute, the base narrowly cuneate, the margin entire, the upper and lower surfaces glabrous, sessile. Flowers few to many in an umbelliform cyme, this terminal or in the axil of the upper leaves, the peduncle ca. 1 cm long, the pedicel 1.5–2.5 cm long; bracts subulate, minute; sepal lobes greenish, lanceolate, ca. 2 mm long, reflexed; corolla lobes lavender with a white margin or whitish basally with lavender veins and a white margin distally, oblanceolate to obovate, 5–7 mm long, reflexed, the corona hoods variously suffused with lavender, ca. 3 mm long, much shorter than the gynostegium, involute, each with 2 lateral lobes extending beyond the truncate apex, the horn subequaling the hood; gynostegium pale lavender or white. Fruit fusiform, 6–8 cm long, 5–10 mm wide, glabrous; seeds ovate, laterally compressed, the margin narrowly winged, apically comose.

Sandhills, wet flatwoods, and bogs. Frequent; northern counties, Levy and Marion Counties. South Carolina south to Florida, west to Mississippi. Spring–summer.

Asclepias connivens Baldwin [Converging, in reference to the corona hoods arching over the gynostegium and meeting above it.] LARGEFLOWER MILKWEED.

Asclepias connivens Baldwin, in Elliott, Sketch Bot. S. Carolina 1: 320. 1817. *Acerates connivens* (Baldwin) Decaisne, in A. de Candolle, Prodr. 8: 521. 1844. *Anantherix connivens* (Baldwin) Feay ex A. W. Wood, Class-Book Bot., ed. 1861. 594. 1861.

Erect perennial herb, to 8 dm; stem slightly pubescent above. Leaves opposite, the blade elliptic, narrowly obovate, or lanceolate, 3–8 cm long, 0.5–2 cm wide, reduced and bract-like toward the base, the apex acute to obtuse, the base truncate, rounded, or cuneate, the margin entire, minutely scabrid, the upper surface glabrous, the lower surface sparsely short-pubescent, sessile or rarely the petiole to 2 mm long. Flowers few to many in an umbelliform cyme, this terminal or in the axil of the upper leaves, the peduncle 1.5–2 cm long, the pedicel 1–1.5 cm long; bracts lanceolate, minute; sepal lobes green, lanceolate, 6–7 mm long, reflexed; corolla lobes greenish white, oblong, 11–15 mm long, reflexed, the corona hoods strongly involute, arching, incurved, and meeting well above the gynostegium, the horns absent. Fruit fusiform, (8)12–15 cm long, 1–1.5 cm wide, smooth, short pubescent; seeds ovate, laterally compressed, the margin narrowly winged, apically comose.

Flatwoods and bogs. Occasional; peninsula, central and western panhandle. South Carolina south to Florida, west to Mississippi. Summer.

Asclepias curassavica L. [Of Curaçao.] SCARLET MILKWEED; BLOODFLOWER.

Asclepias curassavica Linnaeus, Sp. Pl. 215. 1753. *Asclepias bicolor* Moench, Methodus 717. 1794, nom. illegit. *Asclepias aurantiaca* Salisbury, Prodr. Stirp. Chap. Allerton 150. 1796, nom. illegit. *Asclepias nivea* Linnaeus var. *curassavica* (Linnaeus) Kuntze, Revis. Gen. Pl. 1: 418. 1891.

Erect perennial herb, to 1 m; stem pubescent, soon becoming glabrous; leaves opposite, the blade lanceolate or elliptic-lanceolate, 5–12 cm long, 1–3 cm wide, the apex acute, the base cuneate, the margin entire, pubescent, becoming glabrous, the upper surface glabrous, the lower surface minutely pubescent on the veins, becoming glabrate, the petiole 5–10 mm long. Flowers numerous in an umbelliform cyme, this terminal or in the axil of the upper leaves, the peduncle 1.5–3 cm long, the pedicel 1.5–3 cm long; bracts lanceolate, minute; sepal lobes green, lanceolate or triangular-lanceolate, 2–3 mm long, reflexed; corolla bright crimson or yellow, the lobes lanceolate or elliptic-lanceolate, 5–8 mm long, reflexed, the hoods spatulate, orange, erect, 4–5 mm long, longer than the gynostegium, the horn flat, subulate, orange, longer than the gynostegium. Fruit fusiform, 6–10 cm long, glabrous; seeds ovate, laterally compressed, the margin narrowly winged, apically comose.

Disturbed sites. Occasional; central and southern peninsula, panhandle. Escaped from cultivation. Tennessee, Florida, Louisiana, Texas, and California; West Indies, Mexico, Central America, and South America; Africa, Asia, Australia, and Pacific Islands. Native to tropical America. All year.

Asclepias curtissii A. Gray [Commemorates Allen Hiram Curtiss (1845–1907), American botanist.] CURTISS' MILKWEED.

Asclepias curtissii A. Gray, Proc. Amer. Acad. Arts 19: 85. 1883. *Oxypteryx curtissii* (A. Gray) Small, Man. S.E. Fl. 1072, 1507. 1933. TYPE: FLORIDA: s.d., *Curtiss s.n.* (holotype: GH).
Asclepias aceratoides Nash, Bull. Torrey Bot. Club 22: 154. 1895; non M. A. Curtis, 1849. *Asclepias arenicola* Nash, Bull. Torrey Bot. Club 23: 252. 1896. *Oxypteryx arenicola* (Nash) Greene, Pittonia 3: 235. 1897. TYPE: FLORIDA: Lake Co.: vicinity of Eustis, 16–30 Jun 1894, *Nash 1092* (holotype: NY).

Erect perennial herb, to 7 dm; stem puberulent. Leaves opposite, the blade broadly elliptic to ovate, 3–5 cm long, 1–3 cm wide, the apex obtuse to rounded, the base rounded to broadly cuneate, the margin entire, the upper surface glabrous, the lower surface puberulent, the petiole 5–10 mm long. Flowers numerous in an umbelliform cyme, this terminal or in the axil of the upper leaves, the peduncle 1.5–2 cm long, pubescent, the pedicel ca. 1 cm long, pubescent; bracts lanceolate, minute; sepal lobes green, triangular, 2–3 mm long, the outer surface puberulent, reflexed; corolla lobes greenish white, lanceolate, 5–7 mm long, reflexed, the corona hood white, lanceolate, distinctly stalked, involute and dorsiventrally flattened, 4–5 mm long, about twice as long as the gynostegium, erect, the horn adnate to the midvein of the hood up to the middle, horizontally protruding from the slit between the hood margins and curved upward at the apex. Fruit fusiform, 8–11 cm long, ca. 1 cm wide; seeds ovate, laterally compressed, the margin narrowly winged, apically comose.

Scrub. Occasional; peninsula. Endemic. Spring–fall.

Asclepias curtissii is listed as endangered in Florida (Florida Administrative Code, Chapter 5B-40).

Asclepias feayi Chapm. ex A. Gray [Commemorates William T. Féay, physician, teacher, and botanist of Savannah, Georgia.] FLORIDA MILKWEED.

> *Asclepias feayi* Chapman ex A. Gray, Proc. Amer. Acad. Arts 12: 72. 1877. *Acerates feayi* (Chapman ex A. Gray) Chapman, Bot. Gaz. 3: 12. 1878. *Asclepiodora feayi* (Chapman ex A. Gray) Chapman, Fl. South. U.S., ed. 3. 349. 1897. *Asclepiodella feayi* (Chapman ex A. Gray) Small, Man. S.E. Fl. 1073, 1507. 1933. TYPE: FLORIDA: s.d., *Feay s.n.* (holotype: GH?; isotypes: NY).
>
> *Asclepias simpsonii* Chapman, Fl. South. U.S., ed. 2, suppl. 2. 693. 1892. TYPE: FLORIDA.

Erect perennial herb, to 4 dm; stem glabrous. Leaves opposite, the blade linear-filiform, 3–10 cm long, 1–2 mm wide, the apex acute, the base cuneate, the margin entire, the upper and lower surfaces glabrous, sessile. Flowers few in an umbelliform cyme, this terminal or in the axil of the upper leaves, the peduncle 2–5 mm long, the pedicel ca. 2 cm long; bracts subulate, minute; sepal lobes green, lanceolate, ca. 2 mm long, spreading; corolla lobes white to pinkish or pale purple, lanceolate, 6–7 mm long, spreading, slightly upward curved apically, the corona hoods erect, involute, subequaling the gynostegium, open at the apex, the horn absent; gynostegium with the anthers curving over the top. Follicle cylindric-fusiform, 3–4.5 cm long, ca. 5 mm wide, glabrous; seeds ovate, laterally compressed, the margin narrowly winged, apically comose.

Sandhills and scrubby flatwoods. Occasional; peninsula. Endemic. Spring–fall.

Asclepias humistrata Walter [*Humi*, on the ground, and *stratosus*, in layers, in reference to the growth habit.] PINEWOODS MILKWEED.

> *Asclepias humistrata* Walter, Fl. Carol. 105. 1788.

Spreading-ascending perennial herb, to 7 dm; stem glabrous. Leaves opposite, the blade ovate, 6–10 cm long, 4.5–8.5 cm wide, the apex acute to obtuse, the base rounded to amplexicaul, the margin entire, the upper and lower surfaces glabrous, glaucous, the veins pink to lavender, sessile. Flowers numerous in an umbelliform cyme, this terminal or in the axil of the upper leaves, the peduncle 2–8 cm long, the pedicel ca. 2 cm long; bracts lanceolate, minute; sepal lobes green, lanceolate, 2–3 mm long, reflexed; corolla lobes rose to lavender, lanceolate, 5–6 mm long, reflexed, the corona hoods erect, involute, 3–4 mm long, longer than the gynostegium, the horns longer, shorter than the hoods but longer than the gynostegium. Fruit fusiform, 9–14 cm long, 1.3–1.8 cm wide, glabrous; seeds ovate, laterally compressed, the margin narrowly winged, apically comose.

Sandhills and scrub. Frequent; northern counties, central peninsula. North Carolina south to Florida, west to Louisiana. Spring–summer.

Asclepias incarnata L. [Flesh-colored, in reference to the flower color.] SWAMP MILKWEED.

> *Asclepias incarnata* Linnaeus, Sp. Pl. 215. 1753. *Asclepias verecunda* Salisbury, Prodr. Stirp. Chap. Allerton 150. 1795, nom. illegit.

Asclepias pulchra Ehrhart ex Willdenow, Sp. Pl. 1: 1267. 1797. *Asclepias incarnata* Linnaeus var. *pulchra* (Ehrhart ex Willdenow) Persoon, Syn. Pl. 1: 276. 1805. *Asclepias incarnata* Linnaeus forma *pulchra* (Ehrhart ex Willdenow) Voss, Vilm. Blumergärtn., ed. 3. 1: 664. 1894. *Asclepias incarnata* Linnaeus subsp. *pulchra* (Ehrhart ex Willdenow) Woodson, Ann. Missouri Bot. Gard. 41: 53. 1954.

Erect perennial herb, to 2(2.5) m; stem glabrous or pubescent in decurrent lines from the leaf base. Leaves with the blade linear-lanceolate to lanceolate or ovate, (3)5–15 cm long, (0.5)1–3(4.5) cm wide, the apex acute to acuminate, the base cuneate to rounded or truncate, the margin inconspicuously revolute, the upper and lower surfaces sparsely puberulent, at least so on the veins, the petiole 3–17 mm long. Flowers numerous in an umbelliform cyme, this terminal or in the axil of the upper leaves, the peduncle 1–7 cm long, puberulent, the pedicel 1–1.7(2) cm long, puberulent; bracts lanceolate, minute; sepal lobes green or purple, lanceolate to ovate, 1–2 mm long, the outer surface villous, reflexed; corolla lobes bright pink or rarely white, elliptic to oblanceolate, 5–6 mm long, reflexed, the corona hoods oblong, erect, slightly spreading, 2–3 mm long, shorter than the gynostegium, involute, the apex broadly rounded, flat or nearly so, the horns acicular, adnate to the lower ½ of the hood, arching over the gynostegium; gynostegium pale pink or rarely white, 1.2–1.8 mm long. Fruit fusiform, 5–8 cm long, 8–11 mm wide, smooth, glabrous or sparsely puberulent; seeds ovate, laterally compressed, the margin narrowly winged, apically comose.

Swamps and wet hammocks. Occasional; peninsula, Wakulla County. Quebec south to Florida, west to Manitoba, Idaho, Nevada, and Arizona. Summer.

Asclepias lanceolata Walter [Lance-shaped, in reference to the leaves.] FEWFLOWER MILKWEED.

Asclepias lanceolata Walter, Fl. Carol. 105. 1788.
Asclepias paupercula Michaux, Fl. Bor.-Amer. 1: 118. 1803. *Asclepias lanceolata* Walter var. *paupercula* (Michaux) Fernald, Rhodora 37: 438. 1935.

Erect perennial herb, to 1.2 m; stem glabrous. Leaves opposite, the blade linear-lanceolate to lanceolate, 7–20 cm long, 0.3–1 cm wide, the apex acute, the base cuneate, the margin entire, the upper and lower surfaces glabrous, sessile. Flowers few to many in an umbelliform cyme, this terminal or in the axil of the uppermost leaves, the peduncle ca. 2 cm long, the pedicel ca. 1.5 cm long; bracts subulate, minute; sepal lobes green, triangular, 2–3 mm long, reflexed; corolla lobes dull red or orange-red, lance-elliptic, ca. 10 mm long, about twice as long as the gynostegium, reflexed, the corona hoods orange, ovate, 4–6 mm long, longer than the gynostegium, erect, slightly involute, the apex rounded, outward spreading, the horns narrowly lanceolate, shorter than the hoods; gynostegium shorter than the corona. Fruit 8–10 cm long, ca. 1 cm wide, smooth, glabrous; seeds ovate, laterally compressed, the margin narrowly winged, apically comose.

Wet flatwoods and floodplain forests. Frequent; nearly throughout. New Jersey south to Florida, west to Texas. Summer.

Asclepias longifolia Michx. [Long-leaved.] LONGLEAF MILKWEED.
Asclepias longifolia Michaux, Fl. Bor.-Amer. 1: 116. 1803. *Acerates longifolia* (Michaux) Elliott,

Sketch Bot. S. Carolina 1: 317. 1817. *Gomphocarpus longifolius* (Michaux) Sprengel, Syst. Veg. 1: 849. 1824 ("1825"). *Polyotus longifolius* (Michaux) Nuttall, Trans. Amer. Philos. Soc., ser. 2. 5: 200. 1837. *Oligoron longifolium* (Michaux) Rafinesque, New Fl. 4: 60. 1838 ("1836").

Erect perennial herb, to 7 dm; stem puberulent distally, glabrous proximally. Leaves opposite, the blade linear-lanceolate, 5–15 cm long, 2–12 mm wide, the apex acute, the base cuneate, the margin entire, the upper surface glabrous, the lower surface with scattered trichomes, pubescent on the veins and along the margin, sessile. Flowers few to many in an umbelliform cyme, this terminal or in the axil of the upper leaves, the peduncle 2–4 cm long, pubescent, the pedicel ca. 2 cm long, pubescent; bracts subulate, minute; sepal lobes green, lanceolate to ovate, ca. 2 mm long, the outer surface pubescent, reflexed; corolla lobes whitish on the lower ½, suffused with purple distally, lanceolate, 5–6 mm long, reflexed, the apex arching outward, the corona hoods unevenly suffused with purple, spatulate, ca. 2 mm long, the apex rounded, the horns absent; gynostegium ca. 2 mm long, subequaling the hoods in length. Fruit cylindric-fusiform, ca. 10 cm long, pubescent; seeds ovate, laterally compressed, the margin narrowly winged, apically comose.

Wet flatwoods and bogs. Occasional; peninsula, central and western panhandle. Delaware and Maryland south to Florida and Texas. Spring–summer.

Asclepias michauxii Decne. [Commemorates André Michaux (1746–1802), French botanist and North American explorer.] MICHAUX'S MILKWEED.

Asclepias angustifolia Elliott, Sketch Bot. S. Carolina 1: 325. 1817; non Schweigger, 1812. *Asclepias michauxii* Decaisne, in de Candolle, Prodr. 8: 569. 1844.

Erect or ascending perennial herb, to 4 dm; stem puberulent in weak lines. Leaves opposite, subopposite, or alternate, the blade linear-subulate, 5–10 cm long, 3–6 mm wide, the apex acute to obtuse, the base cuneate, the margin entire, slightly involute, the upper surface glabrous, the lower surface sparsely puberulent along the veins, sessile. Flowers few to many in an umbelliform cyme, this terminal on the main stem or its branches, the peduncle 1–4 cm long, puberulent, the pedicel 1–1.5 cm long, puberulent; bracts subulate, minute; sepal lobes green or purplish, lanceolate, ca. 3 mm long, the outer surface puberulent, reflexed; corolla lobes purple on the outer surface, the inner surface greenish white, dark purple at the apex, lanceolate, 4–6 mm long, reflexed, the apex arching outward and upward; corona lobes spatulate, 5–7 mm long, slightly longer than the gynostegium, erect, inward curved, the apex rounded, purple, the horn subulate, purple, exserted and curving over the top of the gynostegium. Fruit narrowly fusiform, 10–15 cm long, 5–8 mm wide, glabrate; seeds ovate, laterally compressed, the margin narrowly winged, apically comose.

Sandhills, flatwoods, bogs, and marshes. Occasional; northern peninsula, central and western panhandle. South Carolina south to Florida, west to Louisiana. Spring–summer.

Asclepias obovata Elliott [Obovate, in reference to the leaf shape.] PINELAND MILKWEED.

Asclepias obovata Elliott, Sketch Bot. S. Carolina 1: 321. 1817. *Asclepias viridiflora* Rafinesque var.

obovata (Elliott) Torrey, Fl. N. Middle United States 1: 284. 1824. *Acerates obovata* (Elliott) Eaton, Man. Bot., ed. 6. 2: 3. 1833. *Polyotus obovatus* (Elliott) Nuttall, Trans. Amer. Philos. Soc., ser. 2. 5: 201. 1837.

Erect perennial herb, to 1 m; stem tomentose. Leaves opposite, the blade obovate or elliptic to ovate, 3–10 cm long, 2–3.5 cm wide, the apex rounded to obtuse, apiculate, the base cordate, truncate, or rounded, the margin entire, the upper surface glabrous, the lower surface tomentose, sessile or the petiole to 2 mm long. Flowers numerous in an umbelliform cyme, this terminal or in the axil of the upper leaves, the peduncle ca. 1 cm long, tomentose, the pedicel ca. 1.5 cm long, tomentose; bracts small, lanceolate, tomentose, caducous; bracts lanceolate, minute; sepal lobes green, lanceolate, 2–3 mm long, the outer surface tomentose, reflexed; corolla lobes yellow-green, lanceolate, 8–10 mm long, reflexed, the corona hoods 6–8 mm long, erect, involute, the horns arising from the middle of the hood, laterally exserted; gynostegium about ½ as long as the corona. Fruit fusiform, 8–12 cm long, tomentose; seeds ovate, laterally compressed, the margin narrowly winged, apically comose.

Sandhills. Occasional; central and western panhandle. South Carolina south to Florida, west to Oklahoma and Texas. Spring.

Asclepias pedicellata Walter [With evident pedicels.] SAVANNAH MILKWEED.

Asclepias pedicellata Walter, Fl. Carol. 106. 1788. *Podostigma pubescens* Elliott, Sketch Bot. S. Carolina 1: 326. 1817, nom. illegit. *Stylandra pumila* Nuttall, Gen. N. Amer. Pl. 1: 170. 1818, nom. illegit. *Anantherix pumila* Nuttall, Trans. Amer. Philos. Soc., ser. 2. 5: 203. 1837, nom. illegit. *Podostigma pedicellatum* (Walter) Vail, in Small, Fl. S.E. U.S. 939. 1903.

Erect perennial herb, to 4 dm; stem puberulent. Leaves opposite, narrowly lanceolate to linear-elliptic, 2–6 cm long, 2–5 mm wide, the apex acute, the base cuneate, the margin entire, the upper and lower surfaces puberulent, sessile. Flowers few to many in an umbelliform cyme, this terminal or in the axil of the upper leaves, the peduncle 5 mm long, the pedicel ca. 1 cm long; bracts subulate, minute; sepal lobes green, lanceolate, 2–3 mm long, spreading; corolla campanulate, yellowish or greenish yellow, the lobes linear-elliptic, 7–8(12) cm long, the apex abruptly narrowed, the corona hood greenish yellow, ca. ¼ as long as the corolla lobes, much shorter than the gynostegium, greenish yellow, narrowly saclike on the outer side, the apex cuminate and strongly inflexed-reflexed; gynostegium elevated on a slender column, slightly shorter than the corolla lobes. Fruit cylindric-fusiform, 11–15 cm long, 3–5 mm wide; seeds ovate, laterally compressed, the margin narrowly winged, apically comose.

Seasonally wet flatwoods, wet prairies, and depressions in scrub. Frequent; nearly throughout. North Carolina south to Florida. Spring–summer.

Asclepias perennis Walter [Plant perennial.] SWAMP MILKWEED.

Asclepias perennis Walter, Fl. Carol. 107. 1788.
Asclepias parviflora Aiton, Hort. Kew. 1: 307. 1789. *Asclepias pulchella* Salisbury, Prodr. Stirp. Chap. Allerton 150. 1796, nom. illegit. TYPE: "Carolina and East Florida."
Asclepias lancifolia Rafinesque, Autik. Bot. 177. 1840. TYPE: FLORIDA.
Asclepias parviflora Aiton var. *latifolia* Rafinesque, Autik. Bot. 179. 1840. TYPE: FLORIDA.

Erect perennial herb, to 6 dm; stem glabrous or sometimes sparsely puberulent in lines below the nodes. Leaves opposite, the blade lanceolate, lance-ovate, or narrowly elliptic, 6–12 cm long, 1–3 cm wide, the apex acuminate, the base cuneate, the margin entire, the upper and lower surfaces glabrous, the petiole 1–1.5 cm long. Flowers few to many in an umbelliform cyme, this terminal or in the axil of the upper leaves, the peduncle 2–3 cm long, the pedicel 1–1.5 cm long; bracts subulate, minute; sepal lobes green, oblong, ca. 2 mm long, glabrous; green, reflexed; corolla lobes white or pinkish, lanceolate, 3–4 mm long, reflexed, the distal ½ arched outward, the corona hoods spatulate, rounded outward, subequaling the gynostegium, the horns exserted, loosely arching over the gynostegium. Fruit cylindric-fusiform, 6–7 cm long, 1–2.5 cm wide, glabrous; seeds ovate, laterally compressed, the margin narrowly winged, lacking an apical coma.

Marshes, cypress swamps, and floodplain forests. Frequent; northern counties, central peninsula. Indiana and Illinois south to Florida and Texas. Spring–summer.

Asclepias rubra L. [*Ruber*, red, in reference to the flower color.] RED MILKWEED.

> *Asclepias rubra* Linnaeus, Sp. Pl. 217. 1753.
> *Asclepias laurifolia* Michaux, Fl. Bor.-Amer. 1: 117. 1803. *Asclepias periplocifolia* Nuttall, Gen. N. Amer. Pl. 1: 167. 1818, nom. illegit. *Asclepias rubra* Linnaeus var. *laurifolia* (Michaux) R. M. Harper, Bull. Torrey Bot. Club 30: 339. 1903.

Erect perennial herb, to 1 m; stem glabrous or sparsely puberulent in lines below the nodes. Leaves opposite, the blade ovate to lanceolate, 6–14 cm long, 1.5–4 cm wide, the apex acuminate, the base rounded to truncate or subcordate, the upper and lower surfaces glabrous, the lower surface glaucous, sessile or to 2 mm long. Flowers numerous in an umbelliform cyme, this terminal or in the axil of the upper leaves or from leafless nodes, the peduncle 1–3 cm long, the pedicel ca. 2 cm long; bracts lanceolate, minute; sepal lobes green, narrowly triangular, 2–3 mm long, reflexed; corolla lobes dull purple to lavender or pink, oblong, 8–9 mm long, longer than the gynostegium, reflexed, the corona hoods sometimes slightly orange-tinged, lanceolate, the outer side rounded, the horn subulate, exserted about midway from the hood and loosely arching over the gynostegium. Fruit cylindric-fusiform, 8–11 cm long, 1–1.5 cm wide, smooth, glabrous; seeds ovate, laterally compressed, the margin narrowly winged, apically comose.

Bogs. Occasional; Bay County, western panhandle. New York south to Florida, west to Texas. Summer.

Asclepias tomentosa Elliott [Thickly covered with short, matted trichomes.] VELVETLEAF MILKWEED.

> *Asclepias tomentosa* Elliott, Sketch Bot. S. Carolina 1: 320. 1817.
> *Asclepias megalotis* Rafinesque, New Fl. 4: 61. 1838 ("1836"). TYPE: FLORIDA.

Erect perennial herb, to 6 dm; stem pubescent. Leaves opposite, the blade lanceolate to broadly elliptic or obovate, 5.5–8.5 cm long, 1.5–3.5 cm wide, the apex acute to rounded, the base obtuse to rounded, the margin entire, the upper and lower surfaces pubescent, the petiole ca. 1 cm long, pubescent. Flowers numerous in an umbelliform cyme from the axil of the upper leaves, sessile or the peduncle to 3 mm long, the pedicel 1.5–2 cm long, puberulent; bracts subulate, minute; sepal lobes green, lanceolate, 2–3 mm long, the outer surface pubescent, reflexed; corolla lobes yellow-green, lanceolate, 6.5–9 mm long, reflexed, the corona hoods 5–7 mm long, subequaling the gynostegium, involute, the apex truncate, the horns longer than the hoods, exserted apically. Fruit fusiform, 10–12 cm long, 1–1.5 cm wide, smooth, pubescent; seeds ovate, laterally compressed, the margin narrowly winged, apically comose.

Sandhills and coastal dunes. Occasional; peninsula, west to central panhandle. North Carolina south to Florida, also Texas. Spring–summer.

Asclepias tuberosa L. [Swollen, in reference to the roots.] BUTTERFLYWEED; BUTTERFLY MILKWEED.

Asclepias tuberosa Linnaeus, Sp. Pl. 217. 1753.
Asclepias decumbens Linnaeus, Sp. Pl. 216. 1753. *Asclepias tuberosa* Linnaeus var. *decumbens* (Linnaeus) Pursh, Fl. Amer. Sept. 184. 1814. *Asclepias tuberosa* Linnaeus forma *decumbens* (Linnaeus) Voss, Vilm. Blumengärtn., ed. 3. 1: 663. 1894.
Asclepias floridana Lamarck, Encycl. 1: 284. 1783. *Acerates floridana* (Lamarck) Hitchcock, Trans. Acad. Sci. St. Louis 5: 508. 1891. TYPE: FLORIDA.
Asclepias rolfsii Britton ex Vail, in Small, Fl. S.E. U.S. 943, 1336. 1903. *Asclepias tuberosa* Linnaeus subsp. *rolfsii* (Britton ex Vail) Woodson, Ann. Missouri Bot. Gard. 31: 368. 1944. *Asclepias tuberosa* Linnaeus var. *rolfsii* (Britton ex Vail) Shinners, Field & Lab. 17: 89. 1949. TYPE: FLORIDA: Miami-Dade Co.: 1 Apr 1903, *Britton s.n.* (holotype: NY).

Erect perennial herb, to 9 dm; stem villous to hirsute. Leaves alternate, the blade linear or ovate-lanceolate to elliptic or oblanceolate, (2)5–10 cm long, (0.5)1–2.3 cm wide, the apex acuminate to rounded, the base obtuse to truncate or cordate, the margin entire, the upper surface glabrous, the lower surface sparsely to densely villous to hirsutulous, the petiole 1–5 mm long. Flowers few to many in an umbelliform cyme, this terminal or in the axil of the upper leaves, sessile or the peduncle to 3 cm long, pubescent, the pedicel 1–2 cm long, pubescent; bracts subulate, minute; sepal lobes green or purple-tinged, linear to lanceolate, 2–4 mm long, the outer surface villous, reflexed; corolla lobes bright orange, rarely red or yellow, elliptic to lanceolate, 5.5–8.5 mm long, reflexed, the corona hoods lanceolate, 4–6 mm long, slightly longer than the gynostegium, slightly spreading, convolute, the apex rounded, slightly recurved, the horns acicular, adnate to the lower ¼ of the hood, subequaling the hood in length; gynostegium orange or rarely yellow, stipitate. Fruit fusiform, 8–15 cm long, 1–1.5 cm wide, smooth, puberulent; seeds ovate, laterally compressed, the margin narrowly winged, apically comose.

Flatwoods and sandhills. Common; nearly throughout. Quebec south to Florida, west to California. Summer–fall.

Asclepias variegata L. [Variegated, in reference to the flower color.] REDRING MILKWEED.

Asclepias variegata Linnaeus, Sp. Pl. 215. 1753. *Asclepias variegata* Linnaeus var. *major* Hooker, Fl. Bor.-Amer. 2: 52. 1838, nom. inadmiss. *Biventraria variegata* (Linnaeus) Small, Man. S.E. Fl. 1072, 1507. 1933.

Erect or ascending perennial herb, to 1 m; stem usually glabrous below the middle, sparsely short-pubescent distally. Leaves opposite, the blade ovate, suborbicular, or elliptic-lanceolate to obovate, 2–15 cm long, 1–8 cm wide, the apex obtuse to rounded to an abrupt point, the base broadly cuneate to rounded, the margin entire, sometimes undulate, the upper surface glabrous, the lower surface sparsely pubescent, somewhat glaucous, the petiole 1–2 cm long. Flowers numerous in an umbelliform cyme, this terminal or in the axil of the upper leaves, the peduncle 2–2.5 cm long, the pedicel 1–2 cm long; bracts lanceolate, minute; sepal lobes green, lanceolate to triangular-ovate, 2–3 mm long, reflexed; corolla lobes white, occasionally tinged with pale pink, oblong-elliptic, 6–8 mm long, reflexed, the corona hoods 2–3 mm long, longer than the gynostegium, involute, inflated, the apex truncate, the horns medially attached, shorter than the hood, flattened, sickle-shaped, projecting laterally over the gynostegium. Gynostegium white with a purple crown. Fruit fusiform, 10–15 cm long, 1.5–2 cm wide, smooth, minutely puberulent, glaucous; seeds ovate, laterally compressed, the margin narrowly winged, apically comose.

Mesic hammocks. Occasional; panhandle. New York and Ontario south to Florida, west to Oklahoma and Texas. Spring.

Asclepias verticillata L. [Whorled, in reference to the leaf arrangement.] WHORLED MILKWEED.

Asclepias verticillata Linnaeus, Sp. Pl. 217. 1753.

Erect perennial herb, to 9 dm; stem usually puberulent in decurrent lines form the leaf bases. Leaves verticillate or subverticillate, 3–6 at a node, the blade filiform or linear-filiform, 1.5–8 cm long, 1–2(3) cm wide, the apex acute, the base narrowly cuneate, the margin entire, revolute, the upper and lower surfaces glabrous or puberulent, sessile. Flowers many in an umbelliform cyme, this terminal or in the axil of the upper leaves, the peduncle 1–4 cm long, puberulent, the pedicel 5–11 mm long, puberulent; bracts subulate, minute; sepal lobes green to purple-tinged, linear-lanceolate to ovate, 1–3 mm long, sparsely villous on the outer surface, reflexed; corolla lobes white to greenish white or purple-tinged, elliptic, 4–5 mm long, reflexed, the corona hoods oblong, ca. 2 mm long, involute, the apex rounded, open, the horns acicular, adnate to the lower ⅓ of the hood, 1.5–2 times as long as the hood and arching over the gynostegium; gynostegium greenish white. Fruit narrowly fusiform, 8–10.5 cm long, 6–8 mm wide, smooth, sparsely puberulent; seeds ovate, laterally compressed, the margin narrowly winged, apically comose.

Sandhills and flatwoods. Frequent; nearly throughout. Vermont and New York south to Florida, west to Saskatchewan, Montana, Wyoming, and Arizona. Spring–summer.

Asclepias viridiflora Raf. [With green flowers.] GREEN MILKWEED.

Asclepias viridiflora Rafinesque, Med. Repos., ser. 2. 5: 360. 1808.
Asclepias viridiflora Pursh, Fl. Amer. Sept. 181. 1814; non Rafinesque, 1808. *Gomphocarpus viridiflorus* Sprengel, Syst. Veg. 1: 849. 1824. ("1825"). *Acerates viridiflora* (Sprengel) Eaton, Man. Bot., ed. 5. 90. 1829. *Polyotus heterophyllus* Nuttall, Trans. Amer. Philos. Soc., ser. 2. 5: 199. 1837, nom. illegit.

Asclepias lanceolata Ives, Amer. J. Sci. Arts 1: 252. 1819; non Walter, 1788. *Asclepias viridiflora* Rafin-
esque var. *lanceolata* Torrey, Fl. N. Middle United States 1: 284. 1824. *Otanema lanceolata* (Tor-
rey) Rafinesque, New Fl. 4: 61. 1838 ("1836"). *Acerates viridiflora* (Sprengel) Eaton var. *lanceolata*
(Torrey) A. Gray, Syn. Fl. N. Amer. 2(1): 99. 1878. *Acerates viridiflora* (Sprengel) Eaton var. *ivesii*
Britton, Mem. Torrey Bot. Club 5: 265. 1894, nom. illegit. *Acerates ivesii* Wooton & Standley, Contr.
U.S. Natl. Herb. 19: 509. 1915, nom. illegit.

Erect perennial herb, to 1 m; stem puberulent to tomentose or glabrate. Leaves opposite or
subopposite, the blade linear-lanceolate to elliptic or ovate, 4–12 cm long, 0.5–4 mm wide,
the apex acute to obtuse and apiculate, the base cuneate to subrotund, the margin entire, ciliate, the upper surface glabrous, the lower surface sparsely puberulent, especially along the
midvein, sessile or the petiole to 5 mm long. Flowers numerous in an umbelliform cyme, this
terminal or in the axil of the upper leaves, sessile or the peduncle to 3 mm long, the pedicel
to 1 cm long, puberulent; bracts subulate, minute; sepal lobes green, lanceolate, 2–3 mm long,
the outer surface puberulent, reflexed, corolla lobes pale green, elliptic-lanceolate, 5–7 mm
long, reflexed, the corona lobes oblong, 4–5 mm long, erect, convolute, subequaling the gynostegium, the horn absent; gynostegium sessile, pale green. Fruit fusiform, 7–15 cm long, 1.5–2
cm wide, smooth, puberulent; seeds ovate, laterally compressed, the margin narrowly winged,
apically comose.

Open, calcareous hammocks. Rare; Gadsden and Jackson Counties. Connecticut and New
York south to Florida, west to British Columbia, Montana, Wyoming, Colorado, and Arizona.
Spring.

Asclepias viridiflora is listed as endangered in Florida (Florida Administrative Code, Chapter 5B-40).

Asclepias viridis Walter [Green, in reference to the flower color.] GREEN ANTELOPEHORN.

Asclepias viridis Walter, Fl. Carol. 107. 1788. *Podostigma viride* (Walter) Elliott, Sketch Bot. S. Carolina
1: 327. 1817. *Anantherix viridis* (Walter) Nuttall, Gen. N. Amer. Pl. 1: 169. 1818. *Gomphocarpus viridis* (Walter) Sprengel, Syst. Veg. 1: 849. 1824 ("1825"). *Asclepiodora viridis* (Walter) A. Gray, Proc.
Amer. Acad. Arts 12: 66. 1877.
Anantherix paniculata Nuttall, Trans. Amer. Philos. Soc., ser. 2. 5: 203. 1837. *Acerates paniculata* (Nuttall) Decaisne, in de Candolle, Prodr. 8: 521. 1844.

Erect perennial herb, to 6 dm; stem glabrous or sparsely pubescent. Leaves alternate, the blade
lanceolate, oblong, or elliptic to obovate, 5–12 cm long, 1–5.5 cm wide, the apex acute to obtuse
or rounded, the base cuneate to rounded, truncate, or subcordate, the margin entire, the upper
surface glabrous, the lower surface sparsely puberulent, the petiole 0.3–1 cm long. Flowers numerous in an umbelliform cyme, this terminal or in the axil of the upper leaves, the peduncle
1–3 cm long, puberulent, the pedicel 1–2.5 cm long, puberulent; bracts lanceolate, minute; sepal
lobes green or purple-tinged, lanceolate to narrowly ovate, 3–5 mm long, the margin ciliolate,
erect-spreading; corolla lobes pale green, elliptic-lanceolate to ovate, 13–17 mm long, longer
than the gynostegium, arcuate-ascending, the corona hoods purplish, 4–6 mm long, involute
and hooded apically, arcuate-ascending apically, the horn absent; gynostegium pale purple.

Fruit fusiform, 7–13 cm long, 1.3–2 cm wide, smooth, sparsely pubescent; seeds ovate, laterally compressed, the margin narrowly winged, apically comose.

Rockland hammocks and pine rocklands. Occasional; central and southern peninsula, central panhandle. West Virginia south to Florida, west to Nebraska, Kansas, Oklahoma, and Texas. Spring–summer.

Asclepias viridula Chapm. [Greenish, in reference to the flower color.] SOUTHERN MILKWEED; GREEN MILKWEED.

> *Asclepias viridula* Chapman, Fl. South. U.S. 363. 1860. FLORIDA: Franklin Co.: Apalachicola, s.d., *Chapman s.n.* (lectotype: MO; isotype: NY). Lectotypified by Woodson (1954: 192).

Erect perennial herb, to 7 dm, stem pubescent in lines below the upper nodes. Leaves opposite, linear-filiform, 4–10 cm long, 1–3 mm wide, the apex acute, the base cuneate, the upper and lower surfaces glabrous, sessile. Flowers many in an umbelliform cyme, this terminal or in the axil of the upper leaves, the pedicel 2–3 cm long, the pedicel ca. 1 cm long; bracts lanceolate, minute; sepal lobes green, lance-ovate, ca. 2 mm long, reflexed; corolla lobes lanceolate, 4–5 mm long, outward curving distally, the outer surface brownish purple, the inner surface green, the corona hoods basally saclike, ca. 3 mm long, involute, exceeding the gynostegium, erect, the free tip flaring, cream-colored, faintly brownish purple medially, the horns flat, subequaling the hood, slightly exserted and arching over the gynostegium. Fruit fusiform, 8–10 cm long; seeds ovate, laterally compressed, the margin narrowly winged, apically comose.

Wet flatwoods. Occasional; northern peninsula, central and western panhandle. Georgia, Alabama, and Florida. Spring–summer.

Asclepias viridula is listed as threatened in Florida (Florida Administrative Code, 5B-40).

Calotropis R. Br. 1810.

Shrubs or trees, with white latex. Leaves opposite, simple, pinnate-veined, petiolate, estipulate. Flowers in axillary umbelliform cymes, bracteate; sepals 5, basally connate; petals 5, basally connate, with a corona of 5 lobes, these with a tubercle on each side, with a basal, upcurved, dorsal spur; stamens, style, and stigma adnate to form a gynostegium, the pollen in pollinia, the ovary 2-carpellate and -loculate. Fruit a follicle; seeds numerous, comose.

A genus of 4 species; North America, West Indies, Mexico, Central America, South America, Africa, Asia, Australia, and Pacific Islands. [From the Greek *calo*, beautiful, and *tropis*, keel of a ship, in reference to the color and shape of the corolla lobes.]

Calotropis procera (Aiton) W. T. Aiton [*Procerus*, very tall.] ROOSTERTREE.

> *Asclepias procera* Aiton, Hort. Kew. 1: 305. 1789. *Calotropis procera* (Aiton) W. T. Aiton, Hort. Kew. 2: 78. 1811. *Madorius procerus* (Aiton) Kuntze, Revis. Gen. Pl. 2: 421. 1891. *Calotropis gigantea* (Linnaeus) W. T. Aiton var. *procera* (Aiton) P. T. Li, J. S. China Agric. Univ. 12(3): 39. 1991.

Shrub or small tree, to 4 m; stem glabrous and glaucous in age, the bark thick corky, deep furrowed. Leaves with the blade ovate or oblong to obovate, 1–3 dm long, 5–15 cm wide, the apex abruptly acuminate, the base cordate, the margin entire, the upper and lower surfaces grayish green or white pubescent, glaucous, glabrous in age, subsessile. Flowers 3–10 in a terminal or axillary umbelliform cyme, the peduncle 2–5.5 cm long, pubescent, the pedicel 1.5–2.5 cm long, pubescent; sepal lobes elliptic to ovate, ca. 5 mm long, the outer surface pubescent; corolla white on the outer surface, pinkish on the inner, the lobes purple-brown at the apex, broadly ovate or ovate-triangular, 7–10 mm long, upward curved and slightly incurved, the corona erect, as long as the gynostegium, with a tubercle on each side, with a basal, upcurved, dorsal spur. Fruit obliquely ovate, inflated, 6–10 cm long, 3–4 cm wide, the pericarp thick, spongy; seeds ovate, laterally compressed, ca. 6 mm long, the margin winged, with an apical coma.

Disturbed sites. Rare; Hillsborough County. Florida and California; West Indies, Mexico, Central America, and South America; Africa, Asia, Australia, and Pacific Islands. Native to Africa and Asia. Spring–fall.

Carissa L. nom. cons. 1767.

Shrubs or trees, with white latex; stems spiny. Leaves opposite, simple, pinnate-veined, petiolate, estipulate; flowers in terminal or axillary cymes; sepals 5, basally connate; petals 5, basally connate; stamens 5, epipetalous, the filaments and anthers free; ovary 2-carpellate and -loculate, the style and stigma 1. Fruit a berry; seeds numerous.

A genus of about 30 species; North America, West Indies, Mexico, Central America, Africa, Asia, Australia, and Pacific Islands. [From the Greek, beloved, in reference to the fragrant flowers and attractive fruit.]

Carissa macrocarpa (Eckl.) A. DC. [Large-fruited.] NATAL PLUM.

Arduina macrocarpa Ecklon, S. African Quart. J. 1: 372. 1830. *Carissa macrocarpa* (Ecklon) A. de Candolle, in de Candolle, Prod. 8: 336. 1844.

Arduina grandiflora E. Meyer, in E. Meyer & Drege, Comm. Pl. Afr. Aust. 191. 1838. *Carissa grandiflora* (E. Meyer) A. de Candolle, in de Candolle, Prodr. 8: 335. 1844.

Shrub or small tree, to 4(6) m; branches with 1- or 2-forked axillary spines 2–4 cm long. Leaves with the blade broadly ovate, 2.5–7.5 cm long, 2–5 cm wide, coriaceous, the apex obtuse, mucronate, the base rounded to obtuse, the margin entire, the upper and lower surfaces glabrous, the petiole 4–6 mm long. Flowers 1–3 in terminal or axillary cymes, the pedicel 2–3 mm long; sepal lobes narrowly ovate, 3–6 mm long, glabrous; corolla white or pink, the tube 1.1–1.8 cm long, the inner surface pubescent, the lobes oblong, 0.9–2.4 cm long; stamens included. Fruit ovoid, 2–5 cm long, bright red, glabrous; seeds ovate, slightly compressed, brown.

Disturbed sites. Occasional; central and southern peninsula. Escaped from cultivation. Florida and Texas; West Indies, Mexico, and Central America; Africa, Asia, and Pacific Islands.

Catharanthus G. Don 1837. PERIWINKLE

Annual or perennial herbs, with white latex. Leaves opposite or alternate, simple, petiolate, estipulate. Flowers solitary or few, axillary; sepals 5, basally connate, petals 5, basally connate; stamens 5, epipetalous, the filaments and anthers free; ovary 2-carpellate and -loculate, the style and stigma 1. Fruit a follicle; seeds numerous.

A genus of 8 species; nearly cosmopolitan. [From the Greek *cather*, pure, and *anthus*, flower.] *Ammocallis* Small, 1903.

Catharanthus roseus (L.) G. Don [Rose-colored, in reference to the the flower color.] MADAGASCAR PERIWINKLE

> *Vinca rosea* Linnaeus, Syst. Nat., ed. 10. 944. 1759. *Pervinca rosea* (Linnaeus) Gaterau, Descr. Pl. Montauban 52. 1789. *Vinca speciosa* Salisbury, Prodr. Stirp. Chap. Allerton 147. 1796, nom. illegit. *Catharanthus roseus* (Linnaeus) G. Don, Gen. Hist. 4: 95. 1837. *Lochnera rosea* (Linnaeus) Reichenbach ex Endlicher, Gen. Pl. 583. 1838. *Ammocallis rosea* (Linnaeus) Small, Fl. S.E. U.S. 936, 1336. 1903.

Erect annual or perennial herb, to 7 dm; stem puberulent. Leaves with the blade oblong-elliptic, 2–7 cm long, 1.5–3 cm wide, the apex obtuse or rounded, mucronate, the base obtuse or cuneate, the margin entire, the upper and lower surfaces minutely puberulent or glabrate, the petiole 3–10 mm long, with several axillary glands. Flowers solitary or 2–4 in a cyme in the axil or the upper leaves, the pedicel 1–3 mm long, puberulent; sepal lobes linear-lanceolate, 4–7 mm long, puberulent; corolla salverform, the tube 2–3 cm long; the throat closed by dense bristly trichomes, the lobes obovate, 1.5–2.5 cm long, spreading; stamens adnate to the corolla tube in the throat above the stigma, the anthers subsessile; stigma thick, the base umbraculiform, the apex 2-lobed. Fruit narrowly elliptic, 1.5–2.5 cm long, subterete, longitudinally ridged; seeds oblong, the ventral surface with a vertical groove, black, muricate.

Dry, disturbed sites. Frequent; central and southern peninsula, central and western panhandle. Escaped from cultivation. North Carolina south to Florida, west to Texas, also Ohio and California; nearly cosmopolitan. Native to Africa. Spring–summer.

Cryptostegia R. Br. 1820. RUBBERVINE

Woody vines or shrubs, with white latex. Leaves opposite, simple, pinnate-veined, petiolate, estipulate. Flowers in terminal cymes, bracteate; sepal 5, basally connate; petals 5, basally connate, with a 5-lobed corona; stamens 5, epipetalous, the filaments free, the anthers adnate to the stigma head; ovary 2-carpellate and -loculate. Fruit a follicle; seeds numerous, comose.

A genus of two species; North America, West Indies, Mexico, Central America, South America, Africa, Asia, Australia, and Pacific Islands. [From the Greek *crypto*, covered, and *stegos*, shelter, in reference to the stigma covered by the anthers.]

Selected reference: Klackenberg (2001).

1. Corona lobes bifid; corolla light mauve; calyx lobes more than 1.2 mm long .... **C. grandiflora**
1. Corona lobes entire; corolla reddish purple; calyx lobes less than 1.2 mm long ... **C. madagascariensis**

Cryptostegia grandiflora R. Br. [Large-flowered.] PALAY RUBBERVINE.

Cryptostegia grandiflora R. Brown, Bot. Reg. 5: t. 435. 1820.

Decumbent woody vine, to 6 m; stem glabrous or sparsely puberulent. Leaves with the blade elliptic to ovate, 5–11 cm long, 2.5–6 cm wide, the apex acute to acuminate, the base obtuse to cuneate, the margin entire, the upper and lower surfaces glabrous, the petiole 4–6 mm long. Flowers in a terminal cyme, the peduncle ca. 1 cm long, puberulent, the pedicel ca. 5 mm long, puberulent; bracts lanceolate, minute, caducous; sepal lobes broadly lanceolate, 1.2–1.5 cm long, the outer surface puberulent; corolla funnelform, 5–7 cm long, light mauve on the outer surface, the inner surface whitish, the corona lobes subulate, to 1.5 cm long, deeply bifid. Fruit oblong-lanceolate, 10–12 cm long, 3–4 cm wide, with longitudinal ribs or wings; seeds 5–8 mm long, with an apical coma.

Disturbed hammocks. Rare; Monroe County keys. Escaped from cultivation. Florida and Texas; West Indies, Mexico, Central America, and South America; Africa, Asia, Australia, and Pacific Islands. Native to Africa. Spring–summer.

Cryptostegia madagascariensis Bojer ex Decne. [Of Madagascar.] MADAGASCAR RUBBERVINE.

Cryptostegia madagascariensis Bojer ex Decaisne, in de Candolle, Prodr. 8: 432. 1844.

Erect shrub or woody vine, to 3 m; stem with prominent lenticels, glabrous. Leaves with the blade oblong, 4–10 cm long, 3–5 cm wide, the apex acuminate,the base rounded, obtuse, or truncate, the margin entire, the upper and lower surfaces glabrous, the petiole 5–10 mm long. Flowers few to many in a terminal cyme, the peduncle ca. 1 cm long, glabrous, the pedicel ca. 5 mm long, glabrous; bracts lanceolate, minute, caducous; sepal lobes lanceolate, 8–9 mm long, 4–6 mm wide, the outer surface glabrous; corolla funnelform, the tube 1.5–2.5 cm long, the outer surface puberulent, the lobes ovate, 2.5–3 cm long, reflexed, the corona lobes subulate, 8–9 mm long, entire. Fruit woody, 6–8 cm long, 2–3 cm wide, with a dorsal and 2 lateral wings; seeds 5–8 mm long, with an apical coma.

Disturbed hammocks. Rare; Manatee, Martin, and Lee Counties, southern peninsula. Escaped from cultivation. Florida; West Indies and South America; Africa, Asia, and Pacific Islands. Spring.

Cryptostegia madagascariensis is listed as a Category II invasive species in Florida (Florida Exotic Pest Plant Council (FLEPPC, 2015).

Cynanchum L. 1753. SWALLOWWORT

Perennial vines, with white latex. Leaves opposite, simple, pinnate-veined, petiolate, estipulate. Flowers in terminal or axillary umbelliform cymes, bracteate; sepals 5, basally connate; petals 5, basally connate, with a 5-lobed corona; stamens 5, the filaments connate, the anthers adnate to the stigma head, the pollen in pollinia. Fruit a follicle; seeds numerous, comose.

A genus of about 200 species; nearly cosmopolitan. [From the Greek *cynos*, dog, and *anchein*, to choke, hence the vernacular name of "dog-strangling vine" for some species.]

Cynanchum has long been considered a "dustbin genus" until recent molecular studies, which resulted in the segregation of several genera, including *Metastelma*, *Orthosia*, and *Sutera*, which occur in Florida.

Epicion Small, 1913; *Sarcostema* R. Br., 1810.

Selected references: Khanum et al. (2016); Liede and Täuber (2002); Meve and Liede-Schumann (2012).

1. Leaves cordate; corona lobes with 2 slender terminal appendages as long as the blade..............C. laeve
1. Leaves linear-lanceolate or elliptic; corona lobes lacking terminal appendages..............C. northropiae

Cynanchum laeve (Michx.) Pers. [*Laevis*, smooth, in reference to the glabrate stems.] HONEYVINE; BLUEVINE.

Gonolobus laevis Michaux, Fl. Bor.-Amer. 1: 119. 1803. *Cynanchum laeve* (Michaux) Persoon, Syn. Pl. 1: 274. 1805. *Vincetoxicum gonocarpos* (Walter) Britton var. *laeve* (Michaux) Britton, Mem. Torrey Bot. Club 5: 266. 1894. *Ampelamus laevis* (Michaux) Krings, Sida 19: 927. 2001.

Perennial vines, to 10 m; stem pubescent, at least distally and in longitudinal lines, glabrate below. Leaves with the blade cordate, (2)5–7(15) cm long, 3–7(10) cm wide, the apex acute, the base deeply cordate, the margin entire, the upper and lower surfaces glabrous or sparsely pubescent, mostly on the veins, the petiole 2–5 cm long. Flowers few to many in an axillary, umbelliform cyme; bracts lanceolate, minute; sepal lobes ovate, 2–3 mm long, erect, the outer surface pubescent; corolla campanulate-rotate, white to light cream-colored, the lobes narrowly oblong to lanceolate, 4–7 mm long, the corona of 5, erect, petaloid segments, 4–6 mm long, white, the apex with 2 slender appendages as long as the blade, this extending well above the gynostegium. Fruit lanceolate-ovate, 8–14(16) cm long, 2–3 cm wide, smooth, glabrous or minutely puberulent; seeds obovate, 7–9 mm long, with an apical coma.

Riverbanks. Rare; Wakulla, Gadsden, and Jackson Counties. New York and Ontario south to Florida, west to Nebraska, Kansas, Oklahoma, and Texas. Fall.

Cynanchum northropiae (Schltr.) Alain [Commemorates Alice Belle (Rich) Northrop (1863–1922), American botanist who collected in the Bahamas with her husband John Isaiah Northrop.] FRAGRANT SWALLOWWORT.

Metastelma northropiae Schlechter, in Urban, Symb. Antill. 5: 468. 1908. *Epicion northropiae* (Schlechter) Small, Man. S.E. Fl. 1075. 1933. *Cynanchum northropiae* (Schlechter) Alain, Mem. Soc. Cub. Hist. Nat. "Felipe Poey" 22: 118. 1955.

Slender perennial vine, to 2 m; stem glabrous. Leaves with the blade elliptic to ovate, 2–3(4) cm long, 1–1.5 cm wide, the apex obtuse, apiculate, the base rounded, the margin entire, the upper and lower surfaces glabrous, the petiole 3–8 mm long. Flowers few in an axillary umbelliform cyme, the peduncle 3–5 mm long, the pedicel 2–3 mm long; bracts lanceolate, minute; sepal lobes ovate, 1–2 mm long, shorter than the corolla tube, glabrous; corolla campanulate-rotate, white, ca. 5 mm long, the lobes oblong, 2–3 mm long, longer than the tube, the corona segments narrowly lanceolate, 1–2 mm long, the apex slightly outward curved. Fruit linear-filiform, 5–7 cm long, to 7 mm wide, smooth, glabrous; seeds ovate, laterally compressed, with an apical coma.

Hammocks. Rare; Brevard County and Indian River, southern peninsula. Florida; West Indies. All year.

EXCLUDED TAXA

Cynanchum bahamense (Grisebach) Gillis—Reported for Florida by Chapman (1897, as *Metastelma bahamense* Grisebach) and Small (1903, as *Metastelma bahamense* Grisebach; 1913a, 1913b, both as *Epicion bahamense* (Grisebach) Small), the name misapplied to material of *C. northropiae*.

Cynanchum cubense (A. Richard) Woodson—Reported for Florida by Small (1933, as *Metalepis cubensis* (A. Richard) Grisebach). No Florida specimens known.

Cynanchum schlechtendalii (Decaisne) Standley and Steyermark—This Mesoamerican species was reported for Florida by Chapman (1860), as *Metastelma schlechtendalii* Decaisne), the name misapplied to material of *C. northropiae*.

Echites P. Browne 1756.

Woody vines, with white latex. Leaves opposite, simple, pinnate-veined, petiolate, estipulate. Flowers in terminal and axillary umbelliform cymes, bracteate; sepals 5, basally connate; petals 5, basally connate; stamens 5, epipetalous, the anthers subsessile, free; ovary superior, 2-carpellate and -loculate, the style and stigma 1. Fruit a follicle; seeds numerous, comose.

A genus of about 11 species; North America, West Indies, Mexico, and Central America. [From the Greek *echis*, viper, in reference to its smooth shining shoots resembling a snake.]

Echites umbellatus Jacq. [Umbellate, in reference to the inflorescence.] DEVIL'S POTATO; RUBBERVINE.

Tabernaemontana echites Linnaeus, Syst. Nat., ed. 10. 945. 1759. *Echites ovatus* Miers, Apocn. S. Amer. 192. 1878. *Echites echites* (Linnaeus) Britton, in Small, Fl. Miami 147, 200. 1913, nom. inadmiss. *Echites umbellatus* Jacquin, Enum. Syst. Pl. 13. 1760.

Woody vine; stem to 2 m long, glabrous. Leaves with the blade ovate to elliptic, 3–10 cm long, 3–10 cm wide, the apex acute to obtuse or rounded, the base unilaterally obtuse or rounded to subcordate, the margin entire, the upper and lower surfaces glabrous, the petiole 1–3 cm long. Flowers few in a terminal or axillary umbelliform cyme, the peduncle 1–4 cm long, the pedicel 1–2.5 cm long; bracts lanceolate, minute; sepal lobes ovate-lanceolate, 2–3 mm long; corolla salverform, white or greenish white, the tube 4–6 cm long, swollen near the middle at the anther attachment, narrowed above it, the lobes obliquely spatulate-obovate, 1–2.5 cm long, the margin irregular; anthers subsessile, linear-lanceolate. Fruit linear-fusiform, 10–20 cm long, to 1 cm wide, smooth, glabrous; seeds compressed-cylindric, with an apical coma.

Coastal pinelands. Occasional; Brevard County southward along the east coast, southern peninsula. Florida; West Indies, Mexico, and Central America. All year.

Funastrum E. Fourn. 1882. TWINEVINE

Perennial vines, with white latex. Leaves opposite, simple, pinnate-veined, petiolate, estipulate. Flowers in axillary umbelliform cymes, bracteate; sepals 5, basally connate; petals 5, basally connate, with a double corona, the outer a low ring, the inner of 5 fleshy scales; stamens, style, and stigma adnate to form a gynostegium, the pollen in pollinia, the ovary superior, 2-carpellate and -loculate. Fruit a follicle; seeds numerous, apically comose.

A genus of 5 species; North America, West Indies, Mexico, Central America, and South America. [*Funis*, rope, and *astrum*, star, in reference to the character of the stem and flowers.]

Funastrum clausum (Jacq.) Schltr. WHITE TWINEVINE

Asclepias clausa Jacquin, Enum. Syst. Pl. 17. 1760. *Cynanchum clausum* (Jacquin) Jacquin, Select. Stirp. Amer. Hist. 87. 1763. *Sarcostemma brownei* G. Meyer, Prim. Fl. Esseq. 139. 1818, nom. illegit. *Sarcostemma clausum* (Jacquin) Schultes, in Roemer & Schultes, Syst. Veg. 6: 114. 1820. *Philibertella clausa* (Jacquin) Vail, Bull. Torrey Bot. Club 24: 306. 1897. *Philibertia clausa* (Jacquin) K. Schumann, in Engler & Prantl, Nat. Pflanzenfam. 4(2): 229. 1895. *Funastrum clausum* (Jacquin) Schlechter, Repert. Spec. Nov. Regni Veg. 13: 283. 1914.
Asclepias viminalis Swartz, Prodr. 53. 1788. *Sarcostemma swartzianum* Schultes, in Roemer & Schultes, Syst. Veg. 6: 116. 1920. *Philibertia viminalis* (Swartz) A. Gray, Proc. Amer. Acad. Arts 12: 64. 1877.
Sarcostemma crassifolium Decaisne, in de Candolle, Prodr. 8: 540. 1844. *Philibertia crassifolia* (Decaisne) Hemsley, Biol. Cent.-Amer., Bot. 2: 318. 1881. *Philibertella crassifolia* (Decaisne) Vail, Bull. Torrey Bot. Club 24: 306. 1897. *Funastrum crassifolium* (Decaisne) Schlechter, Repert. Spec. Nov. Regi Veg. 13: 284. 1914.

Perennial vine; stem glabrous. Leaves with the blade ovate-elliptic to elliptic-lanceolate, 5–8(10) cm long, 2–4(6) cm wide, the apex acute, the base rounded to cordate, the margin entire, the upper and lower surfaces glabrous, the petiole 1–1.5 cm long. Flowers in an axillary umbelliform cyme, the peduncle 6–13 cm long, the pedicel 1.5–2.5 cm long; bracts lanceolate, minute; sepal lobes elliptic-lanceolate, 2–3 mm long, sparsely puberulent on the outer surface; corolla white, campanulate-rotate, the lobes elliptic to ovate, 4–5 mm long, the outer surface minutely puberulent, the inner surface glabrous, with a double corona, the outer corona a low ring, the inner one of 5 fleshy scales, subequaling the gynostegium; style 2-lobed. Fruit fusiform, 5–7 cm long, ca. 1 cm wide, smooth, glabrous; seeds ovate, laterally compressed, narrowly winged, with an apical coma.

Hammocks. Frequent; central and southern peninsula. Florida and Texas; West Indies, Mexico, Central America, and South America.

Gonolobus Michx. 1803.

Perennial herbaceous vines, with white latex. Leaves opposite, simple, pinnate-veined, petiolate, estipulate. Flowers in axillary umbelliform cymes, bracteate; sepals 5, basally connate; petals 5, basally connate, with a corona consisting of a flattened, irregular, 5-angled disk; stamens, style,and stigmas adnate and forming a gynoecium, the pollen in pollinia, the ovary superior, 2-carpellate and -loculate. Fruit a follicle, 5-angled and winged; seeds comose.

A genus of about 130 species; North America, West Indies, Mexico, Central America, and South America. [From the Greek *gonia*, angle, and *lobos*, lobe, in reference to the angled pod.]

Gonolobus suberosus (L.) R. Br. [Having a somewhat irregular margin, in reference to the slightly fissured lower stems, an apparent misapplication of the Linnaean epithet (see Rosatti, 1989).] ANGULARFRUIT MILKWEED; ANGLEPOD.

Cynanchum suberosum Linnaeus, Sp. Pl. 212. 1753. *Gonolobus suberosus* (Linnaeus) R. Brown, Asclepiadeae 24. 1810. *Vincetoxicum suberosum* (Linnaeus) Britton, Mem. Torrey Bot. Club 5: 266. 1894. *Matelea suberosa* (Linnaeus) Shinners, Field & Lab. 18: 73. 1950.

Vincetoxicum gonocarpos Walter, Fl. Carol. 104. 1788. *Asclepias gonocarpos* (Walter) J. F. Gmelin, Syst. Nat. 2: 446. 1791. *Gonolobus macrophyllus* Michaux, Fl. Bor.-Amer. 1: 119. 1803, nom. illegit. *Gonolobus laevis* Michaux var. *macophyllus* A. Gray, Proc. Amer. Acad. Arts 12: 76. 1877. *Gonolobus gonocarpos* (Walter) L. M. Perry, Rhodora 40: 284. 1938. *Matelea gonocarpos* (Walter) Shinners, Field & Lab. 18: 73. 1950.

Climbing perennial vine; stem to 3(6) m long, sparsely to moderately glandular-puberulent. Leaves with the blade oblong-ovate to cordate, the apex acute, the base cordate, the margin entire, the upper and lower surfaces sparsely pubescent, also with minute glandular trichomes, the petiole 2–2.5 cm long. Flowers few in an axillary umbelliform cyme, the peduncle 5–10 mm long, the pedicel ca. 1 cm long; bracts lanceolate, minute, caducous; sepal lobes lanceolate, 3–6 mm long, spreading, the margin ciliate toward the apex; corolla campanulate-rotate, the lobes lanceolate, 7–15 mm long, the corona a flattened, irregularly 5-angled disk, this shorter than the gynostegium, yellow or orange-brown. Fruit fusiform-lanceolate to -ovate, 7–15 cm long, 5-angled or -winged, glabrous; seeds elliptic-ovate, 6–10 mm long, laterally compressed, the wing irregularly toothed, with an apical coma.

Hammocks. Occasional; northern counties, central peninsula. Maryland south to Florida, west to Illinois, Nebraska, Oklahoma, and Texas. Spring–summer.

Gonolobus suberosa (as *Matelea gonocarpos*) is listed as threatened in Florida (Florida Administrative Code, Chapter 5B-40).

Matelea Aubl. 1775. MILKVINE

Perennial herbaceous vines, with white latex. Leaves opposite, simple, pinnate-veined, petiolate, estipulate. Flowers in simple or compound, axillary, umbelliform cymes, bracteate; sepals 5, basally connate; petals 5, basally connate, with a 5-lobed corona, each lobe with 2 appendages between the lobes on the inner surface; stamens, style, and stigmas adnate and forming a gynostegium, pollen in pollinia, the ovary superior, 2-carpellate and -loculate. Fruit a follicle; seeds numerous, comose.

A genus of about 200 species; North America, West Indies, Mexico, Central America, and South America. [Etymology obscure, perhaps a vernacular name in a native language in French Guiana.]

Cyclodon Small, 1933; *Edisonia* Small, 1933; *Odontostephana* Alexander, 1933.

Selected reference: Drapalik (1970).

1. Corona disk-shaped or saucer-shaped, smooth or only obscurely ridged......................**M. alabamensis**
1. Corona cup-shaped, usually crested or appendaged on the inner surface.
    2. Slender, prostrate, vinelike herb; cymes sessile; corolla campanulate; leaves 2–4 cm long.................
    ....**M. pubiflora**
    2. Stout, twining vine; cymes pedunculate; corolla rotate; leaves 5–11 cm long.
        3. Corolla purplish, ca. 5 mm long; corona predominantly black, appearing 5-lobed, each lobe
        4-toothed..**M. floridana**
        3. Corolla white or yellowish, 9–11 mm long; corona not predominantly black, appearing 10-lobed,
        with 5 narrow lobes, each 2-toothed and alternating with 5 short, broad teeth.
            4. Corolla lobes erect; longer corona lobes divided ⅔ their length into 2 linear teeth.................
            ....**M. baldwyniana**
            4. Corolla lobes spreading; longer corona lobes shallowly 2-toothed......................**M. flavidula**

Matelea alabamensis (Vail) Woodson [Of Alabama.] ALABAMA MILKVINE; ALABAMA SPINY POD.

> *Vincetoxicum alabamensis* Vail, Torrey Bot. Club 30: 178, pl. 9. 1903. *Cyclodon alabamensis* (Vail) Small, Man. S.E. Fl. 1075, 1507. 1933. *Matelea alabamensis* (Vail) Woodson, Ann. Missouri Bot. Gard. 28: 234. 1941.

Perennial vine; stem to 2 m long, glandular-pubescent to glabrate. Leaves with blade elliptic-cordate, 7–15 cm long, 4–12 cm wide, the apex acute, the base cordate, the margin entire, the upper and lower surfaces glandular pubescent, the petiole ca. 2 cm long. Flowers solitary or 2–12 in an axillary umbelliform cyme, the peduncle ca. 1.5 cm long, the pedicel ca. 1.5 cm long; bracts minute; sepal lobes 2–4 mm long, spreading, the outer surface glandular-pubescent; corolla rotate, the lobes green, reticulate with darker green vines, elliptic to broadly lanceolate, 4–10 mm long, the apex rounded, the corona orange yellow, disk-shaped or saucer-shaped, smooth or only obscurely ridged, shorter than the gynostegium, with 5 horn-shaped lobes. Fruit fusiform, spiny-muricate, 6.5–9.5 cm long, 1.3–2 cm wide, glandular-pubescent to glabrate; seeds ovate, laterally compressed, the margin winged, with an apical coma.

Bluff forests. Rare; Gadsden, Liberty, and Walton Counties. Georgia, Alabama, and Florida. Spring.

Matelea alabamensis is listed as endangered in Florida (Florida Administrative Code, Chapter 5B-40).

Matelea baldwyniana (Sweet) Woodson [Commemorates William Baldwin (1779–1819), American physician and botanist.] BALDWIN'S MILKVINE; BALDWIN'S SPINY POD.

> *Gonolobus baldwynianus* Sweet, Hort. Brit., ed. 2. 360. 1830. *Vincetoxicum baldwynianum* (Sweet) Britton, Mem. Torrey Bot. Club 5: 265. 1894. *Odonostephana baldwyniana* (Sweet) Alexander, in Small, Man. S.E. Fl. 1077, 1507. 1933. *Matelea baldwyniana* (Sweet) Woodson, Ann. Missouri Bot. Gard. 28: 277. 1941.

Perennial herbaceous vines; stem to 3 m, moderately pubescent with glandular and nonglandular trichomes. Leaves with the blade lanceolate, 5–16 cm long, the apex acute, the base cordate, the margin entire, the upper and lower surfaces moderately to densely pubescent with short simple trichomes and also with minute glandular trichomes, the petiole 2–4 cm long. Flowers in a simple or compound umbelliform cyme; bracts minute; sepal lobes lanceolate, 2–4 mm long, moderately to densely pubescent on the outer surface with short simple trichomes and minute glandular trichomes; corolla white to cream-colored, campanulate-rotate, the lobes narrowly oblong to oblanceolate or linear, 7–13 mm long, moderately pubescent on outer surface as on the sepals, the corona white to cream-colored to light yellow, the lobes bluntly triangular, the appendages narrow triangular, about twice as long as the lobes. Fruit lanceolate to ovate, 6–10 cm long, 1.5–2 cm wide, spiny-muricate, glabrous; seeds ovate, 7–10 mm long, laterally compressed, slightly winged, apically comose.

Bluff forests. Rare; Gadsden and Jackson Counties. Missouri south to Florida, Arkansas, and Oklahoma. Spring.

Matelea baldwyniana is listed as endangered in Florida (Florida Administrative Code, Chapter 5B-40).

Matelea flavidula (Chapm.) Woodson [Yellowish, in reference to the flower color.] YELLOW CAROLINA MILKVINE; YELLOW-FLOWERED SPINY POD.

Gonolobus flavidulus Chapman, Bot. Gaz. 3: 12. 1878. *Gonolobus hirsutus* Michaux var. *flavidulus* (Chapman) A. Gray, Syn. Fl. N. Amer., ed. 2. 2(1): 404. 1886. *Vincetoxicum flavidulum* (Chapman) A. Heller, Muhlenbergia 1: 2. 1900. *Vincetoxicum hirsutum* (Michaux) Britton var. *flavidulum* (Chapman) Vail, in Small, Fl. S.E. U.S. 955. 1903. *Odontostephana flavidula* (Chapman) Alexander, in Small, Man. S.E. Fl. 1078, 1507. 1933. *Matelea flavidula* (Chapman) Woodson, Ann. Missouri Bot. Gard. 28: 228. 1941. TYPE: FLORIDA: Gadsden Co.

Perennial herbaceous vine; stem to 2 m long, with nonglandular and minute glandular trichomes or glabrate. Leaves with the blade cordate, 7–18 cm long, 5–16 cm wide, the apex acute, the base cordate, the margin entire, the upper and lower surfaces sparsely pubescent with nonglandular and minute glandular trichomes, the petiole 2–4 cm long. Flowers few to many in an axillary umbelliform cyme, the peduncle ca. 1.5 cm long, the pedicel ca. 1 cm long; bracts minute; sepal lobes lanceolate, 2–4 mm long, sparsely pubescent with nonglandular and minute glandular trichomes; corolla rotate, often reflexed, the lobes green to yellow with reticulate dark green veins, sometimes maroon to olive-brown or cream-colored around the corona, broadly lanceolate, 5–10 mm long, the apex obtuse, the corona orange-yellow, sometimes with appendages maroon, if the corolla lobes are maroon-colored, then the corona dark maroon with black appendages, the appendages subequaling the gynostegium. Fruit fusiform, 8–15 cm long, 1–2.5 cm wide, spiny-muricate; seeds ovate, laterally compressed, apically comose.

Bluff forests. Rare; Duval County, central panhandle. North Carolina south to Florida, west to Tennessee and Mississippi. Summer.

Matelea flavidula is listed as endangered in Florida (Florida Administrative Code, Chapter 5B-40).

Matelea floridana (Vail) Woodson [Of Florida.] FLORIDA MILKVINE; FLORIDA SPINY POD.

Vincetoxicum floridanum Vail, Bull. Torrey Bot. Club 26: 428. 1899. *Odontostephana floridana* (Vail) Alexander, in Small, Man. S.E. Fl. 1078, 1507. 1933. *Matelea floridana* (Vail) Woodson, Ann. Missouri Bot. Gard. 28: 229. 1941. TYPE: FLORIDA: "East Florida," s.d., *Leavenworth s.n.* (holotype: NY).

Perennial herbaceous vine; stem to 4.5 m long, sparsely pubescent with nonglandular and minute glandular trichomes. Leaves with the blade cordate, 7–15 cm long, 5–13 cm wide, the apex acute, the base cordate, the margin entire, the upper and lower surfaces sparsely pubescent with nonglandular and minute glandular trichomes, the petiole 3–5 cm long. Flowers in an axillary umbelliform cyme, the peduncle 5–10 mm long, the pedicel 5–10 mm long; bracts minute; sepal lobes lanceolate, 2–4 mm long, the outer surface sparsely pubescent with nonglandular and minute glandular trichomes; corolla rotate, ascending or slightly reflexed, the lobes maroon to green or sometimes dark maroon or with a cream-colored area surrounding the corona, lanceolate, 4–8 mm long, the corona black or black with a dark maroon base, slightly longer than the gynostegium, the appendages subequaling the lobes in length. Fruit fusiform, 6.5–9 cm long, ca. 1 cm wide, spiny-muricate, glabrate; seeds ovate, laterally compressed, the margin winged, with an apical coma.

Hammocks. Occasional; peninsula, west to central panhandle. Endemic. Spring.

Matelea floridana is listed as endangered in Florida (Florida Administrative Code, Chapter 5B-40).

Matelea pubiflora (Decne.) Woodson [With pubescent flowers.] TRAILING MILKVINE; SANDHILL SPINY POD.

Chthamalia pubiflora Decaisne, in de Candolle, Prodr. 8: 605. 1844. *Gonolobus pubiflorus* (Decaisne) A. Gray, Proc. Amer. Acad. Arts 12: 77. 1877. *Vincetoxicum pubiflorum* (Decaisne) A. Heller, Muhlenbergia 1: 2. 1900. *Edisonia pubiflora* (Decaisne) Small, Man. S.E. Fl. 1078, 1507. 1933. *Matelea pubiflora* (Decaisne) Woodson, Ann. Missouri Bot. Gard. 28: 230. 1941. TYPE: FLORIDA.

Prostrate or rarely climbing herbaceous vine; stem to 1 m long, sparsely pubescent with nonglandular and minute glandular trichomes. Leaves with the blade cordate, 2–4 cm long, 2–4 cm wide, the apex acute, the base cordate, the margin entire, the upper and lower surfaces sparsely pubescent with nonglandular and minute glandular trichomes, the petiole 1.5–2 cm long. Flowers in an axillary umbelliform cyme, the peduncle absent, the pedicel 2–8 mm long; bracts minute; sepal lobes lanceolate, 3–4 mm long, the outer surface sparsely pubescent with nonglandular and minute glandular trichomes; corolla campanulate, the lobes olive to red-brown, oblong-lanceolate, 2–4 mm long, the margin obtuse, the inner surface pubescent, the corona olive to red-brown, the apices black, incurved over the gynostegium. Fruit fusiform, 6–10 cm long, 1–2 cm wide, spiny puricate, glabrous; seed ovate, laterally compressed, the margin winged, with an apical coma.

Sandhills. Occasional; northern and central peninsula, eastern panhandle. Alabama and Florida. Spring–summer.

Matelea pubiflora is listed as endangered in Florida (Florida Administrative Code, Chapter 5B-40).

Metastelma R. Br. 1810.

Perennial herbaceous vines, with white latex. Leaves opposite, simple, pinnate-veined, petiolate, estipulate. Flowers in axillary umbelliform cymes, bracteate; sepals 5, basally connate; petals 5, basally connate, with a corona of 5 free segments; stamens, style, and stigmas adnate and forming a gynostegium, the pollen in pollinia, the ovary superior, 2-carpellate and -loculate. Fruit a follicle; seeds numerous, comose.

A genus of about 70 species; North America, West Indies, Mexico, Central America, and South America. [From the Greek *meta*, changed, and *stelma*, crown, in reference to the corona consisting of 5 segments rather than a crown.]

Metastelma usually has been included in the broadly circumscribed *Cynanchum* by many workers.

Selected reference: Liede-Schumann et al. (2014).

Metastelma blodgettii A. Gray [Commemorates John Loomis Blodgett (1809–1853), physician, druggist, and botanist in south Florida.] BLODGETT'S SWALLOWWORT.

Metastelma blodgettii A. Gray, Proc. Amer. Acad. Arts. 12: 73. 1877. *Cynanchum blodgettii* (A. Gray) Shinners, Sida 1: 365. 1964. TYPE: FLORIDA: Monroe Co.: Big Pine Key, s.d., *Blodgett s.n.* (holotype: GH).

Twining vines; stems glabrous. Leaves linear or linear-lanceolate, 1–2 cm long, the apex acute, the base rounded, the margin entire, the upper and lower surfaces glabrous, the petiole 2–3 mm long. Flowers in an axillary umbelliform cyme, the peduncle ca. 2 mm long, the pedicel ca. 2 mm long; bracts minute; sepal lobes ovate, ca. 1 mm long; corolla whitish, campanulate, the lobes lanceolate, 2–3 mm long, the apex slightly spreading, pubescent on the inner surface, the corona segments subulate, subequaling the gynostegium. Fruit linear-fusiform, 4–5 cm long, smooth, glabrous; seeds ovate, laterally compressed, the margin winged, with an apical coma.

Tropical hammocks and pine rocklands. Rare; southern peninsula. Florida; West Indies. Spring–fall.

Metastelma blodgettii is listed as threatened in Florida (Florida Administrative Code, Chapter 5B-40).

EXCLUDED TAXON

Metastelma parviflorum (Swartz) R. Brown ex Schultes—This West Indian species was reported for Florida by Chapman (1860), the name misapplied to material of *C. blodgettii*.

Nerium L. 1753. OLEANDER

Shrub or tree, with white latex. Leaves opposite or whorled, simple, pinnate-veined, petiolate, stipulate. Flowers in terminal cymes, bracteate; sepals 5, basally connate; petals 5, basally connate; stamens 5, epipetalous, the filaments and anthers free; ovary superior, 2-carpellate and -loculate, style and stigma 1. Fruit a follicle; seeds numerous, comose.

A monotypic genus; nearly cosmopolitan. [Apparently from Nereus of Greek mythology, who was the eldest son of Pontus (sea) and Gaia (earth) and who lived in the Aegean Sea, which borders on the area where the plant occurs.]

Nerium oleander L. [*Oliandrum*, obscurely related to *laurea*, laurel, in reference to the leaves, and *rhododendron*, in reference to the flowers.] OLEANDER.

Nerium oleander Linnaeus, Sp. Pl. 209. 1753. *Nerium lauriforme* Lamarck, Fl. Franc. 2: 299. 1779 ("1778"), nom. illegit. *Oleander vulgaris* Medikus, Hist. & Commentat. Acad. Elect. Sci. Theod.-Palat. 6(Phys.): 381. 1790. *Nerium floridum* Salisbury, Prodr. Stirp. Chap. Allerton 147. 1796, nom. illegit. *Nerium oleander* Linnaeus var. *roseum* Stokes, Bot. Mat. Med. 1: 498. 1812, nom. inadmiss.

Shrub or small tree to 6 m; stem puberulent when young, glabrous in age. Leaves opposite or in whorls of 3–4, the blade oblong-lanceolate to linear-lanceolate, to 20 cm long, to 2.5 cm wide, the apex acute or acuminate, the base narrowly cuneate, the margin entire, the upper surface glabrous, the lower surface pubescent when young, glabrous in age, the petiole 1–1.5 cm long. Flowers in a thyrse to 20 cm long, the peduncle 2–10 cm long, pubescent, the pedicel ca. 5 mm long, pubescent; bracts lanceolate, minute; sepal lobes green or pinkish, lanceolate, 4–7 mm long, the outer surface pubescent; corolla funnelform, the tube cylindrical, 8–12 mm long, becoming conic-campanulate at the throat, 9–10 mm long, the lobes obliquely ovate to obovate-oblong, 10–15 mm long, spreading; stamens inserted at the throat, the filaments short, the anthers adhering to the stigma but easily separated, with 2 scalelike appendages at the base, the apex with a filiform pilose appendage, the 5 appendages twisted together; style linear-subclavate, the stigma fusiform, with a reflexed basal collar. Fruit linear, 8–20 cm long, ca. 1 cm wide; seeds laterally compressed, densely pubescent, the apex comose.

Dry, disturbed sites. Occasional; peninsula, central and western panhandle. Escaped from cultivation. North Carolina south to Florida, west to California; nearly cosmopolitan. Native to Europe, Asia, and Africa. Spring–fall.

Ochrosia Juss. 1789. YELLOWWOOD

Shrubs or trees, with white latex. Leaves opposite or whorled, simple, pinnate-veined, petiolate, estipulate. Flowers in subterminal cymes, bracteate; sepals 5, basally connate; petals 5, basally connate; stamens 5, epipetalous, the filaments and anthers free; ovary superior, 2-carpellate and -loculate, the style and stigma 1. Fruit a drupe; seeds 2–4 per locule.

A genus of about 40 species; North America, West Indies, Asia, Australia, and Pacific Islands. [From the Greek, *ochros*, pale yellow, in reference to fruit color of some of the species.]

Ochrosia elliptica Labill. [Elliptic, in reference to the leaf shape.] ELLIPTIC YELLOWWOOD.

> *Ochrosia elliptica* Labillardière, Sert. Austro-Caledon. 25, t. 30. 1824. *Lactaria elliptica* (Labillardière) Kuntze, Revis. Gen. Pl. 1: 415. 1891. *Bleekeria elliptica* (Labillardière) Koidzumi, Bot. Mag. (Tokyo) 37: 52. 1923. *Excavatia elliptica* (Labillardière) Markgraf, Bull. Bishop Mus. Honolulu 141: 128. 1936.

Shrub or tree to 3 m; stem glabrous. Leaves opposite or in a whorl of 3–4, the blade obovate to broadly elliptic, 8–15(17) cm long, 3–5(7) cm wide, the apex obtuse or short acuminate, the base cuneate, the margin entire, the upper and lower surfaces glabrous, the petiole 5–11 mm long. Flowers in a subterminal cyme, sessile or the peduncle to 4 cm in fruit, the pedicel absent; bracts ovate, minute; sepal lobes ovate, ca. 2 mm long, glabrous; corolla white, salverform, slightly dilated above the middle, the tube 1–1.5 cm long, cylindric, the lobes linear, 1–2 cm long; stamens inserted in the widening corolla, the anthers narrowly oblong; style connate at the apex, free below, the stigma globular with 2 small lobes at the apex. Fruit ellipsoid, (2)5–6 cm long, red, glabrous, the endocarp thick; seeds suborbicular, laterally compressed, narrow-margined.

Disturbed sites. Rare; Sarasota and Palm Beach Counties, southern peninsula. Escaped from cultivation. Florida; West Indies, Asia, Australia, and Pacific islands. Native to Asia, Australia, and Pacific Islands. Summer–fall.

Orthosia Decne., 1844.

Perennial herbaceous vines, with white latex. Leaves opposite, simple, pinnate-veined, petiolate, estipulate. Flower in umbelliform cymes, bracteate; sepals 5, basally connate; petals 5, basally connate, with a cupulate 5-lobed corona adnate to the column base; stamens, style, and stigma adnate to form a gynostegium, the pollen in pollinia, the ovary superior, 2-carpellate and -loculate. Fruit a follicle; seeds numerous, comose.

A genus of about 40 species; North America, West Indies, Central America, and South America. [From Greek mythology, the surname of Artemis, goddess of the moon, this in turn derived from Mount Orthosium, in reference to its nocturnal blooming.]

Orthosia has commonly been included in the broadly defined *Cynanchum*.

Amphistelma Griseb., 1862.

Selected references: Liede-Schumann & Meve (2008, 2013).

Orthosia scoparia (Nutt.) Liede & Meve [*Scoparie*, in the form of a broom, in reference to its many clustered filiform stems.] LEAFLESS SWALLOWWORT.

> *Cynanchum scoparium* Nuttall, Amer. J. Sci. Arts 5: 291. 1822. *Lyonia scoparia* (Nuttall) Rafinesque, Autik. Bot. 182. 1840. *Cynoctonum scoparium* (Nuttall) Chapman, Fl. South. U.S. 367. 1860. *Vincetoxicum scoparium* (Nuttall) A. W. Wood, Amer. Bot. Fl. 274. 1870. *Metastelma scoparium* (Nuttall) Vail, in Small, Fl. S.E. U.S. 950. 1903. *Amphistelma scoparium* (Nuttall) Small, Fl. Miami 149, 200. 1913. *Orthosia scoparia* (Nuttall) Liede & Meve, Novon 18: 207. 2008. TYPE: FLORIDA: Oct–Nov 1821, *Ware s.n.* (holotype: PH).
>
> *Lyonia cuspidata* Rafinesque, New Fl. 4: 59. 1838. TYPE: FLORIDA.

Diffuse, scrambling perennial herbaceous vine; stem sparsely puberulent or glabrate. Leaves with the blade narrowly linear-lanceolate, 2–5 cm long, 2–5 mm wide, the apex acute, the base cuneate, the margin entire, the upper and lower surfaces sparsely puberulent or glabrate, the petiole 2–8 mm long. Flowers few in an axillary umbelliform cyme, sessile or the peduncle to 2 mm long, the pedicel 2–3 mm long; bracts minute; sepal lobes deltoid, ca. 1 mm long, sparsely puberulent; corolla rotate-campanulate, greenish white, the lobes elliptic-lanceolate, 1–2 mm long, the corona cuplike, the lobes deltoid, erect, ca. 0.5 mm long, shorter than the gynostegium; stigma flat. Fruit linear-fusiform, 3.5–4.5 cm long, smooth, glabrous; seeds ovate, laterally compressed, with an apical coma.

Hammocks. Frequent; peninsula, central panhandle. South Carolina south to Florida, west to Mississippi; West Indies, Central America, and South America. Spring–fall.

Pentalinon Voigt 1845

Woody vines or shrubs, with white latex. Leaves opposite, simple, pinnate-veined, petiolate, estipulate. Flowers in axillary cymes, bracteate; sepals 5, basally connate; petals 5, basally connate; stamens 5, epipetalous, the stamens each with a threadlike apical appendage; ovary superior, 2-carpellate and -loculate. Fruit a follicle; seeds many, comose.

A genus of 2 species; North America, West Indies, Mexico, and Central America. [*Penta*, five, and *linum*, flax, in reference to the narrow threadlike appendage from the apex of each of the five anthers.]

Urechites Müll.-Arg., 1860.

Pentalinon luteum (L.) B. F. Hansen & Wunderlin [Yellow, in reference to the flower color.] WILD ALLAMANDA; HAMMOCK VIPERSTAIL.

> *Vinca lutea* (Linnaeus) Cent. Pl. 2: 12. 1756. *Echites truncatus* Lamarck, Encycl. 2: 340. 1786, nom. illegit. *Echites catesbaei* G. Don, Gen. Hist. 4: 74. 1837. *Urechites luteus* (Linnaeus) Britton, Bull. New York Bot. Gard. 5: 316. 1907. *Pentalinon luteum* (Linnaeus) B. F. Hansen & Wunderlin, Taxon 35: 167. 1986.
>
> *Urechites pinetorum* Small, Addisonia 4: 21. 1919. TYPE: FLORIDA: Miami-Dade Co.: near Coconut Grove, 9 May 1904, *Small & Wilson 1714* (holotype: NY).
>
> *Urechites luteus* (Linnaeus) Britton var. *sericeus* R. W. Long, Rhodora 72: 31. 1970. *Pentalinon luteum* (Linnaeus B. F. Hansen & Wunderlin var. *sericeum* (R. W. Long) D. B. Ward, Novon 11: 361. 2001.

Twining vines or scrambling shrubs; stem pubescent or glabrous. Leaves with the blade oblong, obovate, to elliptic or suborbicular, 3–8 cm long, the apex obtuse to acute, the base rounded to cuneate, the margin entire, the upper surface glabrous, the lower surface pubescent or glabrescent, the petiole 1–1.5 cm long. Flowers solitary or few to many in an axillary cyme, the peduncle 2–5 cm long, the pedicel 1–1.5 cm long; bracts lanceolate, foliaceous, ca. 1 cm long; sepal lobes lanceolate-subulate, 8–11 mm long, glabrous or pubescent; corolla yellow, salverform, the throat dilated and campanulate, the lobes rotate; stamens attached about the middle of the throat, each with a narrow apical appendage; style linear, slightly clavate, the stigma subglobose, slightly 2-lobed, with a reflexed rim below. Fruit linear-fusiform, 8–20 cm long,

4–5 mm wide, smooth, glabrous or pubescent; seeds elliptic, with a narrow beak, with an apical coma.

Mangroves and coastal hammocks. Occasional; St. Lucie, Martin, Palm Beach, and Lee Counties, southern peninsula. Florida; West Indies and Central America. Spring–summer.

Plumeria L. 1753. FRANGIPANI

Shrubs or trees, with white latex. Leaves alternate, simple, pinnate-veined, petiolate, estipulate. Flowers in terminal cymes, bracteate; sepals 5, basally connate; petals 5, basally connate; stamens 5, epipetalous, the filaments and anthers free; ovary partly inferior, 2-carpellate and -loculate, the stigma 1, the style bicapitate. Fruit a follicle; seeds numerous.

A genus of 7 species; North America, West Indies, Mexico, Central America, South America, and Asia. [Commemorates Charles Plumier (1646–1704), French Franciscan monk and botanist who conducted 3 expeditions to the Caribbean.]

Plumeria obtusa L. [Obtuse, in reference to the leaf apex.] FRANGIPANI.

Plumeria obtusa Linnaeus, Sp. Pl. 210. 1753.

Shrub or small tree, to 10(15) m; branchlets thick, somewhat fleshy, with prominent leaf scars, glabrous. Leaves obovate to ovate-oblong, 5–25 cm long, 2–9 cm wide, the apex rounded or emarginate to short acuminate, the base cuneate, the margin entire, the upper surface glabrous, the lower surface glabrous or slightly pubescent, the petiole 2–4 cm long. Flowers few to many in a subumbellate cyme, the pedicel 7–10 mm long; sepal lobes ovate-triangular, 1–2 mm long, the apex rounded or truncate, the outer surface glabrous or pilosulous; corolla white with a yellow center, salverform, the tube 1–2 cm long, the lobes ovate-oblong or obovate, 1.5–4.5 cm long, the apex obtuse or rounded; stamens included in the tube; style short, the stigma slenderly bicapitate. Fruit 7–24 cm long, 1–2 cm wide, smooth, glabrous; seeds elliptic, 2–4 cm long, slightly laterally compressed, basally winged about half its length.

Disturbed sites. Rare; Monroe County keys. Escaped from cultivation. Florida, West Indies, Mexico, and Central America. Native to West Indies, Mexico, and Central America. Summer–fall.

Rauvolfia L. DEVIL'S-PEPPER

Shrubs, with white latex. Leaves whorled, simple, pinnate-veined, petiolate, estipulate. Flowers in axillary cymes, bracteate; sepal lobes 5, basally connate; petals 5, basally connate; stamens 5, epipetalous, the filaments and anthers free; ovary superior, 2-carpellate and -loculate, the style 1 and stigma 1. Fruit a drupe; seeds 2.

A genus of about 110 species; North America, West Indies, Mexico, Central America, South America, Africa, Asia, Australia, and Pacific islands. [Commemorates Leonhard Rauwolf (1535–1596), German physician and botanist.]

Selected reference: Rao (1956).

Rauvolfia tetraphylla L. [With leaves often in whorls of 4.] BE-STILL-TREE.

Rauvolfia tetraphylla Linnaeus, Sp. Pl. 208. 1753.

Shrub, to 2 m; stem pubescent when young, glabrescent in age. Leaves in whorls of 3–5, the blade ovate or oblong, 1–15 cm long, 0.8–4 cm wide, often of unequal size, the apex acute or obtuse, the base cuneate to rounded, the margin entire, the upper and lower surfaces glabrous or sparsely pubescent, the petiole 2–5 mm long, with glandular teeth on the upper side. Flower in simple or compound axillary cymes, the peduncle 1–4 cm long, the pedicel 4–6 mm long; bracts minute; sepal lobes ovate, ca. 1 mm long, the outer surface glabrous or sparsely pubescent; corolla white, salverform, the tube urceolate, 2–3 mm long, the inner surface pubescent on the distal half, the lobes ovate or suborbicular; stamens inserted in the corolla throat; ovaries connate; anthers sessile; style filiform, the stigma drum-shaped, with pendulous ring of tissue, the apex 2-cleft. Fruit subglobose, 5–10 mm long, red, smooth, glabrous; seeds 2.

Disturbed sites. Rare; Palm Beach County. Escaped from cultivation. Florida; West Indies, Mexico, Central America, and South America; Asia and Australia. Native to tropical America. Summer–fall.

Rhabdadenia Müll. Arg. 1860.

Woody vines, with white latex. Leaves opposite, simple, pinnate-veined, petiolate, estipulate. Flowers in axillary reduced, dichasial cymes, bracteate; sepals 5, basally connate; petals 5, basally connate; stamens 5, epipetalous, the filaments free, the anthers connivent and adnate to the stigma; nectaries 5; ovary superior, 2-carpellate and -loculate, the stigma and style 1. Fruit a follicle; seeds numerous, comose.

A genus of 3 species; North America, West Indies, Mexico, Central America, and South America. [From the Greek *rhabdos*, rod or stick, and *adenos*, gland, apparently in reference to the slender fruit.]

Rhabdadenia biflora (Jacq.) Müll. Arg. [Two-flowered, in reference to the flowers frequently appearing in pairs.] RUBBERVINE; MANGROVEVINE.

> *Echites biflorus* Jacquin, Enum. Syst. Pl. 13. 1760. *Rhabdadenia biflora* (Jacquin) Müller Argoviensis, in Martius, Fl. Bras. 6(1): 175. 1860. *Chariomma scandens* Miers, Apocyn. S. Amer. 114. 1878, nom. illegit.
>
> *Echites paludosus* Vahl, Eclog. Amer. 2: 19. 1798. *Exothostemon paludosum* (Vahl) G. Don, Gen. Hist. 4: 83. 1837. *Rhabdadenia paludosa* (Vahl) Miers, Apocyn. S. Amer. 119. 1878.

Woody vine; stem to 8 m long, glabrous. Leaves obovate-oblong to oblong or lanceolate, 5–15 cm long, 1.5–5 cm wide; the apex rounded, mucronate, the base obtuse to cuneate, the margin entire, the upper and lower surfaces glabrous, the petiole ca. 1 cm long. Flowers solitary or 2–5 in an axillary reduced, dichasial cyme, the peduncle 3–5 cm long, the pedicel 10–13 mm long; bracts minute; sepal lobes ovate-oblong, 5–9 mm long, mucronate; corolla white, sometimes tinged with pink, funnelform, the tube 1.5–2 cm long, conic, 2–3 mm long, the lobes oblique-obovate, 2–2.5 cm long, spreading; anthers connivent and agglutinated to the stigma, the connective enlarged, narrowly bilobate, the style filiform, the stigma apex conic, the base umbraculiform. Fruit linear-fusiform, 8–14 cm long, terete, striate, glabrous; seeds linear-elliptic, laterally compressed, with an apical coma.

Mangroves and coastal hammocks. Occasional; central and southern peninsula. Florida; West Indies, Mexico, Central America, and South America.

Seutera Rchb., nom. cons. 1828.

Perennial herbaceous vines, with white latex. Leaves opposite, simple, pinnate-veined, epetiolate, estipulate. Flowers in axillary umbelliform cymes, bracteate; sepals 5, basally connate; petals 5, basally connate, with a 5-lobed corona; stamens, style, and stigma adnate and forming a gynostegium, the pollen in pollinia, the ovary superior, 2-carpellate and -loculate. Fruit a follicle; seeds numerous, comose.

A genus of 2 species; North America, West Indies, Mexico, and Central America. [Commemorates Bartholomäus Seuter (1678–1754), German engraver of botanical prints.]

Seutera has commonly been included in the broadly circumscribed *Cynanchum* by many workers, following Woodson (1941).

Lyonia Elliott, nom. rej., 1817.

Selected reference: Fishbein & Stevens (2005).

Seutera angustifolia (Pers.) Fishbein & W. D. Stevens [With narrow leaves.] GULF COAST SWALLOWWORT.

Cynanchum angustifolium Persoon, Syn. Pl. 1: 274. 1805. *Funastrum angustifolium* (Persoon) Liede & Meve, Nordic J. Bot. 22: 587. 2003 ("2002"). *Seutera angustifolia* (Persoon) Fishbein & W. D. Stevens, Novon 15: 532. 2005.

Ceropegia palustris Pursh, Fl. Amer. Sept. 184. 1814. *Lyonia maritima* Elliott, Sketch Bot. S. Carolina 1: 316. 1817, nom. illegit. *Cynanchum angustifolium* Nuttall, Gen. N. Amer. Pl. 1: 164. 1818, nom. illegit.; non Persoon, 1905. *Macbridea maritima* Rafinesque, Amer. Monthly Mag. & Crit. Rev. 3: 99. 1818, nom. illegit. *Seutera maritima* Decaisne, in de Candolle, Prodr. 8: 590. 1844, nom. illegit. *Cynanchum maritimum* Maximowicz, Bull. Acad. Imp. Sci. Saint-Pétersbourg 9: 800. 1865, nom. illegit; non Jacquin, 1763; nec Salisbury, 1796. *Vincetoxicum palustre* (Pursh) A. Gray, Syn. Fl. N. Amer. 2(1): 102. 1878. *Cynanchum palustre* (Pursh) A. Heller, Cat. N. Amer. Pl. 6. 1898. *Metastelma palustre* (Pursh) Schlechter, in Urban, Symb. Antill. 1: 258. 1899. *Seutera palustris* (Pursh) Vahl, in Small, Fl. S.E. U.S. 952. 1903. *Lyonia palustris* (Pursh) Small, Fl. Miami 149, 200. 1913.

Perennial herbaceous vine; stem glabrous. Leaves linear, 4–8 cm long, 1–5 mm wide, the apex acute, the base cuneate, the margin entire, the upper and lower surfaces glabrous, sessile. Flowers numerous in an axillary umbelliform cyme, the peduncle 1–5 cm long, glabrous or sparsely puberulent, the pedicel 4–8 mm long, glabrous or sparsely puberulent; bracts linear, minute; sepal lobes lanceolate, 2–3 mm long, sparsely puberulent, erect; corolla rotate-campanulate, greenish white, sometimes tinged with purple, the lobes lanceolate, 3–4 mm long, the apex spreading, the corona 5-lobed, the lobes oblong, 1–2 mm long, longer than the gynostegium, retuse or emarginated; stigma apex conic. Fruit fusiform to narrowly lanceolate, 4–6 cm long, 6–7 mm wide, smooth, glabrous; seeds ovate, laterally compressed, with an apical coma.

Coastal hammocks and salt marshes, rarely inland. Frequent; nearly throughout. North Carolina south to Florida, west to Texas; West Indies, Mexico, and Central America. Spring–fall.

Tabernaemontana L. 1753. MILKWOOD

Shrubs or trees, with white latex. Leaves opposite, simple, pinnate-veined, petiolate, estipulate. Flowers in terminal or axillary cymes, bracteate; sepals 5, basally connate; petals 5, basally connate; stamens 5, epipetalous, the filaments and anthers free; ovary superior, 2-carpellate and -loculate, the style and stigma 1; fruit a follicle; seeds numerous.

A genus of about 100 species; North America, West Indies, Mexico, Central America, South America, Africa, Asia, Australia, and Pacific Islands. [Commemorates Jakob Theodor von Bergzaben (Jacobus Theodorus Tabernaemontanus) (1520–1590), German physician and botanist.]

Selected reference: Leeuwenberg (1994).

1. Flowers 1–1.3 cm long .... T. alba
1. Flowers 2–4 cm long .... T. divaricata

Tabernaemontana alba Mill. [With white flowers.] WHITE MILKWOOD.

Tabernaemontana alba Miller, Gard. Dict., ed. 8. 1768.

Shrub or small tree, to 8 m; stem glabrous, the bark gray. Leaves with the blade narrowly elliptic to ovate-oblong, 7–15(20) cm long, 2–6(7) cm wide, the apex acuminate, the base cuneate, the margin entire, the upper and lower surfaces glabrous, the petiole 1–3 cm long. Flowers in a compound axillary cyme to 14 cm long; bracts ovate, lanceolate, minute; corolla cream-yellow, the tube 1 cm long. Fruit elliptic ovate, ca. 5 cm long, ca. 3 cm wide.

Disturbed pinelands. Rare; Miami-Dade County. Escaped from cultivation. Florida; West Indies, Mexico, Central America, and South America. Native to tropical America. Summer.

Tabernaemontana divaricata (L.) R. Br. ex Roem. & Schult. [Divaricate, in reference to the branches.] CAPE JASMINE; PINWHEEL FLOWER.

Nerium divaricatum Linnaeus, Sp. Pl. 306. 1753. *Tabernaemontana divaricata* (Linnaeus) R. Brown ex Roemer & Schultes, Syst. Veg. 4: 427. 1819. *Ervatamia divaricata* (Linnaeus) Burkill, Rec. Bot. Surv. India 10: 320. 1925.

Shrub or small tree, to 5 m; stem glabrous. Leaves with the blade elliptic to oblanceolate or ovate, 7–13 cm long, 2–3.5 cm broad, the apex acuminate, the base cuneate, the margin entire, the upper and lower surfaces glabrous, the petiole ca. 1 cm long. Flowers in a terminal or axillary cyme, the peduncle 2–4(8) cm long, the pedicel ca. 1 cm long; bracts ovate-lanceolate, minute; calyx ca. 1.5 cm long, the sepal lobes ovate, more than ½ as long, glabrous; corolla white, salverform, the tube narrow, 1.5–2 cm long, the lobes obovate, ca. 1.5 cm long, spreading; stamens inserted below the middle of the tube; stigma umbraculiform below. Fruit obliquely and narrowly ellipsoid, 2–7 cm long, 0.5–1.5 cm wide, 3-ribbed, glabrous; seeds ovate, embedded in a fleshy aril.

Disturbed sites. Rare; Osceola County. Escaped from cultivation. Florida; West Indies; Asia. Native to Asia. Summer.

Thevetia L., nom. cons. 1758.

Trees or shrubs, with white latex. Leaves alternate, simple, pinnate-veined, petiolate or epetiolate, estipulate. Flowers in terminal and axillary cymes, bracteate; sepals 5, basally connate, with a corona of 5 scales; petals 5, basally connate; nectaries present; stamens 5, epipetalous, the filaments and anthers free; ovary superior, 2-carpellate and -loculate, the style and stigma 1. Fruit a berrylike drupe; seeds 2 per locule.

A genus of 8 species; North America, West Indies, Mexico, Central America, South America, Asia, Australia, and Pacific Islands. [Commemorates André Thévet (1516–1590), French Franciscan priest, explorer, and cosmographer, who traveled to Brazil.]

It is inconclusive whether *Thevetia* and *Casabella* should be recognized as separate genera or subgenera of *Thevetia*. We adopt the latter, following Williams and Stutzman (2008).

Selected references: Alvarado-Cárdinas and Ochoterena (2007); Williams and Stutzman (2008).

Thevetia peruviana (Pers.) K. Schum. [Of Peru.] LUCKYNUT.

Cerbera thevetia Linnaeus, Sp. Pl. 209. 1753. *Thevetia linearis* Rafinesque, Sylva Tellur. 91. 1838. *Thevetia thevetia* (Linnaeus) H. Karsten, Deut. Fl. 1035. 1883; (Linnaeus) Millspaugh, Publ. Field Columb. Mus., Bot. Ser. 2: 83. 1900, nom. inadmiss. *Ahouai thevetia* (Linnaeus) M. Gómez de la Maza y Jiménez, Fl. Habanera 357. 1897. *Cascabela thevetia* (Linnaeus) Lippold, Feddes Repert. 91: 52. 1980.

Cerbera peruviana Persoon, Syn. Pl. 1: 267. 1809. *Cascabela peruviana* (Persoon) Rafinesque, Sylva Tellur. 162. 1838. *Thevetia peruviana* (Persoon) K. Schumann, in Engler & Prantl, Nat. Pflanzenfam. 4(2): 159. 1895.

Shrub or small tree, to 6(10) m; branchlets greenish gray, glabrous, the bark gray-brown, lenticellate. Leaves with the blade linear-lanceolate, 5–15 cm long, 5–13 mm wide, the apex acuminate, the base narrowly cuneate, the margin entire, the upper and lower surfaces glabrous, the midvein prominent, the lateral veins obscure, sessile or the petiole ca. 3 mm long. Flowers few in a terminal or axillary cyme 7–10 cm long, the peduncle 1–2 cm long, the pedicel 2.5–5 cm long; bracts subulate, minute; sepal lobes ovate or ovate-lanceolate, 4–7 mm long, with numerous squamelae at the base on the inner surface, glabrous; corolla yellow, funnelform, 6–7 cm long, the tube 1–1.5 cm long, with a corona of 5 scales, these connected by a row of long white trichomes, the lobes obliquely obovate, 3.5–4.5 cm long, spreading; nectary annular; stamens inserted in the distal narrow part of the corolla tube, the filaments short, the anthers small, the style filiform, the stigma umbraculiform below, the apex shortly 2-cleft. Fruit compressed triangular-globose, 2.5–4 cm long, green turning red to blackish; seeds 3–4.5, lenticular, 2–3.5 cm long, light gray.

Disturbed sites. Occasional; central and southern peninsula. Escaped from cultivation. Florida and Texas; West Indies, Mexico, Central America, and South America; Asia, Australia, and Pacific Islands. Native to tropical America. All year.

Thyrsanthella Pichon 1948. CLIMBING DOGBANE

Woody vines, with white latex. Leaves opposite, simple, pinnate-veined, petiolate, estipulate. Flowers in axillary cymes, bracteate; sepals 5, basally connate; petals 5, basally connate; stamens 5, epipetalous, the filaments free, the anthers adherent to the stigma; nectar glands 5; ovary superior, 2-carpellate and -loculate, the style and stigma 1. Fruit a follicle; seeds numerous, comose.

A monotypic genus; North America. [From the Greek *thyrsos*, a thyrse, *anthos*, flower, and the Latin *ellus*, the diminutive, in reference to the flowers.]

Thyrsanthella difformis (Walter) Pichon [Different-formed, in reference to the different-shaped upper and lower leaves.] CLIMBING DOGBANE.

> *Echites difformis* Walter, Fl. Carol. 98. 1788. *Forsteronia difformis* (Walter) A. de Candolle, in de Candolle, Prodr. 8: 437. 1844. *Trachelospermum difforme* (Walter) A. Gray, Syn. Fl. N. Amer. 2: 85. 1878. *Thyrsanthus difformis* (Walter) Miers, Apocyn. S. Amer. 99. 1878. *Thyrsanthella difformis* (Walter) Pichon, Bull. Mus. Natl. Hist. Nat., sér. 2. 20: 192. 1948.
>
> *Echites salicifolius* Rafinesque, New Fl. 4: 59. 1838 ("1836"); non Roemer & Schultes, 1819. TYPE: FLORIDA.

Climbing woody vine; stem to 5 m, glabrous or with sparse reddish trichomes when young. Leaves with the blade ovate, lanceolate, or oblong-elliptic, the apex acuminate to rounded and apiculate, the base cuneate to rounded, the margin entire, the upper surface glabrous, the lower surface sparsely to densely pubescent, the petiole 4–5 mm long. Flowers in compound cymes; bracts subulate, 3–4 mm long, caducous; sepal lobes narrowly ovate, 3–4 mm long, the outer surface glabrous or sparsely pubescent; corolla greenish yellow, the inner surface reddish near the top of the tube, the tube 5–7 mm long, sparsely pubescent on the inner surface, the lobes shorter than the tube, spreading or slightly recurved; stamens attached near the midpoint of the tube, the anthers adhering to the stigma; nectar glands around the ovary base and alternating with the stamens; style slender, the stigma club-shaped. Fruit 10–23 cm long, linear, smooth, glabrous; seeds triangular, slender, 8–11 mm long, slightly laterally compressed, with an apical coma.

Floodplain forests; northern counties south to Hernando County. Maryland south to Florida, west to Illinois, Missouri, Oklahoma, and Texas. Spring.

Trachelospermum Lem., nom. cons. 1851.

Woody vines, with white latex. Leaves opposite, simple, pinnate-veined, petiolate, estipulate. Flowers in terminal or axillary cymes, bracteate; sepals 5, basally connate; petals 5, basally connate; stamens 5, epipetalous, the filaments free, the anthers connivent and adherent to the stigma; disc scales 5; ovary superior, 2-carpellate and -loculate. Fruit a follicle; seeds numerous, comose.

A genus of about 15 species; North America and Asia. [From the Greek *trachelos*, neck, and *sperma*, seed, in reference to the seed shape.]

Trachelospermum jasminoides (Lindl.) Lem. [Resembling *Jasminum* (Oleaceae), in reference to the floral odor.] CONFEDERATE JASMINE.

Rhyncospermum jasminoides Lindley, J. Hort. Soc. London 1: 74. 1846. *Trachelospermum jasminoides* (Lindley) Lemaire, Jard. Fleur. 1: t. 61. 1851.

Woody vine; stem to 10 m, pubescent when young, glabrous in age, lenticellate. Leaves with the blade ovate to oblong or narrowly elliptic, 2–10 cm long, 1–4.5 cm wide, the apex acute, the base cuneate, the margin entire, the upper surface glabrous, the lower surface glabrous or sometimes glabrous, the petiole 2–4 mm long. Flowers in terminal or axillary cymes, the peduncle 2–6 cm long, puberulent or glabrous, the pedicel 4–7 mm long, puberulent or glabrous; bracts ca. 2 mm long; sepal lobes oblong, 2–5 mm long, spreading or reflexed, with basal glands, the outer surface pubescent, the margin ciliate; corolla white, salverform, the tube 5–10 mm long, dilated near the middle, the throat glabrous or pilose on the inner surface near the stamens, the lobes obovate, as long as the tube, spreading; stamens inserted at the middle of the tube; the anthers connivent and adherent to the stigma; style short, the head conical. Fruit linear, 10–25 cm long, smooth, glabrous; seeds oblong, ca. 2 mm long, minutely tuberculate, with an apical coma.

Disturbed sites. Rare; Escambia, Hillsborough, Lee, Miami-Dade, and Escambia Counties. Escaped from cultivation. Florida, Louisiana, and Texas; Europe and Asia. Native to Asia. Spring–fall.

Vallesia Ruiz & Pav. 1794.

Shrubs or trees, with white latex. Leaves alternate, simple, pinnate-veined, petiolate, estipulate. Flowers in axillary cymes, bracteate; sepals 5, basally connate; petals 5, basally connate; stamens 5, epipetalous, the filaments and anthers free; ovary superior, 2-carpellate and -loculate. Fruit a drupe; seed 1 per locule.

A genus of about 12 species; North America, West Indies, Mexico, Central America, and South America. [Commemorates Francisco Valles (1524–1592), physician to Philip II of Spain.]

Vallesia antillana Woodson [Of the Antilles]. TEARSHRUB.

Vallesia antillana Woodson, Ann. Missouri Bot. Gard. 24: 13. 1937. TYPE: FLORIDA: Monroe Co.: thickets on Key West, 13 Apr 1896, *Curtiss 5620* (holotype: MO; isotypes: F, UC, US).

Shrub or small tree, to 4 m; stem glabrous. Leaves with the blade ovate or obovate-elliptic, 2–8 cm long, 1–3 cm wide, the base acuminate to obtuse, the base obtuse to cuneate, the margin entire, the upper and lower surfaces glabrous, the petiole 4–7 mm long. Flowers in an axillary dichasial cyme; pedicel 4–5 mm long; bracts ovate-deltoid, minute, caducous; sepal lobes ovate-deltoid, ca. 1 mm long, glabrous; corolla salverform, white, the tube 6–7(10) mm long, the lobes oblong-elliptic, 4–5 mm long, spreading; stamens subsessile, included in the tube; stigma capitate, the apex 2-lobed. Fruit oblong ovoid, ca. 1 cm long; seed narrowly ovoid.

Tropical hammocks. Rare; Monroe and Miami-Dade Counties. Florida; West Indies and Mexico. All year.

Vallesia antillana is listed as endangered in Florida (Florida Administrative Code, Chapter 5B-40).

EXCLUDED TAXA

Vallesia chiococcoides Kunth—Reported for Florida by Chapman (1860), the name misapplied to material of *V. antillana*.

Vallesia glabra (Cavanilles) Link—Reported for Florida by Chapman (1897), Small (1903, 1913a, 1913d, 1933), and Long and Lakela (1971). Florida material is all of *V. antillana*.

Vinca L. 1753. PERIWINKLE

Perennial herbs, sometimes woody basally, with colorless latex. Leaves opposite, simple, pinnate-nerved, petiolate, estipulate. Flowers solitary, axillary; sepals 5, basally connate; petals basally connate; stamens epipetalous, the filaments and anthers free; nectar glands present; ovary superior, 2-carpellate and -loculate, the style and stigma 1. Fruit a follicle; seeds few.

A genus of 7 species; nearly cosmopolitan. [*Vincire*, to bind, fetter, in reference to the long creeping stems used to make garlands.]

1. Leaves broadly lanceolate, ovate to suborbicular, the margin ciliate, the petiole 1–2 cm long; sepal lobes ca. 9 mm long, the margin ciliate; corolla 1.2–1.5 cm long, the limb 3–5 cm wide .... **V. major**
1. Leaves oblong, elliptic, or ovate, the margin glabrous, the petiole 1–2 mm long; sepal lobes 3–5 mm long, the margin glabrous; corolla 0.8–1.2 mm long, the limb 2.5–3 cm wide .... **V. minor**

Vinca major L. [Greater, in reference to its comparison with V. minor.] BIGLEAF PERIWINKLE.

Vinca major Linnaeus, Sp. Pl. 209. 1753. *Pervinca major* (Linnaeus) Scopoli, Fl. Carniol., ed. 2. 1: 170. 1771. *Vinca grandiflora* Salisbury, Prodr. Stirp. Chap. Allerton 146. 1796, nom. illegit. *Vinca ovatifolia* Stokes, Bot. Mat. Med. 4: 497. 1812, nom. illegit.

Prostrate perennial herb, sometimes woody at the base, to 3 dm; stem trailing and rooting at the nodes, to 5 m long, glabrous. Leaves with the blade lanceolate, ovate to suborbicular, 2–9 cm long, 2–6 cm wide, the apex acute, the base cordate to truncate, the margin entire, ciliate, the upper and lower surfaces glabrous, the petiole 1–2 cm long. Flowers solitary, axillary, the pedicel 3–5 cm long; sepal lobes narrowly triangular, ca. 9 mm long, the margin ciliate; corolla bluish purple, funnelform, the tube 1.2–1.5 cm long, the limb 3–5 cm wide, the lobes obliquely obovate, the apex truncate, spreading; stamens attached near the middle of the tube, the anthers free, incurved and positioned over the stigma, puberulent at the apex; nectar glands 2, on opposite sides of the ovary bases; style filiform, gradually broadened to the apex, the stigma subconic, the base umbraculiform, the apex pubescent. Fruit fusiform, 2–7 cm long, striate; glabrous; seeds 1(2–5) per follicle, 3–7 mm long, narrowly elliptic-ovate, laterally compressed, with a deep longitudinal groove, the surface finely wrinkled, dark brown to black.

Disturbed sites. Rare; Jackson County. Escaped from cultivation. Massachusetts and New York south to Florida, west to California and north to British Columbia; Central America and South America; Europe, Asia, Australia, and Pacific Islands. Native to Europe and Asia. Spring–fall.

Vinca minor L. [Smaller, in reference to its comparison with *V. major.*] COMMON PERIWINKLE.

> *Vinca minor* Linnaeus, Sp. Pl. 209. 1753. *Pervinca minor* (Linnaeus) Scopoli, Fl. Carniol., ed. 2. 170. 1760. *Vinca humilis* Salisbury, Prodr. Stirp. Chap. Allerton 146. 1796, nom. illegit. *Vinca ellipticifolia* Stokes, Bot. Mat. Med. 1: 495. 1812, nom. illegit.

Prostrate perennial herb, sometimes somewhat woody at the base, to 2 dm; stem trailing and rooting at the nodes, to 20 m long, glabrous. Leaves with the blade oblong, ovate, or elliptic, 1–4.5 cm long, 0.5–2.5 cm wide, the apex acute, the base cuneate to rounded, sometimes slightly asymmetric, the margin entire, eciliate, the upper and lower surfaces glabrous, the petiole 1–2 mm long. Flowers solitary, axillary, the pedicel 1–1.5 mm long; sepal lobes narrowly elliptic, 3–5 mm long, glabrous; corolla lilac-blue, funnelform, the tube 8–12 mm long, the limb 2.5–3 cm wide, the lobes obliquely obovate, the apex truncate, spreading; stamens attached near the middle of the tube, the anthers free, incurved and and positioned above the stigma, puberulent at the apex; nectar glands 2, on opposite sides of the ovary bases; style filiform, gradually broadened to the apex, the stigma subconic, the base umbraculiform, the apex pubescent. Fruit fusiform, 2–7 cm long, striate, glabrous; seeds 1(2–5) per follicle, 3–7 mm long, narrowly elliptic-ovate, laterally compressed, with a deep longitudinal groove, the surface finely wrinkled, dark brown to black.

Disturbed sites. Rare; Escambia County. Escaped from cultivation. Nearly cosmopolitan. Native to Europe. Summer.

BORAGINACEAE Juss., nom. cons. 1789. BORAGE FAMILY

Herbs. Leaves alternate, simple, pinnate-veined, petiolate or epetiolate, estipulate. Flowers in terminal cymes or solitary and axillary, actinomorphic or zygomorphic, bisexual, homostylous or heterostylous, sometimes cleistogamous, bracteate or ebracteate; sepals 5, basally connate; petals 5, basally connate; stamens 5, epipetalous, the anthers basifixed, introrse, longitudinally dehiscent; ovary superior, 4-carpellate and -loculate, the style 1, gynobasic or terminal, the stigmas 1, 2, or 4. Fruit of four 1-seeded or two 2-seeded nutlets, drupes, or loculicidal capsules.

A family of about 124 genera and about 2,850 species; nearly cosmopolitan.

The Boraginaceae has traditionally been treated in the broad sense. Takhtajan (1987) recognized the Boraginales with seven families, including Boraginaceae, Cordiaceae, Ehretiaceae, and Hydrophyllaceae, which occur in Florida. Later, Takhtajan (1997) revised the Boraginaceae with five subfamilies, including Boraginoideae, Cordioideae, and Ehretioidese of Florida. This understandably caused confusion. The taxonomy of the Boraginales has since been much revised by Refulio-Rodriquez and Olmstead (2014) and Weigend et al. (2014) using molecular and morphology data, again expanding the number of families. Most recently, the Namaceae was added. This proliferation of families has been accepted by some workers, but others strongly reject this as verified by an online survey (Christenhusz et al., 2015). The Angiosperm Phylogeny Group (2016) recommended that the Boraginaceae s.l. be the sole family in the Boraginales, which we have accepted here. It seems the best practice in a flora such as this is to recognize a large familiar family rather than many smaller families unfamiliar to the user.

Cordiaceae R. Br. ex Dumort., 1829; *Ehretiaceae* Mart., nom. cons., 1817; *Heliotropiaceae* Schrad., nom. cons., 1819; *Hydrophyllaceae* R. Br., nom. cons., 1817; *Namaceae* Molinari, 2016.

Selected reference: Al-Shehbaz (1991).

1. Trees, shrubs, or woody vines; fruit a drupe.
    2. Woody vines (rarely shrubs, *H. verdcourtii*).
        3. Leaves 3–7 cm long, pubescent; fruit strongly lobed .... **Myriopus**
        3. Leaves 10–15 cm long, short-hirsute; fruit not lobed .... **Heliotropium**
    2. Trees or shrubs.
        4. Leaves linear-spatulate, densely silky-tomentose; style obsolete (stigma subsessile), not bifid....
          .... **Heliotropium**
        4. Leaves various, but not as above; style elongate, once or twice bifid.
            5. Style once bifid.
                6. Leaves entire.. **Bourreria**
                6. Leaves serrate or crenate... **Ehretia**
            5. Style twice bifid.
                7. Trees; inflorescence corymbose-cymose .... **Cordia**
                7. Shrubs; inflorescence spicate or globose .... **Varronia**
1. Herbs; fruit of 2 or 4 nutlets or a capsule.
    8. Fruit a capsule.
        9. Leaves pinnate-lobed; styles united to near the middle; fruit 1- or 2-seeded .... **Nemophila**
        9. Leaves not lobed; styles free; fruit many-seeded.. **Nama**
    8. Fruit of four 1-seeded or two 2-seeded nutlets.
        10. Fruit with hooked bristles.
            11. Leaves hirsute; flowers white; fruit with the persistent style inconspicuous............................
              .... **Andersonglossum**
            11. Leaves appressed silky-pubescent; flowers blue; fruit with the persistent style prominent....
              .... **Cynoglossum**
        10. Fruit without hooked bristles.
        12. Calyx with uncinate trichomes.. **Myosotis**
        12. Calyx with straight trichomes.
            13. Cymes short.
                14. Corolla white to bluish.. **Buglossoides**
                14. Corolla pale yellow to bright orange-yellow... **Lithospermum**
            13. Cymes elongate.
                15. Flowers bracteate; anthers coherent at the apex; fruit of four 1-seeded nutlets............
                  .... **Euploca**
                15. Flowers ebracteate; anthers free; fruit of two 2-seeded nutlets .... **Heliotropium**

Andersonglossum J. I. Cohen 2015.

Perennial herbs. Leaves alternate, simple, pinnate-veined, petiolate, estipulate. Flowers in terminal paniculiform, scorpioid cymes, actinomorphic, bracteate; sepals 5, basally connate; petals 5, basally connate, with scalelike appendages in the throat; stamens 5, epipetalous; ovary 4-carpellate and -loculate, the style gynobasic, the stigma capitate. Fruit of four 1-seeded nutlets, the surface covered with hooked bristles.

A genus of 3 species; North America. [Commemorates William Russell Anderson (1942–2013), professor and curator of the herbarium at the University of Michigan, and *glossum*, to note its segregation from the genus *Cynoglossum*.]

Selected reference: Cohen (2015).

Andersonglossum virginianum (L.) J. I. Cohen [Of Virginia.] WILD COMFREY.

Cynoglossum virginianum Linnaeus, Sp. Pl. 134. 1753. *Cynoglossum lucidum* Stokes, Bot. Mat. Med. 1: 277. 1812, nom. illegit. *Andersonglossum virginianum* (Linnaeus) J. I. Cohen, Syst. Bot. 40: 618. 2015.

Erect perennial herb, to 8 dm; stem hirsute. Leaves with the blade of the basal leaves elliptic-oblong, 10–20 cm long, 5–10 cm wide, the apex obtuse, the base cuneate, the margin entire or obscurely dentate, the upper and lower surfaces hirsute, the petiole 6–10 cm long, the cauline leaves progressively smaller upward, sessile, some clasping the stem. Flowers in a paniculiform, scorpioid cyme, the pedicel 5–15 mm long; calyx 3–5 mm long; corolla funnelform, white, 8–12 mm long, the throat constricted with scalelike appendages at the base of each lobe; stamens inserted below the appendages. Fruit of four 1-seeded nutlets, these ovoid, 6–8 mm long, densely covered with hooked bristles.

Bluff forests. Rare; Gadsden and Liberty Counties. Newfoundland south to Florida, west to Yukon, British Columbia, North Dakota, South Dakota, Iowa, Illinois, Oklahoma, and Texas. Spring.

Andersonglossum virginianum (as *Cynoglossum virginianum*) is listed as endangered in Florida (Florida Administrative Code, Chapter 5B-40).

Bourreria P. Browne, nom. Cons. 1756. STRONGBARK

Trees or shrubs. Leaves alternate, simple, pinnate-veined, petiolate, estipulate. Flowers in terminal and upper axillary, racemiform cymes or solitary, actinomorphic or slightly zygomorphic; sepals (4)5, basally connate; petals 5, basally connate; stamens 5, epipetalous; ovary 4-carpellate and -loculate; style terminal, the stigma 2-lobed. Fruit a drupe with four 1-seeded pyrenes.

A genus of about 30 species; North America, West Indies; Mexico, Central America, and South America. [Commemorates Johann Ambrosius Beurer (1716–1754), German apothecary and natural history promoter.]

Bourreria is sometimes placed in the Ehretiaceae by various authors.

1. Upper leaf surface glabrous ....**B. succulenta**
1. Upper leaf surface conspicuous hispid or strigose.
    2. Fruit 6–9 mm long; petiole 1–2(3) mm long ....**B. cassinifolia**
    2. Fruit 1–1.5 cm long; petiole (2)3–7 mm long... **B. radula**

Bourreria cassinifolia (A. Rich.) Griseb. [With leaves like *Ilex cassine* (Aquifoliaceae).] SMOOTH STRONGBARK; LITTLE STRONGBARK.

> *Ehretia cassinifolia* A. Richard, in Sagra, Hist. Fis. Cuba, Bot. 11: 113. 1850. *Bourreria cassinifolia* (A. Richard) Grisebach, Mem. Amer. Acad. Arts, ser. 2. 8: 528. 1863. *Morelosia cassinifolia* (A. Richard) Kuntze, Revis. Gen. Pl. 2: 439. 1891.

Shrub or small tree, to 5 m; branchlets sparsely pubescent or glabrescent, reddish, the bark dark gray, striate, rugulose. Leaves with the blade spatulate to oblanceolate or narrowly obovate, (0.6)1–2 cm long, (3)6–9 mm wide, subcoriaceous, the apex acute to rounded, the base cuneate, the margin entire, slightly revolute, the upper surface hispid, the lower surface glabrate, the petiole 1–2(3) mm long. Flowers few in a racemiform cyme or solitary, the pedicel 2–4 mm long; calyx campanulate, 4–6 mm long, triangular, irregular, acute; corolla salverform, white, ca. 1 cm long, the lobes suborbicular; stamens borne on the corolla tube below the throat, exserted, glabrous; ovary 2-loculate or incompletely 4-loculate, the stigma lobes flattened. Fruit subglobose, 6–9 mm long, orange-red.

Pinelands; Rare; Miami-Dade County, Monroe County keys. Florida; West Indies. All year.

Bourreria cassinifolia is listed as endangered in Florida (Florida Administrative Code, Chapter 5B-40).

Bourreria radula (Poir.) G. Don [Rasp, rough, in reference to the leaf surface.] ROUGH STRONGBARK.

> *Ehretia radula* Poiret, in Lamarck, Encycl., Suppl. 2: 2. 1811. *Bourreria radula* (Poiret) G. Don, Gen. Hist. 4: 390. 1838. *Bourreria havanensis* (Roemer & Schultes) Miers var. *radula* (Poiret) A. Gray, Syn. Fl. N. Amer. 2(1): 1878. *Morelosia radula* (Poiret) Kuntze, Revis. Gen. Pl. 2: 439. 1891.
> *Cordia floridana* Nuttall, N. Amer. Sylv. 3: 107. 1849. TYPE: FLORIDA.

Shrub or small tree; branchlets glabrous or minutely pubescent, reddish, the bark dark gray, longitudinally fissured. Leaves with the blade obovate, 2.5–5 cm long, 1–3 cm wide, subcoriaceous, the apex obtuse or retuse, the base cuneate, the margin entire, the upper surface tuberculate-scabrous or hispidulous-papillose, the lower surface glabrous, the petiole 4–6 mm long. Flowers few to many in a cyme 3–9 cm long, the pedicel 2–3 mm long; calyx campanulate, 6–8 mm long, the lobes lanceolate, irregular, acute; corolla salverform, white, ca. 1 cm long, the lobes suborbicular; stamens borne on the corolla tube below the throat, exserted, glabrous. Fruit subglobose, 1–1.5 cm long, orange-red.

Hammocks. Rare; Monroe County keys. Florida; West Indies. All year.

Bourreria radula is listed as endangered in Florida (Florida Administrative Code, Chapter 5B-40).

Bourreria succulenta Jacq. [Fleshy.] BAHAMA STRONGBARK; BODYWOOD.

Bourreria succulenta Jacquin, Enum. Syst. Pl. 14. 1760.
Bourreria revoluta Kunth, in Humboldt et al., Nov. Gen. Sp. 3: 67. 1818. *Ehretia revoluta* (Kunth) de Candolle, Prodr. 9: 507. 1845. *Crematomia revoluta* (Kunth) Miers, Ann. Mag. Nat. Hist., ser. 4. 3: 309. 1869; Contr. Bot. 2: 251. 1869. *Morelosia revoluta* (Kunth) Kuntze, Revis. Gen. Pl. 2: 439. 1891. *Bourreria succulenta* Jacquin var. *revoluta* (Kunth) O. E. Schulz, in Urban, Symb. Antill. 7: 59. 1911.
Ehretia havanensis Roemer & Schultes, Syst. Veg. 4: 805. 1819. *Bourreria tomentosa* (Lamarck) G. Don var. *havanensis* (Roemer & Schultes) Grisebach, Fl. Brit. W.I. 482. 1861. *Bourreria havanensis* (Roemer & Schultes) Miers, Ann. Mag. Nat. Hist., ser. 4. 3: 207. 1869; Contr. Bot. 2: 238. 1869. *Ehretia tomentosa* Lamarck var. *havanensis* (Roemer & Schultes) M. Gómez de la Maza y Jiménez, Anales Soc. Esp. Hist. Nat. 19: 256. 1890. *Morelosia havanensis* (Roemer & Schultes) Kuntze, Revis. Gen. Pl. 2: 439. 1891.
Bourreria ovata Miers, Ann. Mag. Nat. Hist., ser. 4. 3: 203. 1869; Contr. Bot. 2: 234. 1869. *Morelosia ovata* (Miers) Kuntze, Revis. Gen. Pl. 2: 439. 1891.

Shrub or small tree, to 6(10) m; branchlets glabrous or slightly pubescent, the bark reddish brown. Leaves with the blade oblanceolate to ovate, suborbicular, or elliptic, 4–12 cm long, 2.5–7.5(10) cm wide, subcoriaceous, the apex acute to rounded or emarginate, the base cuneate to obtuse, the margin entire, the upper and lower surfaces glabrous or slightly pubescent, the petiole 1–4 cm long. Flowers numerous in a cyme 5–10 cm long and wide, the pedicel 2–8 mm long; calyx campanulate, 5–6 mm long, the lobes 5, ovate, irregular, acute; corolla salverform, white, ca. 1 cm long, the tube ca. 6 mm long, the lobes suborbicular; stamens borne on the corolla tube below the throat, exserted, glabrous; ovary 2-loculate or incompletely 4-loculate, the stigma lobes flattened. Fruit subglobose, 1–1.5 cm long, orange-red, the 4 pyrenes ridged on the dorsal side.

Hammocks. Rare; Miami-Dade County, Monroe County keys. Florida; West Indies, Mexico, Central America, and South America. All year.

Bourreria succulenta is listed as endangered in Florida (Florida Administrative Code, Chapter 5B-40).

EXCLUDED TAXON

Bourreria baccata Linnaeus—This Jamaican species was reported for Florida by Chapman (1860, as *Ehretia bourreria* (Linnaeus) Linnaeus), the name misapplied to material of *B. succulenta*.

Buglossoides Moench 1794.

Annual herbs. Leaves alternate, simple, pinnate-veined, petiolate or epetiolate, estipulate. Flowers solitary, axillary, actinomorphic or slightly zygomorphic, bracteate; sepals 5, basally connate; petals 5, basally connate; stamens 5, epipetalous; ovary 4-carpellate and -loculate, the style gynobasic, the stigma 2-lobed. Fruit of four 1-seeded nutlets.

A genus of about 7 species; North America, Europe, Asia, and Africa. [From the Greek *buglossum*, ox tongue, and the Latin *oides*, resembling, in reference to the broad, rough leaves resembling an ox tongue.]

Buglossoides arvensis (L.) I. M. Johnston [Of fields.] CORN GROMWELL.

Lithospermum arvensis Linnaeus, Sp. Pl. 132. 1753. *Buglossoides arvensis* (Linnaeus) I. M. Johnston, J. Arnold Arbor. 35: 42. 1954.

Erect annual herb, to 4(6) dm; stem strigose. Leaves with the blade linear to oblanceolate, 1–3 cm long, 4–6 mm wide, the apex acute to obtuse, the upper and lower surfaces strigose, the margin entire, sessile or short-petiolate. Flowers solitary, axillary, eventually appearing as a spiciform raceme, the pedicel 1–2 mm long; bracts foliose; calyx 5–6 mm long, the lobes linear, the outer surface pubescent; corolla funnelform, white to bluish, 6–8 mm long, slightly exceeding the calyx, the tube 4–5 mm long, the outer surface pubescent, the inner surface pubescent to about the middle, the lobes oblong-ovate; stamens subsessile, attached on the lower half of the tube, the anthers ca. 1 mm long, prominently apiculate. Fruit of four 1-seeded nutlets, these angular-ovoid, 2–3 mm long, brownish gray or tan, rugose and pitted with darker depressions.

Disturbed sites. Occasional; northern counties, Hillsborough County. Nearly throughout North America; Europe, Asia, and Africa. Native to Europe, Asia, and Africa. Spring.

Cordia L., nom. cons. 1753.

Shrubs or trees. Leaves alternate, simple, pinnate-veined, petiolate, estipulate. Flowers in terminal or axillary scorpioid cymes, actinomorphic or slightly zygomorphic, bisexual, homostylous or heterostylous; sepals (4)5(6), basally connate; petals (4)5(6), basally connate; stamens (4)5(6), epipetalous; ovary 4-carpellate and -loculate, the style terminal, twice bifid, the stigmas 4. Fruit a drupe.

A genus of about 250 species; nearly cosmopolitan. [Commemorates Valerius Cordus (1515–1544), German physician and botanist.]

Cordia is sometimes placed in the Cordiaceae or Ehretiaceae by various authors. *Sebesten* Adans., 1763.

1. Upper leaf surface glabrous; corolla tube less than 0.5 cm long, white .... **C. dichotoma**
1. Upper leaf surface scabrous; corolla tube 2–3 cm long, bright orange ....**C. sebestena**

Cordia dichotoma G. Forst. [Branched in pairs, in reference to the style.] FRAGRANT MANJACK.

Cordia dichotoma G. Forster, Fl. Ins. Austr. 18. 1786.

Tree, to 4 m; branchlets glabrous. Leaves with the blade ovate or elliptic, 6–13 cm long, 4–9 cm wide, the apex obtuse, the base rounded, the margin undulate-dentate, the upper and lower surfaces glabrous, the petiole 2–5 cm long. Flowers in a corymbose cyme, heterostylous, the pedicel 2–3 mm long, the peduncle 2–3 cm long; calyx campanulate, 5–6 mm long, 5-lobed, the lobes triangular, small, unequal, glabrous; corolla funnelform, white, 5–6 mm long, the lobes elliptic, subequaling the tube; stamens 1–2 mm long, included; style included or exserted. Fruit subglobose, 1–1.5 cm long, yellow or red, enclosed at the base by the accrescent calyx.

Disturbed sites. Palm Beach and Miami-Dade Counties. Escaped from cultivation. Florida; Asia, Australia, and Pacific Islands. Native to Asia, Australia, and Pacific Islands. All year.

Cordia sebestena L. [*Sebestan*, an Arabic vernacular name originally applied to a different plant.] LARGELEAF GEIGERTREE.

Cordia sebestena Linnaeus, Sp. Pl. 190. 1753. *Cordia speciosa* Salisbury, Prodr. Stirp. Chap. Allerton 111. 1796, nom. illegit. *Lithocardium sebestena* (Linnaeus) Kuntze, Revis. Gen. Pl. 2: 976. 1891. *Sebesten sebestena* (Linnaeus) Britton ex Small, Fl. Miami 158, 200. 1913.

Shrub or small tree, to 8 m; branchlets brown-pubescent. Leaves with the blade ovate to ovate-elliptic, 8–20 cm long, 4–15 cm wide, the apex acute, the base rounded to subcordate, the margin entire or undulate to somewhat dentate distally, the upper surface scabrous, the lower surface somewhat scabrous to pubescent. Flowers in a pedunculate paniculiform or dichotomous, few-flowered corymb or cyme, heterostylous, the pedicel usually 3–6(15) mm long, the peduncle 3–7 cm long; calyx tubular-campanulate, 1–1.8 cm long, rupturing irregularly and usually 2- to 3-lobed at anthesis, the outer surface strigose-pubescent, the inner surface glabrous; corolla funnelform, bright orange or red-orange, (2)3–4 cm long, the tube 2–3 cm long, the outer and inner surfaces glabrous; stamens included or slightly exserted; style to ca. 2 cm long, included or exserted. Fruit ovoid, 2–4 cm long, dry, hard, enclosed in the white, fleshy, accrescent calyx.

Coastal hammocks. Rare; Lee County, southern peninsula. Escaped from cultivation. Florida; West Indies, Mexico, Central America, and South America. Native to West Indies. Summer–fall.

EXCLUDED TAXON

Cordia myxa Linnaeus—Reported for Florida by Wunderlin (1998), based on a misidentification of material of *C. dichotoma*.

Cynoglossum L. 1753. HOUND'S TONGUE

Biennial herbs. Leaves alternate, simple, pinnate-veined, petiolate or epetiolate, estipulate. Flowers in terminal racemiform or paniculiform, scorpioid cymes, actinomorphic, bracteate; sepals 5, basally connate; petals 5, basally connate, with scalelike appendages in the throat; stamens 5, epipetalous; ovary 4-carpellate and -loculate, the style gynobasic, the stigma capitate. Fruit of four 1-seeded nutlets, the surface covered with hooked bristles.

A genus of about 70 species; nearly cosmopolitan. [From the Greek *cynos*, dog, and *glossum*, tongue, in reference to the shape and texture of the leaves.]

Cynoglossum zeylanicum (Lehm.) Brand [Of Ceylon.] CEYLON HOUND'S TONGUE.

Anchusa zeylanica Vahl ex Hornemann, Hort. Bot. Hafn. 1: 176. 1813; non J. Jacquin, 1812. *Myosotis zeylanica* Lehmann, Neue Schriften Naturf. Ges. Halle 3(2): 21. 1817. *Echinospermum zeylanicum* (Lehmann) Lehmann, Pl. Asperif Nucif. 116. 1818. *Cynoglossum denticulatum* A. de Candolle var. *zeylanicum* (Lehmann) C. B. Clarke, in Hooker f., Fl. Brit. Ind. 4: 157. 1883. *Cynoglossum zeylanicum* (Lehmann) Brand, in Engler, Pflanzenr. 4(Heft 78): 134. 1921.

Erect biennial herb, to 8 dm; stem strigose. Basal leaves with the blade oblong to elliptic, 15–20 cm long, 3–5 cm wide, the apex obtuse, the base cuneate, the margin entire or obscurely dentate, the upper and lower surfaces densely appressed silky-pubescent, the petiole to 12 cm long, the cauline leaves like the basal but smaller and sessile. Flowers in a paniculiform scorpioid cyme, the pedicel 1–2 mm long, densely strigose; calyx with the outer surface strigose, the lobes ovate, 3–4 mm long; corolla funnelform, blue, ca. 5 mm long, the throat constricted by scale-like appendages at the base of each lobe; stamens inserted below the appendages. Fruit of four 1-seeded nutlets, these ovoid, 3–4 mm long, densely covered with hooked bristles.

Disturbed sites. Rare; Palm Beach and Lee Counties. Florida west to Texas; West Indies and South America; Asia, Australia, and Pacific Islands. Native to Asia. Spring–summer.

EXCLUDED TAXON

Cynoglossum furcatum Wallich—Reported for Florida by Wunderlin (1982), the name misapplied to material of *C. zeylanicum*.

Ehretia P. Browne 1756.

Trees. Leaves alternate, simple, pinnate-veined, petiolate, estipulate. Flowers in paniculiform cymes, actinomorphic, bisexual; sepals 5, basally connate; petals 5, basally connate; stamens 5, epipetalous; ovary 2-carpellate and -loculate, the style terminal, the stigmas 2. Fruit a drupe.

A genus of about 50 species; North America, West Indies, Africa, Asia, and Australia. [Commemorates G. D. Ehret, eighteenth-century, German-born botanical artist.]

Ehretia is sometimes placed in the Ehretiaceae by various authors.

1. Leaf margin regularly serrate.. E. acuminata
1. Leaf margin coarsely crenate .... E. microphylla

Ehretia acuminata R. Br. [Tapering to a point, in reference to the leaf apex.] KODO.

Ehretia acuminata R. Brown, Prodr. 497. 1810.

Tree, to 8 m; branchlets brown, glabrous, the bark gray-black, laciniate. Leaves with the blade elliptic to obovate, 5–13 cm long, 4–6 cm wide, the apex acute, the base broadly cuneate, the margin regularly upward serrate, the upper and lower surfaces glabrous, the petiole 1.5–2.5 cm long. Flowers in a paniculiform cyme 8–15 cm long, short-pubescent to subglabrous; calyx ca. 2 mm long, the lobes ovate, ciliate; corolla campanulate, white, 3–4 mm long, the lobes oblong, longer than the tube; stamens 3–4 mm long, exserted; style 2–3 mm long. Fruit subglobose, 3–4 mm long, yellow or orange, the endocarp divided into two 2-seeded pyrenes, the endocarp surface wrinkled.

Disturbed mesic hammocks. Rare; Alachua County. Escaped from cultivation. Florida; Asia and Australia. Native to Asia and Australia. Summer.

Ehretia microphylla Lam. [With small leaves.] FUKIEN TEA.

Ehretia microphylla Lamarck, Tabl. Encycl. 1: 425. 1792. *Carmona microphylla* (Lamarck) G. Don, Gen. Hist. 4: 391. 1837. *Ehretia buxifolia* Roxbury var. *microphylla* (Lamarck) de Candolle, Prodr. 9: 509. 1845.

Shrub, to 3 m; branchlets hirtellous, soon glabrescent, the bark dark brown. Leaves with the blade obovate to spatulate, 1.5–3.5 cm long, 1–2 cm wide, coriaceous, the apex rounded to acute, the base cuneate, the margin entire or 1- to 4-toothed distally, the upper surface hirtellous, the lower surface glabrous, the petiole 2–3 mm long. Flowers 2–6 in a compact scorpioid cyme 0.5–1.5 cm wide, elongating in fruit, the peduncle 1–1.5 cm long, pubescent, the pedicel obsolete or to 2 mm long; calyx 4–6 mm long, lobed nearly to the base, the lobes linear or linear-oblanceolate, 0.5–1 mm long, attenuate below the middle, the outer surface hirtellous, the inner surface pubescent; corolla campanulate, white or slightly reddish, 4–6 mm long, the lobes oblong, spreading, longer than the tube; stamens inserted near the corolla base, 4–5 mm long, exserted; style 4–6 mm long, glabrous, bifid to below the middle, the stigmas capitate. Fruit subglobose, 3–4 mm long, red or yellow, short-beaked, the endocarp bony, 2–3 mm long, 4-seeded, reticulate-wrinkled.

Disturbed sites. Rare; Miami-Dade County. Escaped from cultivation. Florida; Asia, Australia, and Pacific Islands. Native to Asia and Australia. Spring.

Euploca Nutt. 1836.

Annual or perennial herbs. Leaves alternate, simple, pinnate-veined, petiolate or epetiolate, estipulate. Flowers in terminal or axillary scorpioid cymes, actinomorphic, bracteate; sepals 5, basally connate; petals 5, basally connate; stamens 5, epipetalous, the anthers apically connate; ovary 4-carpellate and -loculate, the style terminal, obsolete, the stigma with a sterile appendage. Fruit of four 1-seeded nutlets.

A genus of about 120 species; North America, West Indies, Mexico, Central America, South America, Africa, Asia, Australia, and Pacific Islands. [From the Greek *eu*, true, and *ploce*, to plait, in reference to the corolla.]

Euploca is sometimes placed in the Heliotropaceae by various authors.

1. Leaves sessile or the petiole to 2 mm long..**E. polyphylla**
1. Leaves distinctly petiolate.
  2. Cymes with foliose bracts ....**E. fruticosa**
  2. Cymes with small bracts ....**E. procumbens**

Euploca fruticosa (L.) J.I.M. Melo & Semir [Shrubby.] KEY WEST HELIOTROPE.

Heliotropium fruticosum Linnaeus, Syst. Nat., ed 10. 913. 1759. *Pioctonum antillanum* Rafinesque, Sylva Tellur. 88. 1838, nom. illegit. *Euploca fruticosa* (Linnaeus) J.I.M. Melo & Semir, Kew Bull. 64: 288. 2009.
Heliotropium phyllostachyum Torrey, in Emory, Rep. U.S. Mex. Bound. 2(1): 137. 1858 ("1859").
Heliotropium myosotoides Chapman, Fl. South. U.S. 330. 1860; non Lehmann, 1817. TYPE: FLORIDA.

Erect annual herb, to 3 dm; stem sparsely strigose. Leaves with the blade oblanceolate or lanceolate, 1–2 cm long, 2–7 mm wide, the apex acute, the base cuneate, the margin entire, the upper and lower surfaces strigose, the petiole 2–5 mm long. Flowers in a terminal scorpioid cyme 1–5 cm long, the pedicel ca. 1 mm long, strigose; bracts irregularly developed and foliose; calyx ca. 1 mm long, the lobes lanceolate, unequal; corolla salverform, white, 3–4 mm long, the outer surface slightly hispid. Fruit subglobose, ca. 1 mm long, separating into four 1-seeded nutlets, the nutlets ovate, the sides concave, the apex villous.

Hammocks and disturbed sites. Rare; Monroe County keys. Florida and Texas west to Arizona; West Indies, Mexico, Central America, and South America. All year.

Euploca fruticosa (as *Heliotropium fruticosum*) is listed as endangered in Florida (Florida Administrative Code, Chapter 5B-40).

Euploca polyphylla (Lehm.) J.I.M. Melo & Semir [Many-leaved.] PINELAND HELIOTROPE.

> *Heliotropium polyphyllum* Lehmann, Neue Schriften Naturf. Ges. Halle 3(2): 9. 1817. *Schleidenia polyphylla* (Lehmann) Fresenius, in Martius, Fl. Bras. 8(1): 36. 1857. *Heliotropium polyphyllum* Lehmann var. *genuinum* I. M. Johnston, Contr. Gray Herb. 81: 63. 1928, nom. inadmiss. *Euploca polyphylla* (Lehmann) J.I.M. Melo & Semir, Kew Bull. 64: 289. 2009.
>
> *Lithospermum floridanum* Rafinesque, New. Fl. 4: 18. 1838 ("1836"). TYPE: FLORIDA.
>
> *Heliotropium polyphyllum* Lehmann var. *leavenworthii* A. Gray, Proc. Amer. Acad. Arts 10: 49. 1874. *Heliotropium leavenworthii* (A. Gray) Torrey ex Small, Fl. S.E. U.S. 1006. 1903. TYPE: FLORIDA: Miami-Dade Co.
>
> *Heliotropium horizontale* Small, Bull. New York Bot. Gard. 3: 435. 1905. *Heliotropium polyphyllum* Lehmann var. *horizontale* (Small) R. W. Long, Rhodora 72: 33. 1970. TYPE: FLORIDA: Miami-Dade Co.: between Cutler and Camp Longview, Nov 1903, *Small & Carter 742* (holotype: NY; isotype: US).

Perennial herb, sometimes suffrutescent, to 6 dm; stem strigose pubescent. Leaves with the blade linear-lanceolate to narrowly elliptic, 1–2 cm long, 2–5 mm wide, the apex acute, the base obtuse, the margin entire, the upper and lower surfaces strigose-pubescent, subsessile or the petiole to 2 mm long. Flowers in a terminal scorpioid cyme to 10 cm long; bracts small; calyx 2–3 mm long, the lobes ovate-lanceolate, the outer surface strigose-pubescent; corolla salverform, white with a yellow eye or all yellow, the tube 3–5 mm long. Fruit ovoid, 2–3 mm long, 4-lobed, separating into four 1-seeded nutlets, short-strigose.

Wet flatwoods, pond margins, and wet disturbed sites. Frequent; peninsula, eastern panhandle, Escambia County. Florida; West Indies and South America. All year.

Euploca procumbens (Mill.) Diane & Hilger [Prostrate.] FOURSPIKE HELIOTROPE.

> *Heliotropium procumbens* Miller, Gard. Dict., ed. 8. 1768. *Euploca procumbens* (Miller) Diane & Hilger, Bot. Jahrb. Syst. 125: 48. 2003.
>
> *Heliotropium canescens* Lehmann, Nov. Acta Phys.-Med. Acad. Caes. Leop.-Carol. Nat. Cur. 9: 88. 1818; Pl. Asperif. Nucif. 38. 1818; non Moench, 1794.

Erect to decumbent, annual or short-lived perennial herb, somewhat suffrutescent at the base; stem to 5 dm long, strigose or appressed pilose. Leaves with the blade narrowly oblanceolate or obovate to elliptic, 1–4 cm long, 0.5–2 cm wide, the apex rounded to obtuse or acute, the base cuneate, the margin entire, flat or revolute, the upper and lower surfaces strigose or appressed-pilose, the petiole 0.5–1 cm long. Flowers in a terminal or axillary solitary or paired scorpioid cyme 2–10 cm long, sessile or subsessile, the peduncle 1–3 cm long; bracts small; calyx ca. 1 mm long, the lobes linear to lanceolate, unequal, strigose or appressed-pilose; corolla salverform, white, the tube ca. 1 mm long, the outer surface sparsely strigose, the inner surface villous in the throat. Fruit subglobose, 1–2 mm long, strigillose, separating into four 1-seeded nutlets.

Floodplain forests. Rare; southern peninsula, Liberty and Calhoun Counties. Maryland, Florida west to California; West Indies, Mexico, Central America, and South America. Native to tropical America. Spring–summer.

Heliotropium L. 1753. HELIOTROPE

Shrubs, woody vines, or herbs. Leaves alternate, simple, pinnate-veined, petiolate or epetiolate, estipulate. Flowers in terminal or axillary scorpioid cymes, actinomorphic, ebracteate; sepals 5, basally connate; petals 5, basally connate; stamens 5, epipetalous; ovary 4-carpellate and -loculate, the style terminal, sometimes obsolete, the stigma with a sterile apical appendage. Fruit of two 2-seeded nutlets or a drupe.

A genus of about 140 species; nearly cosmopolitan. [From the Greek *helios*, sun, and *tropos*, turn, in the belief that the flowers of various species turn toward the sun.]

Heliotropium is sometimes placed in the Heliotropiaceae by various authors.

Argusia Boehm., 1760; *Cochranea* Miers, 1868; *Mallotonia* (Griseb.) Britton, 1915; *Schobera* Scop., 1777; *Tiaridium* Lehm., 1818; *Tournefortia* L., 1753.

1. Shrubs or woody vines.
  2. Leaves linear-spatulate, succulent, sessile, the surfaces densely silky-tomentose; plant a shrub.......
  .... **H. gnaphalodes**
  2. Leaves lanceolate, ovate, or elliptic, petiolate, the surface various, but not silky-tomentose; plant a woody vine or rarely a shrub...**H. verdcourtii**
1. Herbs.
  3. Leaves glabrous, succulent...**H. curassavicum**
  3. Leaves variously with trichomes, chartaceous.
    4. Leaves sessile or subsessile (petiole to 1 mm long) ....**H. amplexicaule**
    4. Leaves distinctly petiolate.
      5. Corolla blue or violet, rarely white; fruit with sharp longitudinal ridges.............**H. indicum**
      5. Corolla white; fruit without longitudinal ridges.
        6. Corolla with the outer surface glabrous, the inner surface bearded in the throat; fruit covered with minute scales, smooth..**H. angiospermum**
        6. Corolla with the outer surface short-pilose, the inner surface glabrous; fruit glabrous, tuberculate ....**H. europaeum**

Heliotropium amplexicaule Vahl [With the leaves clasping the stem.] CLASPING HELIOTROPE.

Heliotropium amplexicaule Vahl, Symb. Bot. 3: 21. 1794. *Heliophytum amplexicaule* (Vahl) Britton & P. Wilson, Bot. Porto Rico 2: 560. 1930.

Heliotropium anchusifolium Poiret, in Lamarck, Encycl., Suppl. 3: 23. 1813. *Heliophytum anchusifolium* (Poiret) de Candolle, Prodr. 9: 554. 1845. *Cochranea anchusifolia* (Poiret) Gürke, in Engler & Prantl, Nat. Pflanzenfam. 4(3a): 97. 1893. *Heliophytum anchusifolium* (Poiret) de Candolle var. *latifolium* de Candolle, Prodr. 9: 554. 1845, nom. inadmiss. *Heliotropium anchusifolium* Poiret var. *latifolium* (de Candolle) Kuntze, Revis. Gen. Pl. 3(2): 205. 1898, nom. inadmiss. *Cochranea anchusifolia* (Poiret) Gürke var. *latifolia* (de Candolle) Hicken, Chlor. Plat. Argent. 194. 1910, nom. inadmiss.

Decumbent perennial herb, sometimes suffrutescent at the base; stem to 4 dm long, hirsute and glandular-pubescent. Leaves with the blade oblong to oblong-lanceolate, 2–6 cm long, 0.4–2 cm wide, the apex acute, the base cuneate, the margin entire, repand, the upper and lower surfaces rugose, hirsute, and glandular-pubescent, sessile or subsessile. Flowers in a terminal or axillary scorpioid cyme 4–8 cm long, the rachis hirsute and glandular-pubescent; ebracteate; calyx ca. 2 mm long, the lobes lanceolate, the outer surface hirsute and glandular-pubescent; corolla salverform, purple with a yellow eye, ca. 3 mm long, the outer surface puberulent, the inner surface pubescent with white, antrorse trichomes in the tube. Fruit ca. 2 mm long, longitudinally constricted and 2-lobed, somewhat laterally compressed, the two 2-seeded nutlets sometimes separating.

Disturbed sites. Occasional; northern counties, central peninsula. Massachusetts, Pensylvania, New Jersey, and Virginia south to Florida, west to Texas, also California; South America; Europe and Australia. Native to South America. Spring–summer.

Heliotropium angiospermum Murray [With covered seeds, in reference to the nutlets conspicuously covered by epidermal scales.] SCORPION TAIL.

Heliotropium angiospermum Murray, Prodr. Stirp. Gott. 217. 1770. *Schobera angiosperma* (Murray) Britton, in Britton & P. Wilson, Bot. Porto Rico 2: 134. 1925.

Heliotropium parviflorum Linnaeus, Mant. Pl. 201. 1771. *Heliophytum parviflorum* (Linnaeus) de Candolle, Prodr. 9: 553. 1845.

Erect or decumbent annual or short-lived perennial herb, sometimes suffrutescent at the base, to 8(15) cm; stem hispidulous. Leaves with the blade ovate, elliptic, or broadly lanceolate, 2–10(12) cm long, 1–3.5(5) cm wide, the apex acute, the base cuneate to rounded, sometimes decurrent, the margin entire, the upper and lower surfaces sparsely hispidulous, the petiole 5–10 mm long. Flowers solitary or paired in a terminal or subterminal, erect, scorpioid cyme to 15 cm long, the peduncle 1–5 cm long, the pedicel obsolete; ebracteate; calyx with the lobes lanceolate to oblong, 1–2 mm long, equal or subequal; corolla salverform, white, 2–3 mm long, bearded in the throat. Fruit ca. 2 mm long, laterally compressed, covered with minute scales, separating into two 2-seeded nutlets.

Hammocks, shell middens, and disturbed sites. Frequent; central and southern peninsula. Florida and Texas; West Indies, Mexico, Central America, and South America. All year.

Heliotropium curassavicum L. [Of Curaçao.] SEASIDE HELIOTROPE; SALT HELIOTROPE.

Heliotropium curassavicum Linnaeus, Sp. Pl. 130. 1753. *Heliotropium glaucum* Salisbury, Prodr. Stirp. Chap. Allerton 113. 1796, nom. illegit. *Heliotropium glaucophyllum* Moench, Suppl. Meth. 147. 1802, nom. illegit. *Heliotropium angustifolium* Rafinesque, Herb. Raf. 79. 1803, nom. illegit. *Heliotropium curassavicum* Linnaeus var. *genuinum* I. M. Johnston, Contr. Gray Herb. 81: 14. 1928, nom. inadmiss.

Prostrate or decumbent annual or perennial herb; stem to 4 dm long, glabrous, somewhat glaucous. Leaves with the blade narrowly oblanceolate, linear-oblong to linear, succulent, 2–4(5) cm long, 4–8(10) mm wide, the apex obtuse to rounded, the base narrowly cuneate, the margin entire, the upper and lower surfaces glabrous, somewhat glaucous, sessile. Flowers in a terminal or axillary scorpioid cyme 1–5 cm long, the rachis somewhat broad and flattened, glabrous, sessile or subsessile; ebracteate; calyx lobed nearly to the base, the lobes lanceolate to oblong, 1–2 mm long, equal or subequal, glabrous, succulent; corolla salverform, white, 2–3 mm long, the tube shorter than the lobes, glabrous. Fruit subglobose, 4-lobed, 2–3 mm long, slightly laterally compressed, smooth, glabrous, the nutlets with a thick, firm exocarp, separating at maturity into two 2-seeded nutlets ca. 2 mm long.

Coastal dunes, rarely inland. Frequent; peninsula, Franklin County. Nearly throughout North America; West Indies, Mexico, Central America, and South America; Australia. Native to North America and tropical America. Spring–summer.

Heliotropium europaeum L. [Of Europe.] EUROPEAN HELIOTROPE.

Heliotropium europaeum Linnaeus, Sp. Pl. 130. 1753. *Heliotropium canescens* Moench, Methodus 415. 1794, nom. illegit. *Heliotropium humile* Salisbury, Prodr. Stirp. Chap. Allerton 113. 1796, nom. illegit.; non Lamarck, 1792.

Erect or ascending annual herb, to 5 dm; stem strigose or hirtellous. Leaves with the blade elliptic or elliptic-ovate, 1.5–4 cm long, 1–2.5 cm wide, the apex obtuse to acute, the base cuneate, the margin entire, the upper surface sparsely hirtellous, the lower surface densely hirtellous, the petiole 0.5–1.5 cm long. Flowers in a terminal or axillary, scorpioid, simple or dichotomously branched cyme 2–4 cm long, sessile; ebracteate; calyx with the lobes ovate to ovate-lanceolate, 2–3 mm long, the outer surface strigose; corolla salverform, white, 4–5 mm long, the throat slightly contracted, the outer surface short-pilose, the inner surface glabrous; stigma apex deeply 2-cleft, short-strigose. Fruit subglobose, ca. 2 mm long, separating into two 2-seeded nutlets, tuberculate, glabrous.

Disturbed sites. Rare; Escambia County. Not recently collected. Massachusetts and New York south to Florida, west to Illinois, Arkansas, and Texas, also California; nearly cosmopolitan. Native to Europe, Africa, and Asia. Spring–summer.

Heliotropium gnaphalodes L. [Resembling *Gnaphalium* (Asteraceae).] SEA ROSEMARY; SEA LAVENDER.

Heliotropium gnaphalodes Linnaeus, Syst. Nat., ed. 10. 913. 1759. *Tournefortia gnaphalodes* (Linnaeus) R. Brown ex Roemer & Schultes, Syst. Veg. 4: 538. 1819. *Mallotonia gnaphalodes* (Linnaeus) Britton, Ann. Missouri Bot. Gard. 2: 47. 1915. *Messerschmidia gnaphalodes* (Linnaeus) I. M. Johnston, J. Arnold Arbor. 16: 165. 1935. *Argusia gnaphalodes* (Linnaeus) Heine, Fl. Nouv. Caled. 7: 108. 1976.

Shrub, to 3 m; branchlets white-sericeous. Leaves with the blade linear or linear-spatulate, 4–8 cm long, 0.5–1 cm wide, succulent, the apex obtuse or rounded, the base cuneate, the margin entire, the upper and lower surfaces densely white-sericeous, sessile or subsessile. Flowers in a short, dense-branched scorpioid cyme, the peduncle 3–5 cm long; calyx 3–5 mm long, the outer surface hispid-tomentose, the lobes broadly lanceolate to ovate, the apex acute; corolla tubular, white, tinged pink in the throat, the tube base green, 5–6 mm long, the tube a little longer than the calyx, the lobes lanceolate, 2–3 mm long, the apex acute, the outer surface tomentose; stamens inserted below the tube throat, sessile, the anthers ca. 1 mm long; style shorter than the anthers. Fruit ovoid or subglobose, 5–6 mm long, brown or black, glabrous, becoming dry, bony, and separating into two 2-seeded nutlets with a corky exocarp.

Coastal dunes. Rare; Brevard County southward along the east coast, southern peninsula. Florida; West Indies, Mexico, Central America, and South America. Winter–spring.

Heliotropium gnaphalodes (as *Argusia gnaphalodes*) is listed as endangered in Florida (Florida Administrative Code, Chapter 5B-40).

Heliotropium indicum L. [Of India.] INDIAN HELIOTROPE.

Heliotropium indicum Linnaeus, Sp. Pl. 130. 1753. *Heliotropium cordifolium* Moench, Methodus 415. 1794, nom. illegit. *Heliotropium foetidum* Salisbury, Prodr. Stirp. Chap. Allerton 112. 1796, nom. illegit. *Tiaridium indicum* (Linnaeus) Lehmann, Pl. Asperif. Nucif. 14. 1818. *Eliopia serrata* Rafinesque, Sylva Tellur. 90. 1838, nom. illegit. *Heliophytum indicum* (Linnaeus) de Candolle, Prodr. 9: 556. 1845.

Erect annual herb, to 1 m; stem villous to hispid or hispidulous. Leaves with the blade ovate to elliptic, 4–8(15) cm long, 2–6(10) cm wide, the apex acute, the base obtuse to subcordate, the margin entire, repand or undulate, the upper and lower surfaces sparsely hispid, the petiole 2–8 cm long, winged. Flowers in two ranks in a simple terminal or upper axillary scorpioid cyme to 3 dm long, sessile; ebracteate; calyx ca. 2 mm long, the lobes linear or linear-lanceolate, unequal; corolla salverform, blue or violet (rarely white), the tube 3–4 mm long, constricted at the throat, the outer surface puberulent or strigose, the lobes ca. 1 mm long. Fruit mitreform, 2–3 mm long, with sharp longitudinal ridges, separating into two 2-seeded nutlets, glabrous.

Disturbed sites. Occasional; central and western panhandle. Massachusetts and New York south to Florida, west to Kansas, Oklahoma, and Texas; West Indies, Mexico, Central America, and South America; Africa, Asia, Australia, and Pacific Islands. Native to South America. All year.

Previously believed to be native to Asia, but Johnston (1928) maintains that it is most likely South American as it is unrelated to any Asian species of the genus and is closely related to *H. elongatum* Hoffmann ex Roemer & Schultes of South America.

Heliotropium verdcourtii Craven. [Commemorates Bernard Verdcourt (1925–2011), British botanist whose publications include the Boraginaceae.] CHIGGERY GRAPES.

Tournefortia hirsutissima Linnaeus, Sp. Pl. 140. 1753, nom. cons. *Messerschmidia hirsutissima* (Linnaeus) Roemer & Schultes, Syst. Veg. 4: 541. 1819. *Heliotropium verdcourtii* Craven, Blumea 50: 378. 2005.

Shrub or woody vine, to 3 m; stem hirsute, glabrescent in age. Leaves with the blade lanceolate, ovate, or elliptic, 7–20 cm long, 3–8 cm wide, the apex acute to acuminate, the base obtuse to rounded, the margin entire, the upper surface scabrous-hirsute, the lower surface scabrous, pubescent, or glabrate, the petiole 0.5–2.5 cm long. Flowers in a terminal or axillary, branched, scorpioid cyme to 5 cm long, sessile; sepal lobes subulate to lanceolate, 2–3 mm long, the outer surface hispidulous; corolla salverform, white, the tube 4–6 mm long, the lobes ca. 2 mm long, the outer surface strigose distally, the base glabrous; stigma appendage short, entire. Fruit globose-pyramidal, 5–6 mm long, yellowish green to white with black dots, blackening on drying, pubescent; pyrenes 4.

Hammocks. Occasional; Hendry County, southern peninsula. Florida; West Indies, Mexico, Central America, and South America. All year.

Heliotropium verdcourtii (as *Tournefortia hirsutissima*) is listed as endangered in Florida (Florida Administrative Code, Chapter 5B-40).

Lithospermum L. 1753. GROMWELL

Annual or perennial herbs. Leaves alternate, simple, pinnate-veined, petiolate or epetiolate, estipulate. Flowers in terminal or axillary scorpioid cymes, actinomorphic or slightly zygomorphic, homostylous or heterostylous, sometimes cleistogamous, bracteate; sepals 5, basally connate; petals 5, basally connate, often with crests at the throat; stamens epipetalous; ovary 4-carpellate and -loculate, the style gynobasic, the stigmas 2. Fruit of four 1-seeded nutlets.

A genus of about 50 species; nearly cosmopolitan. [From the Greek *lithos*, stone, and *sperma*, seed, in reference to the stonelike nutlets.]

Batschia J. G. Gmel., 1791; *Onosmodium* Michx., 1803.

Selected reference: Turner (1995).

1. Corolla white, 4–6 mm long; basal leaves broadly oblanceolate or obovate; cauline leaves reduced......
  ....**L. tuberosum**
1. Corolla yellow, 0.8–14 mm long; basal leaves usually absent at anthesis; cauline leaves numerous, linear to lanceolate-oblong.
    2. Styles exserted 5–15 mm from the corolla at anthesis; nutlets brownish..................**L. virginianum**
    2. Syles included within the corolla at anthesis; nutlets white.
        3. Stem strigose; leaves linear (2–3 mm wide at midstem) ....**L. incisum**
        3. Stem puberulent and pilose; leaves linear-oblong to lanceolate (5–10 mm wide at midstem)......
          .... **L. caroliniense**

Lithospermum caroliniense (J. F. Gmel.) MacMill. [Of Carolina]. HAIRY PUCOON.

Batschia caroliniensis J. F. Gmelin, Syst. Nat. 2: 315. 1791. *Lithospermum caroliniense* (J. F. Gmelin) MacMillan, Metasp. Minnesota Valley 438. 1791. *Lithospermum hirtum* Lehmann, Pl. Asperif. Nucif. 304. 1818, nom. illegit.

Batschia gmelinii Michaux, Fl. Bor.-Amer. 1: 130. 1803. *Lithospermum gmelinii* (Michaux) Hitchcock, Key Spring Fl. Manhattan 30. 1894.

Lithospermum strigosum Rafinesque, New Fl. 4: 18. 1838 ("1836"). TYPE: FLORIDA.

Erect perennial herb, to 4(6) m; stem puberulent and pilose. Basal leaves usually absent at anthesis, the cauline ones with the blade linear to lanceolate, 2–6 cm long, 3–12 mm wide, the apex obtuse to acute, the base broadly cuneate to rounded, the margin entire, the upper and lower surfaces puberulent and pilose, sessile. Flowers in a terminal scorpioid cyme, heterostylous; bracts much longer than the calyx; calyx 9–11 mm long, the lobes lanceolate; corolla salverform, dark yellow, 1–2 cm long, the tube in the short-styled flowers exceeding the calyx, in the long-styled ones subequaling the calyx, the crests weakly invaginate, the lobes entire; stamens attached near the middle of the tube in the long-styled flowers, near the top of the tube in the short-styled ones. Fruit of four 1-seeded nutlets, these ovoid, 3–4 mm long, smooth, white, lustrous, porcelainlike, ventrally keeled.

Sandhills. Occasional; central and western panhandle. New York south to Florida, west to Ontario, Minnesota, South Dakota, Nebraska, Colorado, Oklahoma, and Texas. Spring.

Lithospermum incisum Lehm. [Deeply cut, in reference to the leaves.] NARROWLEAF GROMWELL.

Batschia longiflora Pursh, Fl. Amer. Sept. 132. 1814. *Lithospermum incisum* Lehmann, Pl. Asperif. Nucif. 303. 1818. *Lithospermum longiflorum* (Pursh) Sprengel, Syst. Veg. 1: 544. 1824; non Salisbury, 1796.

Lithospermum linearifolium Goldie, Edinburgh Philos. J. 6: 322. 1822. *Cyphorima linearifolia* (Goldie) Lunell, Amer. Midl. Naturalist 4: 514. 1916. *Batschia linearifolia* (Goldie) Small, Man. S.E. Fl. 1126. 1933.

Erect perennial herb, to 4 dm; stem strigose. Basal leaves usually absent at anthesis, the cauline ones with the blade linear, 2–4.5 cm long, 2–3 mm wide, the apex acute, the base broadly cuneate, the margin entire, the upper and lower surfaces strigose, sessile. Flowers in a short terminal scorpioid cyme; bracts longer than the calyx. Chasmogamous flowers with the calyx 3–5 mm long, the lobes linear; corolla salverform, light yellow, 2–4 cm long, the tube slender, the crests weakly to strongly invaginate, the lobes erose, fimbriate-denticulate, or entire; stamens attached near the top of the tube. Cleistogamous flowers with the calyx much exceeding the corolla; corolla 2–6 mm long or absent. Fruit of four 1-seeded nutlets, these ovoid, 3–4 mm long, smooth, white, lustrous, porcelainlike, sometimes pitted, constricted at the base.

Disturbed sites. Occasional; northern and central peninsula, west to central panhandle. Ontario, Tennessee, Florida, and Louisiana west to British Columbia and California; Mexico. Spring–summer.

Lithospermum tuberosum Rugel ex A. DC. [Tuber-bearing, in reference to the roots.] TUBEROUS GROMWELL.

Lithospermum tuberosum Rugel ex de Candolle, Prodr. 10: 76. 1846.

Erect annual herb, to 7 dm; stem hispid. Basal leaves with the blade oblanceolate to obovate, 8–10 cm long, 3–4 cm wide, the apex rounded, mucronate, the base cuneate, the margin entire, the upper and lower surfaces hirsute, the petiole 1–2 cm long; cauline leaves reduced upward and becoming sessile. Flowers in a short terminal or axillary scorpioid cyme; bracts foliose, equaling or longer than the corolla; calyx 4–6 mm long, the lobes linear to linear-spatulate; corolla salverform, white, 5–7 mm long, the lobes ca. ½ as long as the tube, the crests weakly to strongly invaginate; stamens attached near the middle of the tube. Fruit of four 1-seeded nutlets, these ovoid, ca. 2 mm long, smooth, white, lustrous, porcelainlike.

Bluff forests. Occasional; northern peninsula, central panhandle. Virginia south to Florida, west to Texas. Spring–summer.

Lithospermum virginianum L. [Of Virginia.] FALSE GROMWELL; WILD JOB'S TEARS.

Lithospermum virginianum Linnaeus, Sp. Pl. 132. 1753. *Onosmodium hispidum* Michaux, Fl. Bor.-Amer. 1: 133. 1803, nom. illegit. *Purshia hispida* Lehmann, Pl. Asperif. Nucif. 382. 1818, nom. illegit. *Onosmodium virginianum* (Linnaeus) de Candolle, Prodr. 10: 70. 1846.

Onosmodium virginianum (Linnaeus) de Candolle var. *hirsutum* Mackenzie, Bull. Torrey Bot. Club 32: 499. 1905.

Onosmodium floridanum Gandoger, Bull. Soc. Bot. France 65: 63. 1918. TYPE: FLORIDA: Polk Co.: s.d., *Ohlinger 618* (holotype: P).

Erect perennial herb, to 7 cm; stem strigose or hirsute. Basal leaves with the blade oblanceolate, 8–10 cm long, 2–4 cm wide, sometimes absent at anthesis, the cauline ones oblanceolate to elliptic, 5–8 cm long, 1.2–2 cm wide, the apex obtuse to acute, the base cuneate, the margin entire, the upper and lower surfaces strigose or hirsute, sessile or subsessile. Flowers in a branched terminal scorpioid cyme, the branches 6–14 cm long at anthesis, strigose or hirsute, the pedicel 1–3 mm long; bracts lanceolate to ovate, 4–6 mm long, foliaceous; calyx 4–7 mm long, the lobes linear-lanceolate, 2–3 mm long, pubescent on the outer and inner surfaces; corolla campanulate, yellow, 8–14 mm long, the tube 6–9 mm long, the lobes linear 3–5 mm long, erect, the outer surface sparsely pubescent; stamens included, the anthers 2–3 mm long; style exserted 5–10 mm. Fruit deeply lobed, splitting into four 1-seeded nutlets, these ovoid, 2–3 mm long, brownish, smooth.

Sandhills, flatwoods, and open hammocks. Frequent; northern counties, central peninsula. Massachusetts and New York south to Florida, west to Louisiana. Spring–summer.

Myosotis L. 1753. FORGET-ME-NOT

Annual or biennial herbs. Leaves alternate, simple, pinnate-veined, petiolate or epetiolate, estipulate. Flowers in terminal scorpioid cymes; sepals 5, basally connate, the lobes unequal;

petals 5, basally connate, with scalelike appendages in the throat; stamens 5, epipetalous; ovary 4-carpellate and -loculate, the style gynobasic, the stigma 1, capitate. Fruit separating into four 1-seeded nutlets.

A genus of about 100 species; nearly cosmopolitan. [From the Greek *myos*, mouse, and *ous*, ear, in reference to the soft, short leaves of some species.]

Myosotis macrosperma Engelm. [Large-seeded.] LARGESEED FORGET-ME-NOT.

> *Myosotis macrosperma* Engelmann, Amer. J. Sci. Arts 46: 98. 1844. *Myosotis virginica* (Linnaeus) Britton et al. var. *macrosperma* (Engelmann) Fernald, Rhodora 10: 55. 1908.
>
> *Myosotis verna* Nuttall var. *macrosperma* Chapman, Fl. South. U.S. 333. 1860. TYPE: FLORIDA.

Erect annual or biennial herb, to 6 dm; stem pilose. Basal leaves with the blade oblanceolate, 2–8 cm long, 0.5–2 cm wide, the apex rounded or obtuse, tapering to a margined petiole 1–2 cm long, the margin entire, the upper and lower surfaces pilose, the leaves gradually reduced and sessile upward, the lower cauline ones with the blade oblanceolate, this becoming oblong to linear-oblong upward, the apex round to obtuse or acute, the base rounded. Flowers in a terminal scorpioid cyme, this elongating to 4 dm and becoming racemiform when fully developed, the pedicel 1–3 mm long; calyx 4–6 mm long, the lobes triangular, 2–3 mm long, with 3 lobes distinctly shorter than the other 2, with spreading apically hooked trichomes; corolla salverform, white, the throat partly closed by scalelike appendages; stamens inserted near the base of the tube. Fruit of four 1-seeded nutlets, the pedicel 3–6 mm long, the accrescent calyx 6–9 mm long, inflated, the nutlets ovate, slightly lenticular, ca. 2 mm long, narrowly winged, the surface tan, smooth, lustrous.

Bluffs and floodplain forests. Rare; eastern and central panhandle. Ontario south to Florida and Texas. Spring.

EXCLUDED TAXA

> *Myosotis laxa* Lehmann—Reported for Florida by Chapman (1860). No Florida specimens known.
>
> *Myosotis verna* Nuttall—Reported for Florida by Correll and Johnston (1970). No Florida specimens known.
>
> *Myosotis virginica* (Linnaeus) Britton et al.—Reported for Florida by Small (1903, 1913a, 1933). No Florida specimens known.

Myriopus Small (1933).

Woody vines. Leaves alternate, simple, pinnate-veined, petiolate, estipulate. Flowers in terminal or axillary scorpioid cymes, actinomorphic, ebracteate; sepals 5, basally connate; petals 5, basally connate; stamens 5, epipetalous; ovary 4-carpellate and -loculate, the style terminal, elongate, with a sterile apical appendage. Fruit a drupe.

A genus of about 10 species; North America, West Indies, Mexico, Central America, and South America. [From the Greek *myrios*, numberless, and *pus*, footed, in reference to the many flowering branches.]

Myriopus is sometimes placed in the Heliotropiaceae by various authors.

Myriopus volubilis (L.) Small [Twining.] TWINING SOLDIERBUSH.

Tournefortia volubilis Linnaeus, Sp. Pl. 140. 1753. *Messerschmidia volubilis* (Linnaeus) Roemer & Schultes, Syst. Veg. 4: 541. 1819. *Myriopus volubilis* (Linnaeus) Small, Man. S.E. Fl. 1131, 1508. 1933. *Verrucaria volubilis* (Linnaeus) Medikus, Malvenfam. 104. 1787.

Tournefortia poliochros Sprengel, Syst. Veg. 1: 644. 1824. *Messerschmidia poliochros* (Sprengel) G. Don, Gen. Hist. 4: 371. 1838. *Myriopus poliochros* (Sprengel) Small, Man. S.E. Fl. 1131, 1508. 1933.

Twining or trailing woody vine; stem to 6 m long, branchlets sparsely and minutely strigose or tomentulose. Leaves with the blade lanceolate to ovate or elliptic, 3–10 cm long, 1–4 cm wide, the apex acute to acuminate, the base broadly cuneate to rounded, the margin entire, the upper surface glabrous or minutely strigose to pubescent, the lower surface puberulent to scabrous, the petiole 5–15 mm long. Flowers in a 3- to 10-branched, slender scorpioid cyme 3–8 cm long, sessile or subsessile; calyx lobes linear-lanceolate, ca. 2 mm long, the outer surface strigose or pubescent; corolla salverform, yellowish or greenish white, the tube 2–3 mm long, the lobes subulate, 1–2 mm long, the outer surface strigose; stigma appendage long-conic, bifid. Fruit subglobose, the exocarp white, blackening on drying, (1)2- to 4-lobed, 3–4 mm long, glabrous; pyrenes 1–4.

Coastal hammocks and shell middens. Occasional; central and southern peninsula. Florida and Texas; West Indies, Mexico, Central America, and South America. Spring–summer.

Nama L., nom. cons. 1759. FIDDLELEAF

Annual herbs. Leaves alternate, simple, pinnate-veined, petiolate, estipulate. Flowers axillary, solitary or paired; sepals 5, basally connate; petals 5, basally connate; stamens 5, epipetalous; ovary 2-carpellate and 1-loculate with 2 large parietal placenta, the styles terminal, 2, free. Fruit a loculicidal 2-valved capsule; seeds numerous.

A genus of about 35 species; North America, West Indies, Mexico, Central America, South America, and Pacific Islands. [From the Greek, a river or stream, in reference to the habitat of *Nama Zeylanica* L. (=*Hydrolea zeylanica* (L.) Vahl, Hydroleaceae).]

Nama is sometimes placed in the Hydroleaceae, Hydrophyllaceae, or Namaceae by various authors.

Marilaunidium Kuntze, 1891.

Nama jamaicensis L. [Of Jamaica.] JAMAICANWEED.

Nama jamaicense Linnaeus, Syst. Nat., ed. 10. 950. 1759. *Hydrolea jamaicensis* (Linnaeus) Raeuschel, Nomencl. Bot., ed 3. 3: 76. 1797. *Marilaunidium jamaicense* (Linnaeus) Kuntze, Revis. Gen. Pl. 2: 434. 1891. *Conanthus jamaicensis* (Linnaeus) A. Heller, Cat. N. Amer. Pl. 6. 1898.

Prostrate or ascending annual herb; stem to 4 m long, pilose-hirsute, somewhat winged by the decurrent petiole margins. Leaves with the blade obovate to spatulate, 1–3.5(8) cm long, 3–8(20) mm wide, the apex obtuse or apiculate, the base decurrent on the petiole, the margin entire, the upper and lower surfaces pilose-hirsute, sessile. Flowers 1–2 in the leaf axils, subsessile or the pedicel to 1.5 cm long; calyx lobes linear-lanceolate, 4–7 mm long, hirsute;

corolla salverform, white or lavender, 6–7 mm long, equaling or exceeding the calyx, the lobes suborbicular, 1–2 mm long; stamens unequally inserted on the corolla tube, included, the filament base dilated. Fruit oblong, 6–8 mm long, puberulent distally; seeds subglobose, minute, alveolate.

Disturbed sites. Rare; Leon, Manatee, and Lee Counties, southern peninsula. South Carolina, Florida, Alabama, Louisiana, and Texas; West Indies, Mexico, Central America, and South America. Native to tropical America. Spring–fall.

Nemophila Nutt. 1822. BABY BLUE-EYES

Annual herbs. Leaves alternate, simple, pinnate-veined, petiolate, estipulate. Flowers axillary, solitary; sepals 5, basally connate, with appendages between the lobe bases; petals 5, basally connate; stamens 5, epipetalous; ovary 2-carpellate and 1-loculate with 2 large parietal placenta, the styles 2, terminal, connate to near the middle. Fruit a loculicidal capsule; seeds few.

A genus of 11 species; North America and Mexico. [From the Greek *nemos*, a grove, and *philein*, to love, in reference to the habitat of *Nemophila phacelioides* Nutt.]

Nemophila is sometimes placed in the Hydroleaceae or Hydrophyllaceae by various authors. *Galax* L., nom. rej., 1753.

Nemophila aphylla (L.) Brummitt [From the Greek *a*, without, and *phyllon*, leaf, based by Linnaeus on a floral description of a plant by John Mitchell (1711–1767), American physician and botanist.] SMALLFLOWER BABY BLUE-EYES.

Galax aphylla Linnaeus, Sp. Pl. 200. 1753. *Erythrorhiza rotundifolia* Michaux, Fl. Bor.-Amer. 2: 35. 1803, nom. illegit. *Nemophila aphylla* (Linnaeus) Brummitt, Taxon 21: 315. 1972.

Hydrophyllum trilobum Rafinesque, Fl. Ludov. 33. 1817. *Nemophila triloba* (Rafinesque) Thieret, Rhodora 72: 400. 1970.

Ellisia microcalyx Nuttall, Trans. Amer. Philos. Soc., ser. 2. 5: 191. 1837. *Nemophila microcalyx* (Nuttall) Fischer & C. A. Meyer, Sert. Petrop. 1: t. 8. 1846. *Viticella microcalyx* (Nuttall) Nieuwland, Amer. Midl. Naturalist 3: 158. 1913.

Ascending or prostrate annual herb; stem to 4 dm long, much branched from the base, pilose. Leaves with the blade ovate to obovate, 2–3(4) pinnately and irregularly lobed, these in turn irregularly lobed, 1–2 cm long and wide, the apex rounded to obtuse, mucronate, the base cordate, the upper and lower surfaces pilose, the petiole usually as long as the blade. Flowers solitary, axillary, the pedicel 0.5–1.5 cm long; calyx 2–3 mm long, lobed to about the middle, the lobes lanceolate, elliptic, or oblong, the outer surface pilose, with a minute appendage between the lobes; corolla campanulate, white or pale blue, 3–4 mm long; stamens included; styles connate to near the middle, the stigmas capitate. Fruit subglobose, 3–4(5) mm long, pilose; seeds 1 or 2, pale brown, shallowly pitted.

Floodplain forests. Rare; central panhandle. Delaware and Maryland south to Florida, west to Oklahoma and Texas. Spring.

Varronia P. Browne 1756.

Shrubs. Leaves alternate, simple, pinnate-veined, petiolate, estipulate. Flowers in glomerate or spicate scorpioid cymes, actinomorphic or slightly zygomorphic; sepals 4 or 5, basally connate; petals 4 or 5, basally connate; stamens 4 or 5, epipetalous; ovary 4-carpellate and -loculate, the style terminal, twice bifid, the stigmas 4. Fruit a drupe.

A genus of about 100 species; North America, West Indies, Mexico, Central America, South America, and Asia. [Commemorates Marcus Terentius Varro (116–27 BC), distinguished Roman scholar and writer.]

Varronia is sometimes placed in the Cordiaceae or Ehretiaceae by various authors.

1. Inflorescence spicate...**V. curassavica**
1. Inflorescence globose.
    2. Leaves coarsely serrate .... **V. globosa**
    2. Leaves entire or with only a few teeth .... **V. bahamensis**

Varronia bahamensis (Urb.) Millsp. [Of the Bahama Islands.] BAHAMA MANJACK.

> *Cordia bahamensis* Urban, Symb. Antill. 1: 392. 1900. *Varronia bahamensis* (Urban) Millspaugh, Publ. Field Columb. Mus., Bot. Ser. 2: 310. 1909.

Shrub, to 2(4) m; branchlets appressed-setulose. Leaves with the blade linear, linear-oblong, elliptic, or obovate-elliptic, 2–10 cm long, 1–4 cm wide, subcoriaceous, the apex acute to obtuse or rounded, the base obtuse to subcordate, the margin entire or few-toothed, the upper surface setulose-scabrous, the lower surface pilose, at least on the veins, the petiole 0.5–2 cm long. Flowers several to many in a globose cyme, the peduncle as long as the leaves or shorter; calyx 2–3 mm long, pubescent, the lobes triangular, 2–3 mm long; corolla subcylindrical, white, 3–4 mm long, the lobes ovate, the outer surface glabrous, the inner surface with a ring of trichomes at the throat; stamens included. Fruit ovoid, 3–4 mm long, red or black, glabrous.

Pine rocklands. Rare; Miami-Dade County. Florida, West Indies. All year.

Varronia curassavica Jacq. [Of Curaçao.] BLACK SAGE.

> *Varronia curassavica* Jacquin, Enum. Syst. Pl. 14. 1760. *Cordia curassavica* (Jacquin) Roemer & Schultes, Syst. Veg. 4: 460. 1819.

Shrub, to 5 m; branchlets with stiff, curved trichomes. Leaves with the blade lanceolate, ovate, or oblong-ovate, 7–15 cm long, 3–5 cm wide, the apex acute, the base cuneate, the margin serrate, the upper surface scabrous, the lower surface pubescent, the petiole 5–6 mm long. Flowers in a terminal scorpioid spike, the peduncle 1–2 cm long; calyx 4–5 mm long, the lobes triangular, 1–2 mm long, the outer surface finely pubescent or papillose; corolla tubular, white, 4–5 mm long, white, the lobes ca. 2 mm long, erose, reflexed; stamens included. Fruit globose, ca. 5 mm long, red, enclosed in the accrescent calyx.

Disturbed sites. Rare; Broward County. Escaped from cultivation. Florida; West Indies, Mexico, Central America, and South America; Asia. Native to tropical America. All year.

Varronia globosa Jacq. [Ball-like, in reference to the inflorescence.] CURAÇAO BUSH; BUTTERFLY SAGE.

Varronia globosa Jacquin, Enum. Syst. Pl. 14. 1760. *Cordia globosa* (Jacquin) Kunth, in Humboldt et al., Nov. Gen. Sp. 3: 76. 1818. *Lithocardium globosum* (Jacquin) Kuntze, Revis. Gen. Pl. 2: 438. 1891. *Varronia humilis* Jacquin, Enum. Syst. Pl. 14. 1760. *Cordia humilis* (Jacquin) G. Don, Gen. Hist. 4: 383. 1838. *Lithocardium corymbosum* (Linnaeus) Kuntze var. *humile* (Jacquin) Kuntze, Revis. Gen. Pl. 2: 438. 1891. *Cordia globosa* (Jacquin) Kunth var. *humilis* (Jacquin) I. M. Johnston, J. Arnold Arbor. 30: 98. 1949. *Cordia globosa* (Jacquin) Kunth subsp. *humilis* (Jacquin) Borhidi, Bot. Közlem 58: 176. 1971. *Cordia bullata* (Linnaeus) Roemer & Schultes subsp. *humilis* (Jacquin) Gaviria, Mitt. Bot. Staatssaml. München 23: 189. 1987. *Varronia globosa* Jacquin subsp. *humilis* (Jacquin) Borhidi, Act. Bot. Hung. 34: 385. 1988. *Varronia bullata* Linnaeus subsp. *humilis* (Jacquin) Feuillet, J. Bot. Res. Inst. Texas 2: 837. 2008.

Shrub, to 3 m; branchlets strigose. Leaves with the blade rhomboid-ovate to lanceolate or ovate-oblong, 1.5–7 cm long, 1–3 cm wide, the apex acute to obtuse, the base cuneate to rounded, sometimes abruptly contracted and decurrent, the margin coarsely serrate, the upper surface strigose, somewhat rugose, the lower surface densely strigose, the petiole 1–1.5 cm long. Flowers many in a globose inflorescence, the peduncle 1–2 cm long; calyx campanulate, 3–4 mm long, hispid or hirsute, 5-lobed, the lobes filiform, 1–2 mm long; corolla campanulate, white, 5–9 mm long, the outer surface glabrous, the inner surface hispid at the throat; stamens included. Fruit ovoid to subglobose, 3–4 mm long, red, partly enclosed by the accrescent calyx, glabrous.

Hammocks. Rare; Hendry County, southern peninsula. Florida; West Indies, Mexico, Central America, and South America. All year.

Varronia globosa (as *Cordia globosa*) is listed as endangered in Florida (Florida Administrative Code, Chapter 5B-40).

EXCLUDED TAXON

Varronia bullata Linnaeus—Reported for Florida by Chapman (1860, 1883, 1897, all as *Cordia bullata* (Linnaeus) Roemer & Schultes), who misapplied the name to material of *V. globosa*.

EXCLUDED GENUS

Hackelia virginiana (Linnaeus) I. M. Johnston—Reported for Florida by Radford et al. (1968). No Florida specimens known.

Literature Cited

Almeda, F., and P. W. Fritsch. 2009. Symplocaceae. *In*: Flora of North America Editorial Committee. Flora of North America North of Mexico. 8: 329–31. New York/Oxford: Oxford University Press.

Al-Shehbaz, I. 1991. The genera of Boraginaceae in the southeastern United States. J. Arnold Arbor. Suppl. Ser. 1: 1–169.

Alvarado-Cárdenas, L. O., and H. Ochoterena. 2007. A phylogenetic analysis of the *Cascabela-Thevetia* species complex (Plumerieae, Apocynaceae) based on morphology. Ann. Missouri Bot. Gard. 94: 298–323.

Anderson, E. F. 2001. The Cactus Family. Portland, OR: Timber Press.

Austin, D. F. 1984. Resume of the Florida taxa of *Cereus* (Cactaceae). Florida Sci. 47: 68–72.

Austin, D. F., D. M. Binninger, and D. J. Pinkava. 1998. Uniqueness of the endangered Florida semaphore cactus (*Opuntia corallicola*). Sida 18: 527–34.

Benson, L. D. 1969. The cacti of the United States and Canada—new names and nomenclatural combinations. Cact. Succ. J. (Los Angeles) 41: 124–28, 185–90, 233–34.

———. 1982. The Cacti of the United States and Canada. Stanford, CA: Stanford University Press.

Boetsch, J. R. 2002. The Aizoaceae and Molluginaceae of the southeastern United States. Castanea 67: 42–53.

Bogle, A. L. 1969. The genera of Portulacaceae and Basellaceae in the southeastern United States. J. Arnold Arbor. 50: 566–98.

———. 1974. The genera of Nyctaginaceae in the southeastern United States. J. Arnold Arbor. 51: 431–62.

Britton, N. L., and J. N. Rose. 1919. The Cactaceae. Volume 1. Publ. Carnegie Inst. Washington 248: 1–236.

———. 1920. The Cactaceae. Volume 2. Publ. Carnegie Inst. Washington 248: 1–241.

———. 1922. The Cactaceae. Volume 3. Publ. Carnegie Inst. Washington 248: 1–258.

———. 1923. The Cactaceae. Volume 4. Publ. Carnegie Inst. Washington 248: 1–318.

Burch, D., R. P. Wunderlin, and D. B. Ward. 1975. Contributions to the Flora of Florida: 9, *Psychotria* (Rubiaceae). Castanea 40: 273–79.

Burckhalter, R. E. 1992. The genus *Nyssa* (Cornaceae) in North America: a Revision. Sida 15: 323–42.

Caulkins, D. B., and R. Wyatt. 1990. Variation and taxonomy of *Phytolacca americana* and *P. rigida* in the southeastern United States. Bull. Torrey Bot. Club 117: 357–67.

Channell, R. B., and C. E. Wood. 1959. The genera of the Primulales in the Southeastern United States. J. Arnold Arbor. 40: 268–88.

Chapman, A. W. 1860. Flora of the Southern United States. New York: Ivison, Phinney & Co.

———. 1883. Flora of the Southern United States. 2nd ed. New York: Ivison, Blakeman, Taylor & Co.

———. 1897. Flora of the Southern United States. 3rd ed. New York: American Book Co.

Cholewa, A. F. 2009a. *Samolus*. *In*: Flora of North America Editorial Committee. Flora of North America North of Mexico. 8: 254–56. New York/Oxford: Oxford University Press.

———. 2009b. *Anagallis. In*: Flora of North America Editorial Committee. Flora of North America North of Mexico. 8: 305–8. New York/Oxford: Oxford University Press.

———. 2009c. *Lysimachia. In*: Flora of North America Editorial Committee. Flora of North America North of Mexico. 8: 308–18. New York/Oxford: Oxford University Press.

Christenhusz, M. J. M, M. S. Vorontsova, M. F. Fay, and M. W. Chase. 2015. Results from an online survey of family delimitation in angiosperms and ferns: recommendations to the Angiosperm Phylogeny Group for thorny problems in plant classification. Bot. J. Linn. Soc. 178: 501–28.

Clemants, S. E. 2009. *Bejaria. In*: Flora of North America Editorial Committee. Flora of North America North of Mexico. 8: 449–50. New York/Oxford: Oxford University Press.

Clement, J. S. 2003. *Gaupira. In*: Flora of North America Editorial Committee. Flora of North America North of Mexico. 4: 74. New York/Oxford: Oxford University Press.

Clement, J. S., and R. W. Spellenberg. 2003. *Pisonia. In*: Flora of North America Editorial Committee. Flora of North America North of Mexico. 4: 71–73. New York/Oxford: Oxford University Press.

Clewell, A. F. 1985. Guide to the Vascular Plants of the Florida Panhandle. Tallahassee: University Presses of Florida/Florida State University Press.

Cohen, J. I. 2015. *Adelinia* and *Andersonglossum* (Boraginaceae), two new genera from New World species of *Cynoglossum*. Syst. Bot. 40: 611–19.

Correll, D. S., and M. C. Johnston. 1970. Manual of the Vascular Plants of Texas. Renner: Texas Research Foundation.

Danin, A., and L. C. Anderson. 1986. Distribution of *Portulaca oleracea* L. (Portulacaceae) subspecies in Florida. Sida 11: 318–24.

Danin, A., I. Baker, and H. G. Baker. 1978. Cytogeography and taxonomy of the *Portulaca oleracea* L. polyploidy complex. Israel J. Bot. 27: 177–211.

Darlington, J. 1934. A monograph of the genus *Mentzelia*. Ann. Missouri Bot. Gard. 21: 103–227.

Delprete, P. G. 2007. New combinations and new synonymies in the genus *Spermacoce* (Rubiaceae) for the Flora of Goiás and Tocantins (Brazil) and the flora of the Guianas. J. Bot. Res. Inst. Texas 1: 1023–30.

Diamond, A. R. 2013. New and noteworthy woody vascular plant records from Alabama. Phyroneuron 2013-47: 1–13.

Douglas, N. A., and P. S. Manos. 2007. Molecular phylogeny of Nyctaginaceae: taxonomy, biogeography, and characters associated with a radiation of xerophytic genera in North America. Amer. J. Bot. 94: 856–72.

Drapalik, D. J. 1970. A biosystematic study of the genus *Matelea* in the southeastern United States. PhD dissertation. University of North Carolina, Chapel Hill.

Dwyer, J. D. 1980. Rubiaceae. *In*: R. E. Woodson and R. E. Schery (Editors). Flora of Panama. Ann. Missouri Bot. Gard. 67: 1–522.

Eckenwalder, J. E. 2009. Ebenaceae. *In*: Flora of North America Editorial Committee. Flora of North America North of Mexico. 8: 247–50. New York/Oxford: Oxford University Press.

Elias, T. 1976. A monograph of the genus *Hamelia* (Rubiaceae). Mem. New York Bot. Gard. 26: 81–144.

Elisens, W. J. 2009a. *Mimusops. In*: Flora of North America Editorial Committee. Flora of North America North of Mexico. 8: 233–34. New York/Oxford: Oxford University Press.

———. 2009b. *Ceratiola. In*: Flora of North America Editorial Committee. Flora of North America North of Mexico. 8: 490–91. New York/Oxford: Oxford University Press.

Elisens, W. J., and J. M. Jones. 2009. *Sideroxylon. In*: Flora of North America Editorial Committee. Flora of North America North of Mexico. 8: 236–44. New York/Oxford: Oxford University Press.

Endress, M. E., and P. V. Bruyns. 2000. A revised classification of the Apocynaceae. Bot. Rev. (Lancaster) 66: 1–56.

Endress, M. E., S. Liede-Schumann, and U. Meve. 2014. An updated classification for Apocynaceae. Phytotaxa 159: 174–94.

Ernst, W. R., and H. J. Thompson 1963. The Loasaceae in the southeastern United States. J. Arnold Arbor. 44: 138–42.

Eyde, R. H. 1966. The Nyssaceae in the southeastern United states. J. Arnold Arbor. 47: 117–25.

Ferguson, I. K. 1966. The Cornaceae in the southeastern United States. J. Arnold Arbor. 47: 106–16.

Ferren, W. R. 2003a. *Sesuvium*. *In*: Flora of North America Editorial Committee. Flora of North America North of Mexico. 4: 80–81. New York/Oxford: Oxford University Press.

———. 2003b. *Cypselea*. *In*: Flora of North America Editorial Committee. Flora of North America North of Mexico. 4: 82. New York/Oxford: Oxford University Press.

———. 2003c. *Trianthema*. *In*: Flora of North America Editorial Committee. Flora of North America North of Mexico. 4: 82–83. New York/Oxford: Oxford University Press.

Fishbein, M., and W. D. Stevens. 2005. Resurrection of *Seutera* Reichenbach (Apocynaceae, Asclepiadoideae). Novon 15: 531–33.

Florida Exotic Pest Plant Council (FLEPPC). 2017. Florida Exotic Pest Plant Council's 2017 List of Invasive Plant Species. http://www.fleppc.org/list/list.htm.

Franck, A. R. 2012. Synopsis of *Harrisia* including a newly described species, several typifications, new synonyms, and a key to species. Haseltonia 18: 95–104.

Freeman, C. C. 2009. *Chimaphila*. *In*: Flora of North America Editorial Committee. Flora of North America North of Mexico. 8: 385–87. New York/Oxford: Oxford University Press.

Fritsch, P. W. 2009. Styracaceae. *In*: Flora of North America Editorial Committee. Flora of North America North of Mexico. 8: 339–47. New York/Oxford: Oxford University Press.

Gajdeczka, M. T., K. M. Neubig, W. S. Judd, W. M. Whitten, N. H. Williams, and K. D. Perkins. 2010. Phylogenetic analysis of the *Gaylussacia frondosa* complex (Ericaceae: Vaccinieae) based on molecular and morphological characters. J. Bot. Inst. Texas. 4: 245–60.

Gilbert, M. G. 1993. A review of *Gisekia* (Gisekiaceae). Kew Bull. 48: 343–56.

Gillett, J. M. 1959. A revision of *Bartonia* and *Obolaria* (Gentianaceae). Rhodora 61: 43–63.

Gleason, H. A., and A. Cronquist. 1991. Manual of the Vascular Plants of Northeastern United States and Adjacent Canada. 2nd ed. Bronx: New York Botanical Garden.

Godfrey, R. K., and J. W. Wooten. 1981. Aquatic and Wetland Plants of Southeastern United States: Dicotyledons. Athens: University of Georgia Press.

Goyder, D. J. 2003. A synopsis of *Morrenia* Lindl. (Apocynaceae subfam. Asclepiadoideae). Kew Bull. 58: 713–21.

Gray, A. 1887. Contributions to American Botany. I. Revision of some polypetalous genera and orders precursory to the Flora of North America. Proc. Amer. Acad. Arts. Sci. 22: 270–324.

Hawkes, M. W. 2003a. *Pereskia*. *In*: Flora of North America Editorial Committee. Flora of North America North of Mexico. 4: 100–101. New York/Oxford: Oxford University Press.

———. 2003b. *Hylocereus*. *In*: Flora of North America Editorial Committee. Flora of North America North of Mexico. 4: 175. New York/Oxford: Oxford University Press.

———. 2003c. *Selenicereus*. *In*: Flora of North America Editorial Committee. Flora of North America North of Mexico. 4: 176–78. New York/Oxford: Oxford University Press.

———. 2003d. *Epiphyllum*. *In*: Flora of North America Editorial Committee. Flora of North America North of Mexico. 4: 178–79. New York/Oxford: Oxford University Press.

Hooten, M. L. 1991. A new species of *Harrisia* from south Florida. Cact. Succ. J. (Los Angeles) 63: 64–66.

Johnston, I. M. 1928. Studies in the Boraginaceae—VII. The South American species of *Heliotropium*. Contr. Gray Herb. 81: 3–73.

Judd, W. S. 1981. A monograph of *Lyonia* (Ericaceae). J. Arnold Arbor. 62: 63–209, 315–436.

———. 2009a. *Oxydendrum*. *In*: Flora of North America Editorial Committee. Flora of North America North of Mexico. 8: 496–97. New York/Oxford: Oxford University Press.

———. 2009b. *Pieris*. *In*: Flora of North America Editorial Committee. Flora of North America North of Mexico. 8: 497–98. New York/Oxford: Oxford University Press.

———. 2009c. *Agarista*. *In*: Flora of North America Editorial Committee. Flora of North America North of Mexico. 8: 499. New York/Oxford: Oxford University Press.

———. 2009d. *Lyonia*. *In*: Flora of North America Editorial Committee. Flora of North America North of Mexico. 8: 500–503. New York/Oxford: Oxford University Press.

Judd, W. S., and K. A. Kron. 2009a. *Rhododendron*. *In*: Flora of North America Editorial Committee. Flora of North America North of Mexico. 8: 455–75. New York/Oxford: Oxford University Press.

———. 2009b. *Epigaea*. *In*: Flora of North America Editorial Committee. Flora of North America North of Mexico. 8: 475. New York/Oxford: Oxford University Press.

Judd. W. S., N. C. Melvin, K. Waselkov, and K. A. Kron. 2012. Taxonomic revision of *Eubotrys* (Ericaceae, Gaultherieae). Brittonia 64: 165–78.

———. 2013. A taxonomic revision of *Leucothoë* (Ericaceae; tribe Gaultherieae). Brittonia 65: 417–38.

Kartesz, J. T., and K. N. Gandhi. 1992. A new combination in *Spermacoce* (Rubiaceae) and lectotypification of *Borreria terminalis*. Brittonia 44: 370–71.

Khanum, R., S. Surveswaran, and C. Liede-Schumann. 2016. *Cynanchum* (Apocynaceae: Asclepiadoideae): a pantropical asclepiadoid genus revisited. Taxon 65: 467–86.

Kiger, R. W. 2003a. *Phemeranthus*. *In*: Flora of North America Editorial Committee. Flora of North America North of Mexico. 4: 488–95. New York/Oxford: Oxford University Press.

———. 2003b. *Talinum*. *In*: Flora of North America Editorial Committee. Flora of North America North of Mexico. 4: 502–4. New York/Oxford: Oxford University Press.

Kimnach, M. 1964. *Epiphyllum phyllanthus*. Cact. Succ. J. (Los Angeles) 36: 105–15.

Klackenberg, J. 2001. Revision of the genus *Cryptostegia* R. Br. (Apocynaceae, Periplocoideae). Addansonia, Sér. 3. 23: 205–18.

Kubitzki, K. Editor. 2004. The Families and Genera of Vascular Plants. 6. Flowering Plants. Dicotyledons: Celastrales, Oxidales, Rosales, Cornales, Ericales. Berlin/Heidelberg/New York: Springer-Verlag.

Kurz, H., and R. K. Godfrey. 1962. Trees of North Florida. Gainesville: University of Florida Press.

Le Duc, A. 1995. A revision of *Mirabilis* section Mirabilis (Nyctaginaceae). Sida 16: 613–48.

Leeuwenberg, A.J.M. 1994. A Revision of *Tabernaemontana*, Vol. 2: The New World Species. Kew: Royal Botanical Gardens.

Legrand, C. D. 1962. Las species americanas de *Portulaca*. Anales Mus. Nac. Montevideo, ser. 2. 7(3): 1–147.

Lemke, D. E. 2009. Cyrillaceae. *In*: Flora of North America Editorial Committee. Flora of North America North of Mexico. 8: 367–69. New York/Oxford: Oxford University Press.

Lewis, W. H., and R. L. Oliver. 1974. Revision of *Richardia* (Rubiaceae). Brittonia 26: 271–301.

Liede, S., and A. Täuber. 2002. Circumscription of the genus *Cynanchum* (Apocynaceae—Asclepiadoideae). Syst. Bot. 27: 789–800.

Liede-Schumann, S., and U. Meve. 2008. Nomenclatural novelties and one new species in *Orthosia* (Apocynaceae, Asclepiadoideae). Novon 18: 202–10.

———. 2013. The Orthosiinae revisited (Apocynaceae, Asclepiadoideae, Asclepiadeae). Ann. Missouri Bot. Gard. 99: 44–81.

Liede-Schumann, S., M. Nikolaus, U. C. Soares e Silva, A. Rapini, R. D. Mangelsdorff, and U. Meve. 2014. Phylogenetics and biogeography of the genus *Metastelma* (Apocynaceae-Asclepiadoideae-Asclepiadeae: Metastelmatinae). Syst. Bot. 39: 594–612.

Lima, A. N. 1996. *Epiphyllum phyllanthus* var. *hookeri* (Haw.) Kimn. in Florida. Cact. Succ. J. (Los angeles) 68: 96–97.

Liogier, A. H. [Hermano Alain]. 1962. Flora de Cuba. 5. Río Piedras, P.R.: Editorial Universitaria, Universidad de Puerto Rico.

———. 1997. Descriptive Flora of Puerto Rico and Adjacent Islands. 5. Río Piedras, P.R.: Editorial de la Universidad de Puerto Rico.

Liu, S., K. E. Denford, J. E. Ebinger, J. G. Packer, and G. C. Tucker. 2009. *Kalmia. In*: Flora of North America Editorial Committee. Flora of North America North of Mexico. 8: 480–85. New York/Oxford: Oxford University Press.

Long, R. W. 1970. Additions and nomenclatural changes in the flora of southern Florida—I. Rhodora 72: 17–46.

Long, R. W., and O. Lakela. 1971. A Flora of Tropical Florida. Coral Gables: University of Miami Press.

Maas, P. J. M., and P. Ruyters. 1986. *Voyria* and *Voyriella* (saprophytic Gentianaceae). Flora Neotrop. Monogr. 41: 1–93.

Majure, L. C. 2014. Typifications and a nomenclatural change in some eastern North American *Opuntia* (Cactaceae). Phytoneuron 2014-106: 1–2.

Majure, L. C., W. S. Judd, P. S. Soltis, and D. E. Soltis. 2012a. Cytogeography of the Humifusa clade of *Opuntia* s.s. Mill. 1754. (Cactaceae, Opuntioideae, Opuntieae): correlations with pleistocene refugia and morphological traits in a polyploidy complex. Comp. Cytogenet. 6: 53–77.

Majure, L. C., R. Puente, M. P. Griffith, W. S. Judd, P. S. Soltis, and D. E. Soltis. 2012b. Phylogeny of *Opuntia* s.s. (Cactaceae): clade delineation, geographic origins, and reticulate evolution. Amer. J. Bot. 99: 847–64.

Majure, L. C., D. S. Soltis, P. S. Soltis, and W. S. Judd. 2014. A case of mistaken identity, *Opuntia abjecta*, long-lost in synonymy under the Caribbean species, *O. triacantha*, and a reassessment of the enigmatic *O. cubensis*. Brittonia 66: 118–30.

Matthews, J. F. 2003. *Portulaca. In*: Flora of North America Editorial Committee. Flora of North America North of Mexico. 4: 497–501. New York/Oxford: Oxford University Press.

Matthews, J. F., D. A. Ketron, and S. F. Zane. 1993. The biology and taxonomy of the *Portulaca oleracea* L. (Portulacaceae) complex in North America. Rhodora 95: 166–83.

Matthews, J. F., and P. A. Levins. 1985. The genus *Portulaca* in the southeastern United States. Castanea 50: 96–104.

Matthews, K. G., N. Dunne, E. York, and L. Struwe. 2009. A phylogenic analysis and taxonomic revision of *Bartonia* (Gentianaceae: Gentianeae), based on molecular and morphological evidence. Syst. Bot. 34: 162–72.

Mellichamp, T. L., and F. W. Case. 2009. *Sarracenia. In*: Flora of North America Editorial Committee. Flora of North America North of Mexico. 8: 350–63. New York/Oxford: Oxford University Press.

Menninger, E. A. 1964. Seaside Plants of the World. New York: Hearthside Press.

Meve, U., and S. Liede-Schumann. 2012. Taxonomic dissolution of *Sarcostemma* (Apocynaceae: Asclepiadoideae). Kew Bull. 67: 751–58.

Michaux, A. 1803. Flora Boreali-Americana. 1. Paris/Strasbourg: Caroli Crapelet.

Murrell, Z. E. 1992. Systematics of the genus *Cornus* (Cornaceae). PhD dissertation. Duke University.

———. 1993. Phylogenetic relationships in *Cornus* (Cornaceae). Syst. Bot. 18: 469–95.

Naczi, R. F. C., E. M. Soper, F. W. Case, and R. B. Case. 1999. *Sarracenia rosea* (Sarraceniaceae), a new species of pitcher plant from the southeastern United States. Sida 18: 1183–1206.

Negrón-Ortiz, V. 2005. Taxonomic revision of the Neotropical genus: *Erithalis* (Rubiaceae: Chiococceae). Sida 21: 1565–98.

Negrón-Ortiz, V., and R. J. Hickey. 1996. The genus *Ernodea* (Rubiaceae) in the Caribbean Basin. II. Morphological analysis and systematic. Syst. Bot. 21: 445–58.

Nienaber, M. A., and J. W. Thieret. 2003. Phytolaccaceae. *In*: Flora of North America Editorial Committee. Flora of North America North of Mexico. 4: 3–11. New York/Oxford: Oxford University Press.

Nowicke, J. W. 1968. Palynotaxonomic study of the Phytolaccaceae. Ann. Missouri Bot. Gard. 55: 294–363.

Nuttall, T. 1843. Description and notices of new or rare plants in the natural orders Lobeliaceae, Campanulaceae, Vacciniaceae, Ericaceae, collected in a journey over the continent of North America, and during a visit to the Sandwich Islands, and upper California. Trans. Amer. Philos. Soc., ser. 2. 8: 251–72.

Ornduff, R. 1970. The systematic and breeding system of *Gelsemium* (Loganiaceae). J. Arnold Arbor. 51: 1–17.

Packer, J. G. 2003. Portulacaceae. *In*: Flora of North America Editorial Committee. Flora of North America North of Mexico. 4: 457–58. New York/Oxford: Oxford University Press.

Parfitt, B. D., and A. C. Gibson. 2003a. *Harrisia*. *In*: Flora of North America Editorial Committee. Flora of North America North of Mexico. 4: 152–54. New York/Oxford: Oxford University Press.

———. 2003b. *Acanthocereus*. *In*: Flora of North America Editorial Committee. Flora of North America North of Mexico. 4: 154–55. New York/Oxford: Oxford University Press.

———. 2003c. *Pilocereus*. *In*: Flora of North America Editorial Committee. Flora of North America North of Mexico. 4: 179–81. New York/Oxford: Oxford University Press.

Pennell, F. W. 1919. Notes on plants of the southern United States. Bull. Torrey Bot. Club 46: 183–87.

Pinkava, D. J. 2003a. *Opuntia*. *In*: Flora of North America Editorial Committee. Flora of North America North of Mexico. 4: 123–48. New York/Oxford: Oxford University Press.

———. 2003b. *Nopalea*. *In*: Flora of North America Editorial Committee. Flora of North America North of Mexico. 4: 148–49. New York/Oxford: Oxford University Press.

———. 2003c. *Consolea*. *In*: Flora of North America Editorial Committee. Flora of North America North of Mexico. 4: 149–50. New York/Oxford: Oxford University Press.

Pipoly, J. J., and J. M. Ricketson. 2009a. *Ardisia*. *In*: Flora of North America Editorial Committee. Flora of North America North of Mexico. 8: 318–20. New York/Oxford: Oxford University Press.

———. 2009b. *Myrsine*. *In*: Flora of North America Editorial Committee. Flora of North America North of Mexico. 8: 321. New York/Oxford: Oxford University Press.

Pollard, C. L. 1895. The genus *Zenobia* Donn. Bull. Torrey Bot. Club 22: 231–32.

Prince, L. M. 2009. Theaceae. *In*: Flora of North America Editorial Committee. Flora of North America North of Mexico. 8: 322–28. New York/Oxford: Oxford University Press.

Pringle, J. S. 1967. Taxonomy and distribution of *Gentiana*, section Pneumonanthae in eastern North America. Brittonia 19: 1–32.

Puff, C. (ed.). 1991. The genus *Paederia* L.: a multidisciplinary study. Opera Bot. Belg. 3: 1–376.

Radford, A. H., H. E. Ahles, and C. R. Bell. 1964. Guide to the Vascular Flora of the Carolinas. Chapel Hill: University of North Carolina.

———. 1968. Manual of the Vascular Flora of the Carolinas. Chapel Hill: University of North Carolina Press.

Rao, A. S. 1956. A revision of *Rauvolfia* with particular reference to the American species. Ann. Missouri Bot. Gard. 43: 253–354.

Rapini, A., J. Fontella Pereira, and D. J. Goyder. 2011. Towards a stable generic circumscription in Oxypetalinae (Apocynaceae). Phytotaxa 26: 9–16.

Refulio-Rodriguez, N. F., and R. G. Olmstead. 2014. Phylogeny of Lamidae. Amer. J. Bot. 101: 1–13.

Reveal, J. L. 2009. *Dodecatheon*. *In*: Flora of North America Editorial Committee. Flora of North America North of Mexico. 8: 268–86. New York/Oxford: Oxford University Press.

Reveal, J. L., and M. J. Seldin. 1976. On the identity of *Halesia carolina* L. (Styracaceae). Taxon 25: 123–40.

Ricketson, J. M., and J. J. Pipoly. 1997. Nomenclatural notes and a synopsis of the genus *Myrsine* (Myrsinaceae) in Mesoamerica. Sida 17: 579–89.

Ridsdale, C. E. 1976. A revision of the tribe Cephalantheae (Rubiaceae). Blumea 23: 177–88.

Rogers, G. K. 1985. The genera of Phytolaccaceae in the southeastern United States. J. Arnold Arbor. 66: 1–37.

———. 1986. The genera of Loganiaceae in the southeastern United States. J. Arnold Arbor. 67: 143–85.

———. 1987. The genera of Cinchonoideae (Rubiaceae) in the southeastern United States. J. Arnold Arbor. 68: 137–83.

———. 2005. The genera of Rubiaceae in the southeastern United States, part II. Subfamily Rubioideae, and subfamily Cinchonoideae revisited (*Chiococca*, *Erithalis*, and *Guettardia*). Harvard Papers Bot. 10: 1–45.

Rosatti, T. J. 1989. The genera of suborder Apocynineae (Apocynaceae and Asclepiadaceae) in the south-eastern United States. J. Arnold Arbor. 70: 307–401, 443–514.

Rose, J. P. 2012. The systematics of *Monotropsis* (Ericaceae). MS thesis, Ohio State University.

Serviss, B. E., and J. H. Peck. 2016. *Camellia sasanqua* (Theaceae) in the Arkansas flora. Phytoneuron 2016-28: 1–7.

Small, J. K. 1903. Flora of the Southeastern United States. New York: Published by the author.

———. 1913a. Flora of the Southeastern United States. 2nd ed. New York: Published by the author.

———. 1913b. Flora of Miami. New York: Published by the author.

———. 1913c. Florida Trees. New York: Published by the author.

———. 1913d. Flora of the Florida Keys. New York: Published by the author.

———. 1913e. Shrubs of Florida. New York: Published by the author.

———. 1933. Manual of the Southeastern Flora. New York: Published by the author.

Sorrie, B. A., A. S. Weakley, and G. S. Tucker. 2009. *Gaylussacia*. *In*: Flora of North America Editorial Committee. Flora of North America North of Mexico. 8: 530–35. New York/Oxford: Oxford University Press.

Spellenberg, R. W. 2003a. *Boerhavia*. *In*: Flora of North America Editorial Committee. Flora of North America North of Mexico. 4: 17–28. New York/Oxford: Oxford University Press.

———. 2003b. *Okenia*. *In*: Flora of North America Editorial Committee. Flora of North America North of Mexico. 4: 39. New York/Oxford: Oxford University Press.

———. 2003c. *Mirabilis*. *In*: Flora of North America Editorial Committee. Flora of North America North of Mexico. 4: 40–57. New York/Oxford: Oxford University Press.

Spellman, D. L., and C. R. Gunn. 1976. *Morrenia odorata* and *Araujia sericofera* (Asclepiadaceae): weeds in citrus groves. Castanea 41: 139–48.

Sperling, C. R. 1987. Systematics of the Basellaceae. PhD dissertation. Harvard University, Cambridge, MA.

Spongberg, S. A. 1972. The genera of Saxifragaceae in the southeastern United States. J. Arnold Arbor. 53: 409–98.

Ståhl, B. 1992. On the identity of *Jacquinia armillaris* (Theophrastaceae) and related species. Brittonia 44: 54–60.

———. 2004. Samolaceae. *In*: K. Kubitzki et al., eds. The Families and Genera of Vascular Plants. 6: 387–89. Berlin, Heidelberg, New York: Springer-Verlag.

Takhtajan, A. 1987. Systema Magnoliophytorum. Leningrad: Soviet Sciences Press.

———. 1997. Diversity and Classification of Flowering Plants. New York: Columbia University Press.

Terrell, E. E. 1986. Taxonomic and nomenclatural notes on *Houstonia nigricans* (Rubiaceae). Sida 11: 471–81.

———. 1990. Synopsis of *Oldenlandia* (Rubiaceae) in the United States. Phytologia 68: 125–33.

———. 2001. Taxonomy of *Stenaria* (Rubiaceae: Hedyotideae), a new genus including *Hedyotis nigricans*. Sida 19: 591–614.

Terrell, E. E., and W. H. Lewis. 1990. *Oldenlandiopsis* (Rubiaceae), a new genus from the Caribbean Basin, based on *Oldenlandia callitrichoides* Grisebach. Brittonia 42: 185–90.

The Angiosperm Phylogeny Group. 2016. An update of the Angiosperm Phylogeny Group classification for the orders and families of flowering plants: APG IV. Bot. J. Linn. Soc. 181: 1–20.

Thomas, J. 1960. A monographic study of the Cyrillaceae. Contr. Gray Herb. 186: 1–114.

———. 1961. The genera of the Cyrillaceae and Clethraceae of the southeastern United States. J. Arnold Arbor. 42: 96–106.

Tucker, G. C. 2009a. *Leucothoe*. *In*: Flora of North America Editorial Committee. Flora of North America North of Mexico. 8: 508–10. New York/Oxford: Oxford University Press.

———. 2009b. *Eubotrys*. *In*: Flora of North America Editorial Committee. Flora of North America North of Mexico. 8: 510–12. New York/Oxford: Oxford University Press.

Tucker, G. C., and S. C. Jones. 2009. Clethraceae. *In*: Flora of North America Editorial Committee. Flora of North America North of Mexico. 8: 364–66. New York/Oxford: Oxford University Press.

Turner, B. L. 1995. Synopsis of the genus *Onosmodium* (Boraginaceae). Phytologia 78: 39–60.

Vander Kloet, S. P. 2009. *Vaccinium*. *In*: Flora of North America Editorial Committee. Flora of North America North of Mexico. 8: 515–30. New York/Oxford: Oxford University Press.

Ventenat, E. P. 1800–1803. Description de Plantes Nouvelles et Peu Connues Cultivés dans le Jardin de J. M. Cels. [10 parts]. Paris.

Vincent, M. A. 2003a. Basellaceae. *In*: Flora of North America Editorial Committee. Flora of North America North of Mexico. 4: 505–8. New York/Oxford: Oxford University Press.

———. 2003b. Molluginaceae. *In*: Flora of North America Editorial Committee. Flora of North America North of Mexico. 4: 509–12. New York/Oxford: Oxford University Press.

Vivrette, N. J. 2003a. *Tetragonia*. *In*: Flora of North America Editorial Committee. Flora of North America North of Mexico. 4: 77–78. New York/Oxford: Oxford University Press.

———. 2003b. *Galenia*. *In*: Flora of North America Editorial Committee. Flora of North America North of Mexico. 4: 78–79. New York/Oxford: Oxford University Press.

———. 2003c. *Aptenia*. *In*: Flora of North America Editorial Committee. Flora of North America North of Mexico. 4: 83–84. New York/Oxford: Oxford University Press.

———. 2003d. *Carpobrotus*. *In*: Flora of North America Editorial Committee. Flora of North America North of Mexico. 4: 86–87. New York/Oxford: Oxford University Press.

Wallace, G. D. 2009a. *Monotropa*. *In*: Flora of North America Editorial Committee. Flora of North America North of Mexico. 8: 392–93. New York/Oxford: Oxford University Press.

———. 2009b. *Monotropsis*. *In*: Flora of North America Editorial Committee. Flora of North America North of Mexico. 8: 394. New York/Oxford: Oxford University Press.

Ward, D. B. 2001. New combinations in the Florida flora. Novon 11: 360–65.

———. 2009. Keys to the Flora of Florida: 23, *Opuntia* (Cactaceae). Phytologia 91: 383–93.

———. 2011. Keys to the Flora of Florida: 29, *Spermacoce* (Rubiaceae) Phytologia 93: 275–82.

Ward, D. B., and P. M. Lyrene. 2007. *Vaccinium tenellum* is not a native Florida blueberry. Castanea 72: 45–46.

Weigend, M., F. Luebert, M. Gottschling, T.L.P. Couvreur, H. H. Hilger, and J. S. Miller. 2014. From capsules to nutlets—phylogenetic relationships in the Boraginales. Cladistics 30: 508–18.

Wen, J., and T. F. Stussy. 1993. Phylogeny and biogeography of *Nyssa* (Cornaceae). Syst. Bot. 18: 68–79.

Wherry, E. T. 1955. The genus *Phlox*. Morris Arbor. Monogr. 3: 1–174.

Whetstone, R. D., and R. P. Wunderlin. 2009. *Jacquinia*. *In*: Flora of North America Editorial Committee. Flora of North America North of Mexico. 8: 253–54. New York/Oxford: Oxford University Press.

Wilbur, R. L. 1955. A revision of the North American genus *Sabatia* (Gentianaceae). Rhodora 57: 1–33, 43–71, 78–104.

Wilbur, R. L., and J. L. Luteyn. 1978. Ericaceae. In: R. E. Woodson et al. Flora of Panama, Part VIII. Ann. Missouri Bot. Gard. 65: 27–144.

Wilhelm, G. S. 1984. Vascular Flora of the Pensacola Region. PhD dissertation, Southern Illinois University, Carbondale.

Willdenow, C. L. 1799. Species Plantarum. 2(1). Berlin: G. C. Nauk.

Williams, J. K., and J. K. Stutzman. 2008. Chromosome number of *Thevetia ahouai* (Apocynaceae: Rauvolfoideae: Plumerieae) with discussion on the generic boundaries of *Thevetia*. J. Bot. Res. Inst. Texas 2: 489–93.

Wilson, K. A. 1960. The genera of Hydrophyllaceae and Polemoniaceae in the southeastern United States. J. Arnold Arbor. 41: 197–212.

Wood, C. E. 1959. The genera of Theaceae of the southeastern United States. J. Arnold Arbor. 40: 413–19.

———. 1960. The genera of Sarraceniaceae and Droseraceae in the southeastern United States. J. Arnold Arbor. 41: 152–63.

———. 1961. The genera of Ericaceae in the southeastern United States. J. Arnold Arbor. 42: 10–80.

Wood, C. E., and R. B. Channell. 1959. The Empetraceae and Diapensiaceae of the southeastern United States. J. Arnold Arbor. 40: 161–71.

———. 1960. The genera of the Ebenales in the southeastern United States. J. Arnold Arbor. 41: 1–35.

Wood, C. E., and R. E. Weaver. 1982. The genera of Gentianaceae in the southeastern United States. J. Arnold Arbor. 63: 441–87.

Woodson, R. E. 1930. Studies in the Apocynaceae. I. A critical study of the Apocynoideae (with special reference to the genus *Apocynum*). Ann. Missouri Bot. Gard. 17: 1–212.

———. 1941. North American Asclepiadaceae I. Perspective of the genera. Ann. Missouri Bot. Gard. 28: 193–244.

———. 1954. The North American species of *Asclepias* L. Ann. Missouri Bot. Gard. 41: 1–211.

Wunderlin, R. P. 1982. Guide to the Vascular Plants of Central Florida. Gainesville: University Presses of Florida/University of South Florida Press.

———. 1998. Guide to the Vascular Plants of Florida. Gainesville: University Presses of Florida/University of South Florida Press.

———. 2009. *Bonellia. In*: Flora of North America Editorial Committee. Flora of North America North of Mexico. 8: 252. New York/Oxford: Oxford University Press.

Wunderlin, R. P., and B. F. Hansen. 2003. Guide to the Vascular Plants of Florida. 2nd ed. Gainesville: University Press of Florida.

———. 2011. Guide to the Vascular Plants of Florida. 3rd ed. Gainesville: University Press of Florida.

Wunderlin, R. P., and R. D. Whetstone. 2009a. *Manilkara. In*: Flora of North America Editorial Committee. Flora of North America North of Mexico. 8: 234–35. New York/Oxford: Oxford University Press.

———. 2009b. *Chrysophyllum. In*: Flora of North America Editorial Committee. Flora of North America North of Mexico. 8: 245–46. New York/Oxford: Oxford University Press.

———. 2009c. *Pouteria. In*: Flora of North America Editorial Committee. Flora of North America North of Mexico. 8: 244–45. New York/Oxford: Oxford University Press.

Xiang, Q.-Y., M. L. Moody, D. E. Soltis, C. Z. Fan, and P. S. Soltis. 2002. Relationships within Cornales and circumscription of Cornaceae—*matK* and *rbcL* sequence data and effects of outgroups and long branches. Mol. Phylog. Evol. 24: 35–57.

Xiang, Q.-Y., D. T. Thomas, and Q. P. Xiang. 2011. Resolving and dating the phylogeny of Cornales—effects of taxon sampling, data partitions, and fossil calibrations. Mol. Phylogen. Evol. 59: 123–38.

Xiang, Q.-Y., D. T. Thomas, W. Zhang, S. R. Manchester, and Z. Murrill. 2006. Species level phylogeny of the genus *Cornus* (Cornaceae) based on molecular and morphological evidence—implications for taxonomy and Tertiary intercontinental migration. Taxon 55: 9–30.

Index to Common Names

Index to Scientific Names

Accepted scientific names of plants and plant families are in roman type. Synonyms are in *italics*.

Richard P. Wunderlin, professor emeritus of biology at the University of South Florida, is the author of *Guide to the Vascular Plants of Central Florida* and *Guide to the Vascular Plants of Florida*, and coauthor of *Guide to the Vascular Plants of Florida*, second edition, *Guide to the Vascular Plants of Florida*, third edition, and the *Atlas of Florida Plants* website (www.florida.plantatlas.usf.edu).

Bruce F. Hansen, curator emeritus of biology at the University of South Florida Herbarium, is coauthor of *Guide to the Vascular Plants of Florida*, second edition, *Guide to the Vascular Plants of Florida*, third edition, and the *Atlas of Florida Plants* website (www.florida.plantatlas.usf.edu).

Alan R. Franck, curator of the University of South Florida Herbarium, is coauthor of the *Atlas of Florida Plants* website (www.florida.plantatlas.usf.edu).

CPSIA information can be obtained
at www.ICGtesting.com
Printed in the USA
LVOW09*1219270318
571254LV00003B/3/P

9 780813 056791